PATTERNS IN EXCITABLE MEDIA
Genesis, Dynamics, and Control

PATTERNS IN EXCITABLE MEDIA
Genesis, Dynamics, and Control

Sitabhra Sinha

S. Sridhar

CRC Press
Taylor & Francis Group
Boca Raton London New York

CRC Press is an imprint of the
Taylor & Francis Group, an **informa** business

A CHAPMAN & HALL BOOK

CRC Press
Taylor & Francis Group
6000 Broken Sound Parkway NW, Suite 300
Boca Raton, FL 33487-2742

First issued in paperback 2019

© 2015 by Taylor & Francis Group, LLC
CRC Press is an imprint of Taylor & Francis Group, an Informa business

No claim to original U.S. Government works

ISBN-13: 978-1-4665-5283-8 (hbk)
ISBN-13: 978-0-367-37798-4 (pbk)

Visit the Taylor & Francis Web site at
http://www.taylorandfrancis.com

and the CRC Press Web site at
http://www.crcpress.com

Contents

Preface

Let us start with a gameshow–like question: what is common to the processes happening on catalytic converters in cars, a heart in the throes of a life-threatening arrhythmia, the periodic change of color in an oscillating chemical reaction discovered by a Russian scientist that every reviewer said was impossible, and the aggregation of starved single-celled slime mold into a multi-cellular organism with a fruiting body which eventually bursts open, spreading spores far and wide? The short answer is that they can all be described using a versatile class of models known as excitable media. These, in turn, form a special class of the family of reaction–diffusion (RD) models that, for decades, have been used to explain pattern formation in many natural systems. Possibly the most famous application of the RD paradigm to date was that devised by Alan Turing in 1952 to explain how the inhomogeneities that characterize morphogenesis — the development of form in biological organisms — can arise spontaneously from a homogeneous initial state. The "Turing mechanism," as it is often referred to, involves small fluctuations in concentrations — that may arise from noise or randomness — eventually growing over time and giving rise to patterns characterized by spots or stripes. This specific mechanism of pattern formation necessarily requires a slowly diffusing activating species and a fast diffusing inhibitory species. Although other processes are known to exist that can result in the spontaneous development of heterogeneities, the predominance of the Turing mechanism in the theoretical literature concerned with modeling pattern formation has occasionally led to the misunderstanding that patterns can only arise in RD systems when the specific conditions associated with this mechanism are met.

Excitable media provide another pathway to pattern formation, through the propagation of self-generated waves. Their behavior is characterized by the existence of a threshold of activation such that small fluctuations about the resting state of the system rapidly decay without having any effect on the system. Only a strong stimulus, i.e., one which exceeds the threshold, is able to make the system switch to another state distinct from the rest state, termed the "activated" or "excited" state. The inactivating or inhibitory component of excitable media is not required to diffuse faster than the activator component — in fact, the inactivating component need not diffuse at all. Once an element in this system reaches its excited state, it stays there only for a finite duration before slowly returning back to its equilibrium resting state within a time interval known as the "refractory period." During this time, the element cannot be re-excited even with a supra-threshold stimulus. One can see how this principle operates even in mundane physical systems like the toilet flush, which requires a minimum force (threshold) that needs to be exerted down on the lever in order to start the water flow and has a refractory period during which the cistern is refilled with water before a flush can be repeated.

The somewhat unexpected consequence of the refractory period is that when two waves in excitable media collide head on, they annihilate each other. Another consequence is that a wave with a broken front (i.e., a singularity at the broken end of the wave that has in its neighborhood points belonging to every phase of the cycle, from resting to activated to refractory states) will curl up to form the hook-like shape of a spiral wave that continually moves outward from the central singularity (the wave source). Such spiral waves are ubiquitous in nature and have, for example, been seen on imaging the catalytic reaction taking place on the surface of platinum, where the poisonous carbon monoxide emission of automobiles is converted into relatively harmless carbon dioxide. Another manifestation of this phenomenon is observed in colonies of the cellular slime mold, *Dictyostelium*, where lack of nutrients can trigger the secretion of cyclic adenosine monophosphate (cAMP). This results in spiral waves of cAMP propagating through the colony, which signals the different cells to start aggregating in order to form a multicellular organism.

Probably the simplest experimental setup in which these waves can be observed is the class of inorganic chemical reactions proposed in the early 1950s by the biochemist Boris Pavlovich Belousov of Moscow State University. Belousov was trying to devise an inorganic analog of the citric acid cycle (possibly better known as the Krebs cycle) that is used by cells to generate energy stored in carbohydrates, fats, and proteins. Instead of the organic reactants of the actual cycle, Belousov used cerium to oxidize citric acid in the presence of bromate in a sulfuric acid solution that acted as a catalyst. The ensuing reaction initially appeared to proceed normally, with the cerium ions oxidizing citric acid, causing a change in the color of the solution from yellow to colorless. However, at some point, the process suddenly appeared to reverse, with bromate ions oxidizing the reduced cerium and the solution turning back to the original yellow coloration. This rhythmic change in color occurred for a long time until the reactants were exhausted. For several years, Belousov tried to publish his findings but his manuscript was rejected again and again, with referees and journal editors claiming that the observation appeared to violate the second law of thermodynamics, and therefore, could not be true.* It was only later in the 1960s, when Anatoly Zhabotinsky, then a graduate student at Moscow State University, began investigating this reaction — introducing variations along the way, such as replacing citrate with malonate — that the concept of oscillating chemical reactions began to be accepted in the community. Zhabotinsky also discovered that when the reaction was allowed to take place on a very thin layer of the chemical solution (later, similar experiments would be carried out on gel surfaces) spectacular spiral waves were produced.† While the above systems — viz., catalytic converters, slime molds and the Belousov–Zhabotinsky chemical system — may appear to be very different from each other, it turns out that the phenomena they exhibit, in particular, the dynamics of spiral waves, can be described accurately by

---

*For more details, see A. T. Winfree, *When Time Breaks Down*, Princeton University Press, Princeton, NJ, 1987.

†Zhabotinsky's contributions are summarized elegantly in his obituary by I. R. Epstein, *Nature* 455, 1053 (2008).

closely related excitable media models.

Under certain circumstances, the singularity defining the vortex of the spiral wave can itself drift in space over time. If at any time this meandering motion brings it close to another region of the spiral, the two wavefronts collide and mutually annihilate. This results in multiple free ends of the wavefronts, each of which eventually curls around, forming several coexisting spiral waves. Thus, instabilities in the spiral waves can lead to a state characterized by many coexisting small spirals — which is a manifestation of spatiotemporal chaos — often termed as turbulence. Such a state has not only been observed in real systems, but has also been associated with clinical conditions, such as arrhythmias which are pathological deviations from the normal rhythmic activity of the heart. A certain type of arrhythmia, tachycardia, during which heart tissue is activated at an abnormally rapid rate, is related to the genesis of spiral waves. This was first pointed out in the late 1960s by the biophysicist Valentin Krinsky, who was instrumental in setting up the cardiac biophysics group in the Institute of Biophysics at Puschino near Moscow that produced several internationally renowned scientists in this area. As a result of their research (as well as that of scientists in other countries) it was realized that spiral breakup and the subsequent irregular activity correspond to cardiac fibrillation, during which the heart effectively stops pumping blood and which can be fatal unless controlled within minutes. The absence of any coherent mechanical action of the muscle can be traced to the different regions of the fibrillating heart getting activated in a completely uncoordinated manner. One can illustrate this with an analogy to the Mexican wave seen during many sporting events in a large stadium, where spectators in one section jump up with their hands raised and then sit down again as the people in the next section rise up, repeating the motion, resulting in the wave gradually making its way around the entire stadium.[†] Now imagine that every individual is still standing up and sitting down periodically but without any heed to what their neighbors are doing. The impression of a coherent wave steadily making its way across the rows of spectators will be quickly lost, although at the scale of individuals activity still persists. The only effective treatment of fibrillation to date is to apply a large electrical shock (a process known as defibrillation). One of the goals of understanding the dynamical basis of spiral wave breakup in excitable media is to arrive at more efficient and intelligent methods of controlling fibrillation-like episodes.

So how does one make a start at understanding the mechanisms by which these patterns form? Just as the Ising model is considered to be the paradigmatic model for understanding the transition from disorder to order in systems in thermal equilibrium, where interactions between elements result in collective phenomena, we believe that simple models of excitable dynamics, such as that proposed independently by Richard FitzHugh and Jin-Ichi Nagumo in the early 1960s as a simple

---

[†]For details, see I. Farkas, D. Helbing and T. Vicsek, Mexican waves in an excitable medium, *Nature* 419, 131 (2002).

model for understanding electrical signal propagation along a nerve, and which is discussed extensively in this book, may play a similar role for understanding complex pattern formation in a non-equilibrium setting. The power of such models goes beyond their immediate brief of explaining excitable systems, as they can also be used to understand (i) systems that have no stable equilibria and which are, consequently, characterized by relaxation oscillations, as well as, (ii) those exhibiting "multistable" behavior corresponding to the co-existence of many stable equilibria. An additional source of complexity is the fact that most excitable systems seen in nature possess a significant degree of heterogeneity. This could take the form of gradients in one or more properties, such as conduction or excitability, or the presence of inexcitable regions that could be either localized in a given location or dispersed throughout the medium. The importance of understanding the impact of heterogeneities on the collective dynamics of a medium is underlined by the critical role that scar tissue (or other inexcitable regions, such as valves) plays in the initiation and termination of arrhythmic episodes in the heart. In this book, we have set forth our efforts over the last decade or so toward understanding various facets of excitable media — in particular, how patterns form, the various categories of patterns that can be observed under different conditions and the possible means of controlling them. Some of the chapters are based on research papers that we have published in different journals over this period. Although many of them were written primarily for an audience of physicists and applied mathematicians, the implications of the results reported therein are much broader in scope and should be of interest to a wider audience comprising chemists, biologists, engineers, and clinicians. Along with the introductory material and the supplementary codes for numerical simulations, the material should be accessible to an advanced undergraduate or beginning graduate student wishing to gain an in-depth understanding of the dynamics of patterns in excitable media and the possible means of manipulating them.

We acknowledge, with pleasure, our debt to researchers we have collaborated with over the years (in alphabetical order): Médéric Argentina, Johannes Breuer, C. K. Chan, Christian Cherubini, David J. Christini, Simonetta Filippi, Nicolas G. Garnier, Antina Ghosh, Marcel Hörning, T. Jesan, Kimmo Kaski, Valentin Krinsky, Pik-Yin Lai, Duy-Manh Le, Stefan Luther, Shakti N. Menon, Yun-Chieh Mi, Ashwin Pande, Rahul Pandit, Alexander V. Panfilov, Alain Pumir, Jari Saramäki, Avishek Sen, T. K. Shajahan, Rajeev Singh, Kenneth M. Stein, and Jinshan Xu. We are particularly grateful to Rahul who has been instrumental in getting us involved in this field in the first place. We have also benefitted from discussions with the following: K. R. Balakrishnan, John Cain, Nivedita Chatterjee, Markus Dahlem Milos Dolnik, Igor Efimov, Irv Epstein, Ulrike Feudel, N. Gautam, Neelima Gupte, Sanjay Kharche, Eckehard Schöll, James Sneyd, and Masahiro Toiya. We have been fortunate to be in the very supportive and creative environment provided by our colleagues and students at the Institute of Mathematical Sciences, Chennai. We would especially like to mention Ramesh Anishetty, G. Baskaran, Gautam Menon, Raj K. Pan, and Sudeshna Sinha with whom we have had long, enjoyable discussion sessions, touching on various aspects of the theory and modeling of excitable systems. We also thank the support staff and the systems administrators of the Institute. Access to the high-performance computing facilities — including the Annapurna, Aravali, and Vindhya clusters — at the Institute is gratefully acknowledged. We thank Edward Cox, Wim Lammers, James Lechleiter, Harm Rotermund, and Harry Swinney for graciously providing permission to use figures from their papers in our book. We would also like to thank everyone involved with the production of the book, especially our commissioning editor, Aastha Sharma, who has been extremely supportive throughout the process of writing this book. Last, but certainly not least, we would like to thank our families without whose encouragement and understanding the book certainly could not have been written.

Sitabhra Sinha and S. Sridhar,
Chennai

I

Introducing Excitable Media

Dynamical systems • What is an excitable system? • The heart: A model excitable system for studying pattern formation and control • Outline of the book
Introduction • Hodgkin–Huxley formalism • A phenomenological two-variable model for excitability • "Daughters" of FitzHugh–Nagumo • Cellular automata • Difference equation description of excitability • Modeling natural systems using excitable media models • Physiologically realistic models of cardiac excitation • Spatial propagation • Compartmental models • Numerically solving excitable media models • Role of high performance computation
Introduction • What is a spiral wave? The phase singularity at the spiral core • The genesis of a spiral wave • Excitability of the medium and the nature of the spiral wave • Kinematical approach to spiral wave dynamics • Scroll waves and filament dynamics • Onset of spatiotemporal chaos • Mother rotor and multiple wavelets
Intracellular: Calcium waves • Cellular aggregation: cAMP waves in slime mold • Organ: Waves of spreading depression in the brain and the retina • Organ: Waves of contraction in uterus near term • Chemical medium: Oscillations and spiral waves in Belousov–Zhabotinsky • Solid state devices: Spirals in CO oxidation on Pt(110) surface) • Population: Waves of spreading infection in epidemics • Population: Invasion by parasites of a host population • Ecology: Plankton growth

1

Introducing excitable media

1.1 Dynamical systems

Almost anything around us that evolves over time can be thought of as a dynamical system, the word "dynamic" having the connotation of a force that produces change in its state. The use of mathematics to study motion is a recent venture, usually traced to the development of calculus. Prior to that, the language of mathematics was mostly employed in the study of static patterns and structures, as in geometry that can be traced to the earliest civilizations. One of the first attempts to understand time-evolution is the study of integer sequence generating processes, one of the most famous of which is attributed to the medieval Italian mathematician, Fibonacci (c.1170- c.1250) [Devlin 2011]. In his book *Liber Abaci* (1202), through which the Arabic (or, rather Indian) numerals were introduced in Europe, Fibonacci described a problem concerning a growing rabbit population to illustrate certain aspects of the new system of numbering [Fibonacci 2003]:

> A certain man had one pair of rabbits together in a certain enclosed space, and one wishes to know how many are created from the pair in one year when it is the nature of them in a single month to bear another pair, and in the second month those born to bear also.

Although this may be wanting as a description of how the population of a biological species really changes over time, it is easy to observe that the process described is a simple dynamical system, evolving in discrete time. If at any given time n the number of rabbit pairs is p_n, then the process described above can be represented as

$$p_{n+1} = p_n + p_{n-1}, \tag{1.1}$$

with the initial conditions $p_1 = 1$ and $p_2 = 2$. As the state of the single dynamical variable p at any time depends not only on the state at the previous instant, but also on the state in the instant before that, one can recognize this to be a system with delay. By introducing another dynamical variable $q_n = p_{n-1}$, we can rewrite the above system as a dynamical system without delay (i.e., to determine the state of the system in the immediate future one requires to know only the present state of the system):

$$p_{n+1} = p_n + q_n, \quad q_{n+1} = p_n. \tag{1.2}$$

We now know that the problem in fact is of a much earlier vintage. As Marcus du Sautoy mentions: "... a 6th-century Indian poet called Virahanka was perhaps the first to single them out as significant," in the context of trying to figure out how many possible ways one can combine long and short notes in a sequence of given length. [†]

One obvious drawback of using the above dynamical system as a model for population growth is that it results in an exponential increase. The rate at which the population size diverges i.e., $\phi_n = p_{n+1}/p_n$, asymptotically approaches the golden ratio $(1+\sqrt{5})/2 \sim 1.618$. However, in reality, the increasing population would soon reach a limit set by finite resources available for consumption by the members of the population. Thus, if the population cannot increase beyond a maximum size K (that is a function of the resource available in the surrounding environment) a much more reasonable mathematical description of how population sizes actually change over time is:

$$p_{n+1} = p_n[1 - (p_n/K)], \tag{1.3}$$

or its continuous time analog that was proposed in 1838 by the French mathematician Pierre Francois Verhulst (1804-1849), namely the ordinary differential equation

$$\frac{dP}{dt} = rP\left(1 - \frac{P}{K}\right), \tag{1.4}$$

which is known as the *Logistic equation*. Here, $P(t)$ is the dynamical variable, representing the number of individual members of the population at a given time t, while r and K are *parameters* of the system describing, respectively, the intrinsic reproduction rate and the carrying capacity of the environment (i.e., the maximum number of individuals that can be supported).

As seen in Fig. 1.1 (a), a system following this dynamics will, starting from any initial population size, eventually reach the maximum value allowed by the available resources, K. If the growth rate is varied, the speed with which the population reaches the asymptotic value changes. The population size $P = K$ is therefore a *fixed point* or an *equilibrium* state for the system Eq. (1.4) — where a "fixed point" or an "equilibrium" corresponds to any value of the dynamical variable (P) for which the rate of change (dP/dt) is zero. Moreover, it is a *stable* equilibrium (or an *attracting* fixed point), as any perturbation that takes the system away from K will decay rapidly and the population size will again return to K. This is related to the sign of the rate of change dP/dt around the equilibrium, represented by the arrows indicating the flows in Fig. 1.1 (b). If $dP/dt > 0$ when P smaller than K but less

---

[†]For more details on the relation between rhythm patterns and Fibonacci sequence, see M. du Sautoy, Unlocking the secrets of the sequence, *Times2*, April 15, 2009.

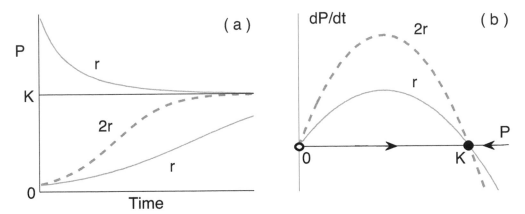

FIGURE 1.1   (a) The time-evolution of the logistic equation, Eq. (1.4), for two different values of the intrinsic reproduction rate (one twice the other). Starting from any initial condition, the population size converges to K, the carrying capacity of the environment. (b) The growth rate dP/dt shown as a function of the population size P at any given time, showing two equilibria: one at $P = 0$ (shown by an unfilled circle to indicate that it is unstable) and the other at $P = N$ (represented by a filled circle to imply that it is stable). The arrows indicate the direction of flow, i.e., whether the population increases or decreases, at any given value of P.

than 0 when P is greater than K, any small deviation from K will trigger changes that will bring it back to K. On inspection it can be seen that the system has another equilibrium at $P = 0$. This means that the population cannot spontaneously grow to finite values out of nothing — there has to be initially a few individuals from which the population will start to grow. However this equilibrium is unstable (i.e., it is a *repelling* fixed point), as a small perturbation, viz., the addition or loss of a few individuals, will result in the system moving away from $P = 0$, as the population starts growing, toward the stable equilibrium at K.

Suppose we want to introduce other factors observed in the growth dynamics of real populations, for example, the dependence of individual fitness on the population size that is referred to as *Allee effect* after the ecologist Warder C. Allee. At very low densities or numbers, a population may become unviable and go extinct once it falls below a critical population size.[†] In order to incorporate this effect, one can augment the logistic equation Eq. (1.4) to:

$$\frac{dP}{dt} = rP(P - \alpha)\left(1 - \frac{P}{K}\right), \tag{1.5}$$

where the earlier quadratic function has been replaced by a cubic one and the new parameter α is the threshold population size below which the population is doomed

---

[†]For more details see F. Courchamp, L. Berec and J. Gascoigne, *Allee Effects in Ecology and Conservation*, Oxford University Press, Oxford, 2008.

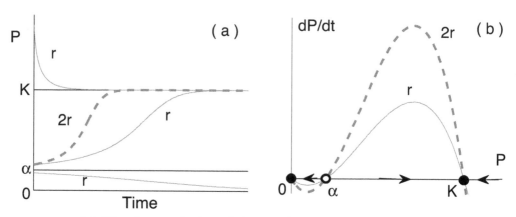

FIGURE 1.2 (a) The time-evolution of the population equation with strong allee effect, Eq. (1.4), for two different values of the intrinsic reproduction rate (one twice the other). If the initial population size is above the critical value α, the population reaches an asymptotic equilibrium corresponding to the carrying capacity K. However, if the initial size is below α, the population is destined for extinction. (b) The growth rate dP/dt shown as a function of the population size P at any given time, showing three equilibria: two stable fixed points at $P = 0$ and $P = K$ (shown by filled circles) and an unstable one at $P = \alpha$ (represented by an unfilled circle). The arrows indicate the direction of flow, i.e., whether the population increases or decreases, at any given value of P.

to go extinct, i.e., go to $P = 0$.

As seen from Fig. 1.2 (a), the state that the population reaches asymptotically now depends on its initial value. If the population size is initially too low (i.e., $P < \alpha$), then it eventually decays to zero resulting in the extinction of the population. However, if the initial size is greater than the critical value α, then the population converges to the finite value K (as in the earlier example). Fig. 1.2 (b) shows that there are now three equilibria, the two stable ones at $P = 0$ and $P = K$ separated by an unstable one at $P = \alpha$. Thus, the population size α acts as a *threshold* for the system dynamics, with the system going to one state or another, depending on whether one begins at a value lower or higher than α. As we shall see, this is an important characteristic feature for excitable systems.

So far we have considered a single population whose growth is affected by factors in the environment, such as availability of nutrients on which the individuals depend for their survival, in which it resides. As the system is described by the dynamics of a single variable, it is referred to as a *one-dimensional* dynamical system. Very often, the environment will also contain other species, whose interaction with the species we are interested in is an important factor governing the growth or decline of the latter population. If we explicitly want to introduce another population comprising, for example, predators that prey on members of the population we originally wished to model, one can use the following pair of coupled differential equations (i.e., a *two-*

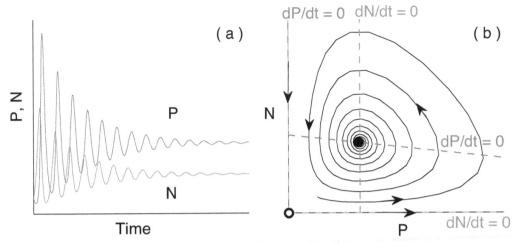

FIGURE 1.3 (a) The time-evolution of the prey (P) and predator (N) populations shows both asymptotically converging to stable equilibria after a series of transient oscillations during which the sizes of both populations rise and fall periodically, out of phase with each other. (b) The system evolution in the (P, N)-phase space. The stable equilibrium to which the population sizes converge eventually is shown with a filled circle, while the unfilled circle at $P = 0, N = 0$ represents the unstable equilibrium located there. The P and N nullclines, corresponding to $dP/dt = 0$ and $dN/dt = 0$, respectively, are shown using broken lines. Each intersection of a P nullcline with a N nullcline corresponds to a fixed point of the system dynamics.

dimensional dynamical system):

$$\frac{dP}{dt} = F_1(P, N) = rP\left(1 - \frac{P}{K}\right) - aPN,$$

$$\frac{dN}{dt} = F_2(P, N) = bPN - dN.$$

(1.6)

Here, P and N represent the population sizes for the prey and predator, respectively. The intrinsic growth dynamics of the prey are described, as in Eq. (1.1), by a logistic term comprising a growth rate r and a carrying capacity K, but that of the predator is represented only by an effective growth rate d. This is because the predator is assumed to be dependent solely on the prey and whose maximum size is limited by the total amount of prey available, so that the explicit introduction of a carrying capacity is considered unnecessary. The interaction between the two species is captured by the terms containing the products of the two dynamical variables, with the coefficients a and b representing the effect predation has on the prey and the availability of prey has on the predator, respectively.

Fig. 1.3 (a) shows the behavior of the system for a particular set of parameter values, for which the populations of both species initially shows a series of oscillations with gradually diminishing amplitude, eventually converging to the stable equilibrium at $P = d/b, N = (r/a)[1 - (d/bK)]$. There are two other equilibria, one at $P = 0, N = 0$ and the other at $P = K, N = 0$, both of which are unstable.

The dynamical behavior of the system can be visually represented in the phase space whose axes correspond to the variables P and N, shown in Fig. 1.3 (b). A qualitative understanding of how the system evolves over time can be simply obtained by considering the *nullclines* of Eq. (1.6), i.e., the curves corresponding to $dP/dt = 0$ and $dN/dt = 0$ (shown in Fig. 1.3 (b) using broken lines). As is easily seen, the P-nullclines are the two lines $P = 0$ and $N = (r/a)[1 - (P/K)]$; similarly, the N-nullclines are the lines $P = d/b$ and $N = 0$. The direction along which the system state will flow at any point in the (P, N)-plane can be obtained from a vector whose direction is governed by the sign of dP/dt and dN/dt at that point. Thus, every time one crosses a P (N) nullcline, the direction of the flow in the horizontal (vertical) direction will reverse. Note that any intersection between a P-nullcline and a N-nullcline is an equilibrium as the rate of change for both the variables will be zero at that point. Thus, we easily determine that the system will have three equilibria: (i) $P = 0, N = 0$ (intersection of the nullclines $P = 0$ and $N = 0$), (ii) $P = K, N = 0$, which is the intersection of the nullclines $N = (r/a)[1 - (P/K)]$ and $N = 0$), and (iii) $P = d/b, N = (r/a)[1 - (d/bK)]$ (intersection of the nullclines $N = (r/a)[1 - (P/K)]$ and $P = d/b$).

As the flow at each point in the (P, N)-plane is now governed by two slopes, dP/dt and dN/dt, the stability of each fixed point (P^*, N^*) is decided by the eigenvalues of the 2×2 *Jacobian matrix*, J, evaluated at the point:

$$J = \left| \begin{array}{cc} \partial F_1/\partial P & \partial F_1/\partial N \\ \partial F_2/\partial P & \partial F_2/\partial N \end{array} \right|_{P=P^*, N=N^*}. \tag{1.7}$$

The eigenvalues λ of this matrix are related to the rate at which small perturbations about the fixed point evolve, such that, if $\delta(0)$ is the magnitude of the initial perturbation, after some time t the deviation from the equilibrium $\delta(t) \sim \delta(0) \exp(\lambda t)$. Thus, if $\lambda < 0$, the perturbation decays rapidly so that the point is stable, while if $\lambda > 0$, the perturbations grow with time, moving the system further and further from the equilibrium. For a two-dimensional system, we need to take into account the sign of both the eigenvalues, to determine how small perturbations about an equilibrium will develop over time. If both the eigenvalues are negative (for a complex eigenvalue, we focus on its real part), then the fixed point is stable and is called an *attractor*, while if both are positive the fixed point is obviously unstable and called a *repeller*. If one is negative and the other positive, the fixed point corresponds to a *saddle* with perturbations decaying along one eigendirection (the *stable manifold*) and growing along the other (the *unstable manifold*). Thus, for a time the system state may seem to approach the fixed point (if it is close to the stable eigendirection) but will eventually move away from the point (along the unstable eigendirection). The existence of saddle points is an important feature of many nonlinear dynamical systems, and in fact is crucial for the control of chaotic systems. When the eigenvalues are complex, the imaginary part of the eigenvalue implies that the solution will have an oscillatory nature. So the occurrence of a complex eigenvalue with a negative real part means that the solution will show transient oscillations with a diminishing amplitude, as seen in Fig. 1.3 (a).

There is, of course, the possibility that the real part of the complex eigenvalues may be zero. This will be seen for Eq. (1.6) when the carrying capacity $K \to 0$ and the system becomes identical to the famous predatory–prey model proposed by Alfred Lotka and Vito Volterra. As is well known in this case, apart from the saddle at $P = 0, N = 0$, the only other equilibrium is the one at $P = d/b, N = r/a$. The eigenvalues of the Jacobian matrix corresponding to this point have no real parts (such fixed points are referred to as *centers*) which implies that it is *neutrally stable*. Thus, a perturbation about this point will not asymptotically move away from it nor will it steadily decay over time. Instead both the populations will show undamped oscillations about $P = d/b, N = r/a$, the amplitude being decided by the initial values of P and N.

Having used the example of population growth to introduce some of the key notions of dynamical systems, we now proceed to explain the basic elements of excitable systems.

1.2 What is an excitable system?

To understand the properties of an excitable element, let us try to delve into what it means to be "excitable." When we say that someone is excitable, we usually mean that if the person is disturbed beyond some threshold, then he gets into a state (*excited* state) that he is normally not in. And once he has been pushed into this excited state, if left to himself he will eventually calm down (return to *resting* state). So the excited state is metastable, in the sense that the system spontaneously returns to the resting state. The resting state, on the other hand, is a stable equilibrium from which the system can be shifted only by a large enough perturbation. Any excitable medium shares these qualities: It has a stable resting state and a metastable excited state; the transition from the former to the latter can only take place if there is an external perturbation whose magnitude is above a certain *threshold*. Another property of an excitable medium is the existence of a *refractory period*. Suppose you anger an excitable person with some stimulus (e.g., snatching away the newspaper he is reading); then when he is exhausted after blowing his temper it requires some time for him to get back to the same state he was in before. If you try to disturb him again while he has still not recovered from the earlier outburst, he will most probably not have the energy to shout at you or do any of the things that he would have done in his excited state. Even if he does, it would be at an intensity much lower than normal. Precisely the same phenomenon is observed in all excitable media. Once the medium has been excited, then it takes a while for it to recover, during which time the system cannot be re-excited by another supra-threshold stimulus.

These two key properties of having (i) a metastable excited state that can be reached only when a stimulation higher than a threshold value is applied, and (ii) a refractory period immediately following excitation during which re-excitation is not possible, even with stimulation much greater than the threshold value, give rise to a number of interesting features. One of these is that traveling waves of excitation

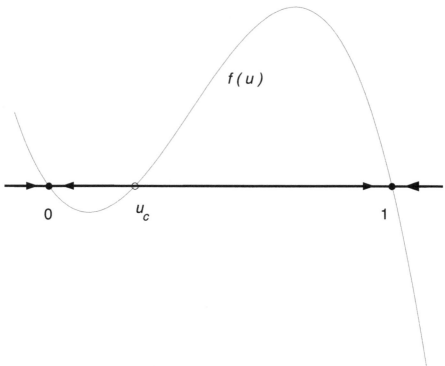

FIGURE 1.4   The function $f(u)$ of Eq. (1.8). The two stable fixed points at $u = 0$ and $u = 1$ are indicated by filled circles, while the unfilled circle at $u = u_c$ indicates that it is an unstable fixed point. Arrows indicate the flow of u at different regions in the u-axis. Arrows converge toward the stable points, while they diverge away from the unstable point.

in such a medium cannot pass through each other. When two such waves meet they annihilate each other, just as in the common technique of fighting fire with fire, when two fire-fronts meet they extinguish each other. The reason is that behind the wavefronts lie regions that are in their refractory periods and cannot be excited. Therefore, the excitation wavefronts cannot cross each other.

To model such a system in terms of equations, we note that we require at least a cubic non-linearity, as there are two equilibrium states of the system. Let us consider the following differential equation:

$$\frac{du}{dt} = f(u) = u(u - u_c)(1 - u), \tag{1.8}$$

which has three equilibria at $u = 0$, $u = u_c$, and at $u = 1$. To check the stability of these equilibria, observe that $f(u) > 0$ if $u < 0$, while $f(u) < 0$ if $u_c > u > 0$; on the other hand, $f(u) > 0$ if $1 > u > u_c$, while $f(u) < 0$ if $u > 1$. This implies that both the equilibria at $u = 0$ and $u = 1$ are stable, while the one at $u = u_c$ is unstable. The location of the initial value of u relative to u_c determines which stable node it will finally converge to (Fig. 1.4). However, such a *bistable* system is not exactly what we are looking for; we need one of the "stable" nodes to eventually become unstable so that the system can return to the resting state, identified with the other

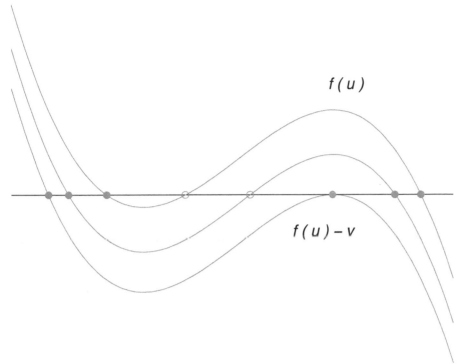

FIGURE 1.5 The function $f(u)$ progressively shifted by v showing a saddle-node bifurcation. The unstable node and one of the stable nodes come closer to each other as v is increased until they collide and disappear, leaving only one stable node.

stable node.

Now, we note that if we subtract a constant u from $f(u)$, the equilibria are displaced along the u-axis (Fig. 1.5). As v is increased from 0, the unstable node (initially at $u = u_c$) and the stable node (initially at $u = 1$) start moving toward each other, until, at a critical value of v they collide with each other. This is an example of a saddle-node bifurcation, in which a stable and unstable node collide and disappear. At higher values of v there is only one (stable) equilibrium, to which the system converges starting from any initial condition.

Eventually, of course, the system should recover so that it can converge to an excited state when starting from a value higher than the threshold. This implies that v is not a constant, but changes with time. Therefore it can be described by another differential equation:

$$\frac{dv}{dt} = \epsilon(ku - v), \tag{1.9}$$

where $k \geq 0$. The parameter $\epsilon(<< 1)$ specifies how slowly v evolves with time compared to u. If $\epsilon = 0$, we recover the system we started with, where v is a constant.

Combining Eqs. (1.8) and (1.9), we now get the complete set of equations for

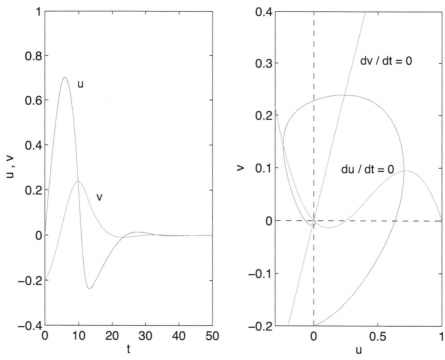

FIGURE 1.6   The time evolution of u and v variables in Eq.(1.10) for $u_c = 0.25$, $k = 1$, $\epsilon = 0.1$ (left) and the corresponding trajectory in the (u, v)-phase plane (right). The nullclines are also indicated. The initial condition (i.e., at $t = 0$) is $u = 0$, $v = -0.2$.

describing an excitable medium:

$$\frac{du}{dt} = u(u - u_c)(1 - u) - v,$$

$$\frac{dv}{dt} = \epsilon(ku - v). \tag{1.10}$$

This model of excitable media is often referred to as the FitzHugh–Nagumo model after the two scientists who had independently developed it in the early 1960s.[†] Fig. 1.6 (left) shows the time evolution of the u and v variables for a particular set of parameter values. In the context of biological cells, this profile is called

---

[†]Richard FitzHugh (1922–2007), a scientist at the Biophysics Laboratory of the National Institutes of Health, Bethesda, USA, proposed the system of equations which he called *Bonhoeffer–van der Pol* model while working out the mathematical basis of the equations proposed by Hodgkin and Huxley for describing the electrical activity in the squid giant axon. See R. FitzHugh, Impulses and physiological states in theoretical models of nerve membrane, *Biophysical J.* 1, 445-466 (1961).

Shortly afterward, Jin-Ichi Nagumo (1926–1999), a faculty member of the Department of Applied Physics in the University of Tokyo, Japan, along with his colleagues, Suguru Arimoto and Shuji Yoshizawa, independently published a similar model. An electronic circuit implementation of the model, called a monostable multivibrator, was built by Nagumo using tunnel diodes that had current-voltage characteristics similar to the cubic non-linearity that characterizes the

the action potential. When plotted in the $u - v$ phase plane (Fig. 1.6, right) we observe clearly the large excursion from the stable state at $u = 0$ that is characteristic of excitation. The dynamics of the model can be qualitatively understood by looking at the nullclines, i.e., the curves representing $du/dt = 0$ and $dv/dt = 0$. The intersection of the two curves at $u = 0, v = 0$ shows that the system has a single equilibrium there. For regions below (above) the cubic u-nullcline, $du/dt > 0(< 0)$, while for regions to the left (right) of the v-nullcline, $dv/dt < 0(> 0)$. So by following the direction of flow given by the above conditions, starting from an initial state located anywhere in the $(u - v)$ plane, one can qualitatively trace the trajectory of the system dynamics.

1.3 The heart: A model excitable system for studying pattern formation and control

In this section, we introduce an excitable biological organ, viz., the heart, that is used throughout the book as a model system for studying the dynamics and control of excitable systems. The heart is an extremely efficient pump that sends deoxygenated blood to the lungs and oxygenated blood to the rest of the body by rhythmic contraction of its two upper chambers (atria) followed by the two lower chambers (ventricles) (Fig. 1.7). The mechanical action of contraction, and subsequent relaxation, is initiated by electrical activity in the excitable cells in the heart wall.

The extracellular and intracellular environment in the heart wall comprises a variety of charged ions such as Na^+, K^+, and Ca^{++}. In the resting state, cardiac cells are in a hyperpolarized state with a membrane potential of ~ -85 mV (Fig. 1.8, right). Small variations in the transmembrane potential that fall below a threshold value are quickly compensated for by the activity of ion pumps in the cell membrane. In addition, the cardiac cell at rest is impervious to charged ions and has an excess of K^+ ions inside compared to the outside; conversely, there is an excess of Na^+ and Ca^{++} ions outside the cell in comparison to inside the cell. However, upon being excited by a sufficiently large stimulus (i.e., a stimulus which is sufficient to increase the membrane potential beyond the excitation threshold, ~ -60 mV), voltage-sensitive ion channels in the cell membrane that are selective for Na^+ ions open up. This allows the passage of a Na^+ ionic current into the cell, increasing the transmembrane potential which, in turn, results in a further increase in the number of open channels until the cell membrane is almost completely depolarized, resulting in a transmembrane potential of ~ 30 mV. This process is extremely fast ($\simeq 10$ ms) and the transmembrane potential shows a steep increase in the *depolarization* phase. Once an equilibrium is reached between the Na^+ concentration outside and inside the cell, the Na^+ current subsides but a slower Ca^{++} ionic current continues to enter

---

FitzHugh–Nagumo equations. See J. Nagumo, S. Arimoto, and S. Yoshizawa, An active pulse transmission line simulating nerve axon, *Proc. IRE* **50**, 2061-2070 (1962).

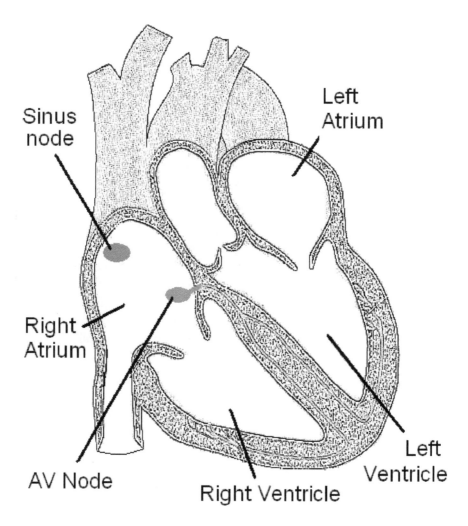

FIGURE 1.7   A cross-sectional view of the human heart showing the main components of electrical excitation: the sino-atrial or sinus node, the source of spontaneous activity and the atrio-vetricular or AV node, a secondary source of spontaneous activation that is normally slaved to the sinus node. Excitation from the sinus node spreads throughout the upper chambers of the heart, viz., the left and right atria, resulting in their contraction. As a result blood from these chambers is pushed into the lower chambers, viz., the left and right ventricles. The AV node is excited by activity emanating from the sinus node (after a time delay). Excitation then spreads from the AV node to the entire ventricle which contracts sending the blood to the lungs and other parts of the body, completing a full cardiac cycle.

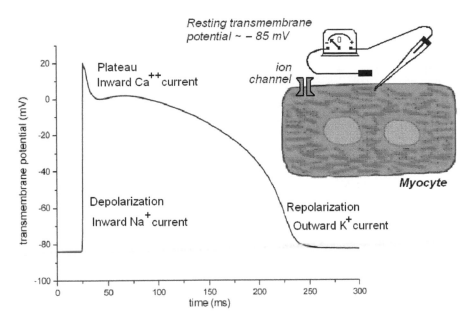

FIGURE 1.8 (right) A schematic diagram for measuring the potential difference across the cell membrane of a myocyte, the principal type of cell constituting the heart wall. At the resting state, the myocyte is impervious to the passage of electrically charged ions and has a transmembrane potential difference ~ -85 mV. On being excited by a suprathreshold stimulus ion channels (selective for specific ions) located on the cell membrane open up allowing ionic currents to pass across the membrane. This results in the characteristic profile of the cardiac action potential (left).

the cell through ion channels selective for Ca^{++} ions. The transmembrane potential is found to decay first and then remain almost constant for some time during this *plateau* phase. Finally, in the *repolarization* phase, ion channels selective for K^+ ions open up and the K^+ ionic current out of the cell causes the transmembrane potential to decrease back to the value corresponding to the resting state. However, the cell can be considered to have fully recovered only when the ion pumps located on the cell membrane restore the original ionic concentration inside the cell. This series of steps constitutes the action potential (AP) (Fig. 1.8, left). In human ventricles, the duration of the AP is ~ 200 ms.

Neighboring cells in the cardiac tissue communicate with each other through low-resistance connections known as gap junctions that allow excitation to propagate from cell to cell through currents due to differences in the cells' membrane potential. Waves of excitation are therefore observed to propagate along the heart wall. Two such waves annihilate each other upon collision because of the existence of a refractory period in cardiac cells. Refractory period refers to the time during which the cardiac cell slowly recovers its resting state properties after an episode of activation. During this time the cell cannot be excited even if a supra-threshold stimulus is applied.

During normal functioning of the heart, the sinus node (a group of cells that

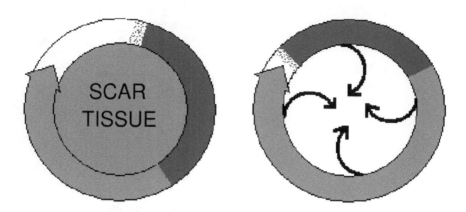

FIGURE 1.9   Reentrant pathways due to a physical obstacle in the core region (left) or due to a temporarily inexcitable zone of refractory tissue in the core region (right). The arrow in the reentry pathway indicates the direction of propagation of the excitation wavefront; the shaded region at the waveback represents refractory tissue that has not yet recovered from the excitation.

are spontaneously excited at periodic intervals) acts as the natural oscillator of the heart, governing its rhythm of activation. The excitation generated by the sinus node propagates throughout the atria from the sinus node. As the ventricles are electrically isolated from the atria, the propagation of excitation to the lower chambers of the heart can occur only via the slow conduction pathway of the AV node. This serves the dual purpose of allowing a time delay between the activation of the atria and the ventricles (thus allowing the blood from the atria to be fully pumped into the ventricles, before contracting the latter), as well as protecting the ventricle from being affected by very rapid excitations in the atria. Once the AV node is excited by the waves propagating from the sinus node, this excitation propagates throughout the ventricles, from the apex (the pointed end of the heart) to the base. Note that the AV node is also capable of spontaneous excitation, but as its natural frequency is lower than that of the sinus node, in normal circumstances it is slaved to the faster rhythm of the sinus node.

The most common reason for disturbances in the natural cardiac rhythm (i.e., arrhythmias) is the formation of a *reentrant pathway*. This involves the sudden occurrence of a feedback loop of excitation around a core region. Reentry can be around a physical obstacle to propagation, with the excitation wave going round and round an existing inexcitable zone, e.g., a region of scar tissue (Fig. 1.9, left). Reentry can also occur around a normal region which was only transiently inexcitable (e.g., during the refractory state while recovering from a premature excitation). These reentrant excitations are typically manifested as spiral waves (Fig. 1.9, right). Once established, spiral waves persist even when the original refractory region has recovered its excitability. The frequency of excitation activity around these reentrant pathways is governed by the perimeter length of the feedback circuit and is, therefore, likely to be much higher than the frequency of excitation arriving from the sinus node. As in a competition between two sources of excitation, the one with the higher frequency always dominates, the heart goes into a state of abnormally rapid activity that is no longer controlled by the sinus node. Even worse, such high frequencies of activation make the excitation wave propagation unstable. Often these waves break up, forming in turn other spiral waves, and this process rapidly multiplies leading to a state of spatiotemporal chaos, that is clinically manifested as ventricular fibrillation.

About half of all cardiac related deaths are due to fibrillation. The natural question is how should we treat a person whose heart is fibrillating. Giving him medicine is not of much use as yet. In fact, designing pharmaceutical drugs meant to prevent cardiac arrest has had rather embarrassing episodes, e.g., the infamous CAST clinical trials where, people being given trial medications were found to be dying at more than twice the normal rate from cardiac arrhythmias! The best option for saving a patient as of now is to give electrical intervention therapy. Familiar to most viewers of medical soap operas on TV, such defibrillation therapy (when applied externally) involves two paddle electrodes connected to a high voltage source being placed on the patient's chest and back. For a very brief duration, hundreds of Joules of electricity are passed across them in an attempt to reset the heart into its resting state, when its normal rhythm can take over once more.

It has become very common to place such devices (which have also become easier to use) in public places in developed nations. Artificial respiration and defibrillation devices can keep a person alive until he can be taken into the ICU of a hospital, enhancing severalfold the survival chances from a cardiac arrest. However, if a person is diagnosed to be at high risk of cardiac disease, it is desirable to do something more to be prepared for the next episode of a cardiac arrest, rather than hoping passively that the next time he suffers a cardiac arrest, a defibrillator will be nearby and that someone with a cool head and courage enough to use it will be close at hand.

So what does a doctor do? The patient gets his own mini defibrillator — a device implanted in his chest that constantly monitors his heart rhythm and delivers a suitable treatment when it detects the onset of cardiac arrest. The only problem with such an internal defibrillator is that it uses extremely high currents so that it is not only extremely painful but also dangerous. Burns caused by such high currents may create scars on the heart wall that can help bring about future arrhythmias.

So the problem that we are faced with is to find a safe defibrillation device that uses much smaller currents. The answer lies in understanding the dynamics of pattern formation in the heterogeneous excitable medium of cardiac muscles.

1.4 Outline of the book

The contents of the book can be broadly divided into three parts. The first part, comprising the first four chapters, including this one, introduces the basic properties of excitable media and their dynamics. The aim of Chapter 1 was to outline the general approach to modeling systems that exhibit excitability. In Chapter 2 we will be looking in detail at the theoretical background of such models starting from the equations proposed by Hodgkin and Huxley for describing nerve signal transmission that forms the foundation to which almost all models of excitable media can be traced back. We introduce the reader to several examples ranging from fairly simple, generic ones to more complicated models that have been proposed to describe in detail specific excitable systems. The strongly non-linear nature of the equations makes it unlikely that such systems can be solved analytically; thus, very often, we need to take the aid of numerical methods to understand how these models behave. A brief outline of the techniques used for solving such equations is given in Chapter 2, as is a brief discussion of the recent widespread use of high-performance computers for this purpose. Chapter 3 discusses the dynamics of patterns seen in excitable media that extend across in space in one or more dimensions. We begin by considering the case of a wave circulating around a ring, which is one of the simplest spatial phenomena associated with excitable media. An important problem associated with this phenomenon is whether it can be terminated by a precisely timed stimulus, which has important implications because of its relation to the clinical use of electrical pacing to treat life-threatening tachyarrhythmias of the heart. We then consider higher dimensional spatial patterns such as spiral waves and scroll waves, and the mechanisms by which they occur in excitable media. We conclude the first part with Chapter 4, which describes many systems occurring in nature that can be characterized as excitable media. Examples as diverse as the intracellular waves of calcium seen in the *Xenopus* oocyte to plankton patches that span across enormous distances in oceans are briefly discussed to give a flavor of the range of scales that excitable media occupy in the natural world.

While in the first part we consider homogeneous media for simplicity of treatment, most systems that we see around us are disordered. Thus, in the second part, we focus on the dynamics of excitable systems displaying various types of heterogeneities. For example, disorder can be manifested in a excitable medium as localized inhomogeneities, e.g., obstacles that do not allow the passage of excitation, or regions which have very different excitation or conduction properties than their neighbors. These are described in Chapter 5, where we show that the evolution of the spatiotemporal patterns can depend sensitively on the precise location of such localized heterogeneities. In Chapter 6, we consider spatially extended disorder such as gradients in local kinetics or conduction. We show that the drift of spiral waves can show unexpected features in the presence of such gradients. These results have important implications for clinical conditions such as "mother rotor" fibrillation characterized by a stationary persistent source of high-frequency excitations giving rise to turbulent activity. Chapter 7 considers another type of spatially distributed disorder where heterogeneities are randomly dispersed through an excitable medium. We show that a variable number of passive cells distributed throughout an excitable medium can result in emergence of collective oscillations when the conduction properties of the system are varied. This can be connected to the sudden transition to coherent contractions in the uterus in the late stages of pregnancy. Finally Chapter 8 considers disorder introduced through altering the topology of connections between the excitable elements constituting the medium. We consider how a complex network of links, comprising both a regular topology of connections between spatially neighboring elements and a set of sparse but long-range connections between a few randomly chosen elements can result in transitions between different pattern regimes as the ratio of these two types of connections is altered. It points the way toward extending the notions of excitable media dynamics to general dynamical systems defined on arbitrary connection topologies.

In the final part, we consider the question of how to control the patterns in excitable media, turning the inherently non-linear nature of such systems to our advantage. The connection of specific patterns with life threatening disorders, for instance, in the case of cardiac arrhythmias, gives special importance to the problem of using weak perturbations to control excitable systems. In Chapter 9 we provide a very brief review of the principle of controlling chaos in dynamical systems, discussing how excitable media differ from other systems in this respect. Chapter 10 looks at how multiperiodic behavior, such as that seen in the spatiotemporal variations in the dynamics of waves propagating in an excitable ring, can be controlled by external signals. In Chapters 11 and 12, we deal with the control of more complex patterns that occur in higher dimensions, spiral waves, and spatiotemporal chaos, respectively. Here we discuss in detail several novel methods proposed recently for terminating activity in a medium, including the use of sub-threshold stimulation as well as generating simulated waves that terminate chaos using an array of spatially distributed electrodes. Several programs that can be used for exploring features of excitable systems written in the MATLAB® programming language are provided at the end.

2

The world of excitable systems

2.1 Introduction

Spatially extended excitable media are a sub-class within the more general framework of reaction-diffusion systems. As suggested by the name, reaction-diffusion models provide a natural description for the dynamics of a chemical system: the reagents are *reacting* with each other and the reactants as well as the products being transported through *diffusion*. Over time, these models have been used to analyze a wide class of spatially extended systems in chemistry, physics, biology, and ecology [Cross and Greenside 2009; Cross and Hohenberg 1993; Murray 2002]. Under coarse-graining, these systems are modeled using partial differential equations (PDEs) having the form:

$$\frac{\partial \mathbf{q}(\mathbf{x}, t)}{\partial t} = \mathbf{D}\nabla^2 \mathbf{q}(\mathbf{x}, t) + \mathbf{R}(\mathbf{q}) \qquad (2.1)$$

where each component of $\mathbf{q}(\mathbf{x}, t)$ represents one of the several variables describing the state of a system (e.g., concentration of a chemical species in case of chemical reactions), $\mathbf{D}$ is the diffusion matrix, and $\mathbf{R}(\mathbf{q})$ represents the different (non-linear) reaction terms. Thus, the term on the right-hand side of Eq. (2.1) represents the transport of the different components while the second term contains details of all the local dynamical processes operating on each of the components including production, decay, etc.

A commonly used analytical tool for understanding the dynamics of non-linear PDEs is to perform linear stability analysis for various solutions. An important example of such analysis carried out for reaction-diffusion systems is that of Alan Turing [Turing 1952]. While trying to understand the mechanisms responsible for morphogenesis, Turing discovered a striking, counter-intuitive effect of diffusion, namely, a homogeneous solution of a reaction-diffusion system can be destabilized by diffusion under certain circumstances. This is surprising as diffusion usually smooths out any spatial fluctuation in a system. This crucial insight of Turing has provided one of the most well-known mechanisms of pattern formation in reaction-diffusion systems and the resulting patterns (see Fig. 2.1) are named after Turing.

Several models have successfully used this mechanism to describe the generation of a wide variety of patterns, e.g., stripes and spots that occur in animal coat patterns [Murray 2002].

Apart from *Turing patterns*, reaction-diffusion systems can exhibit a wide range of other spatiotemporal dynamical behavior such as traveling waves, dissipative solitons, spatiotemporal chaos, etc. While some of these can be explained through an analytical treatment, one has to resort to numerical simulations to study the rest. The first step is the discretization of the Laplacian or diffusion operator for a finite system which turns the space continuum into a discrete lattice. Thus, this process converts a system of PDEs into a large number of coupled ordinary differential equations (ODEs). Diffusion is now represented by the coupling of suitable variables at a given lattice point with those on its nearest neighbors. We will get back to this later in the chapter when we discuss techniques of numerical methods.

Traditionally, the space continuum (and hence the system of PDEs) is assumed to represent reality, while the lattice (correspondingly, the system of coupled ODEs) is considered to be an approximation. However, with modern technology it is possible to investigate systems where the spatially discrete lattice is the more accurate description and the corresponding PDE is an approximation. Examples include recent experiments involving beads containing chemical reactants suspended in a medium within a microfluidic channel or simulations of a system of cells interacting with each other in biological tissue [Singh and Sinha 2013; Toiya et al. 2008]. In these situations it is natural to model the individual beads or cells as a single unit, so that the system is represented as a lattice of dynamical elements. It is important to make this distinction as these recent experiments report observations of phenomena that are natural for a lattice but are difficult to understand in terms of a spatial continuum, e.g., anti-phase oscillations and heterogeneous oscillator death [Singh and Sinha 2013; Toiya et al. 2008].

As mentioned in the Preface, excitable media models form a special class within the framework provided by the reaction-diffusion system. A characteristic feature that distinguishes most excitable systems is that patterns arise through interaction between waves of transient activity that propagates through the medium. One of the first systems where the existence of propagating excitation waves was analyzed in detail is the squid giant axon. Nerve impulse transmission in this cell was studied using electrophysiology techniques by Hodgkin and Huxley. They also provided a basis for describing such activity in terms of mathematical models that forms the basis for all subsequent theoretical work on excitable systems. In the subsequent section, beginning with a discussion of the pioneering work of Hodgkin and Huxley, we present several models that have been used to describe a variety of excitable behavior seen in natural systems.

2.2 Hodgkin–Huxley formalism

Before Hodgkin and Huxley's pioneering work, knowledge about the electrical activity of the cell was fairly limited, primarily due to the fact that it was not possible

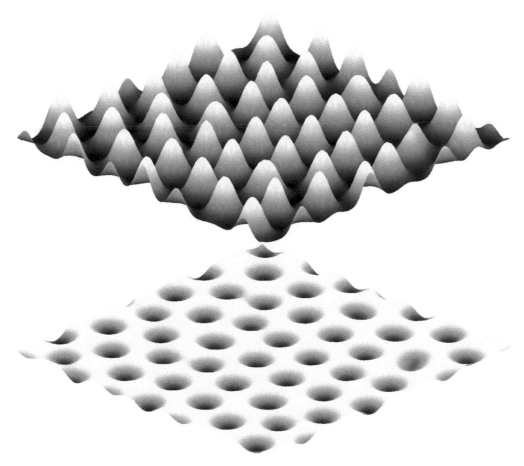

FIGURE 2.1 An example of a two-dimensional Turing pattern produced using the Brusselator model [Eq. (2.19), with parameters $a = 1.5$ and $b = 2.5$] showing the asymptotic spatial distribution of the concentrations for the activator, u (top) and inhibitor, v (bottom) species, the former diffusing much slower compared to the latter (viz., $D_u = 2.8, D_v = 22.4$). The variation of the concentrations across space is manifested as either spots or stripes (depending on parameters and domain geometry) in the dynamical regime that satisfies the conditions for Turing mechanism.

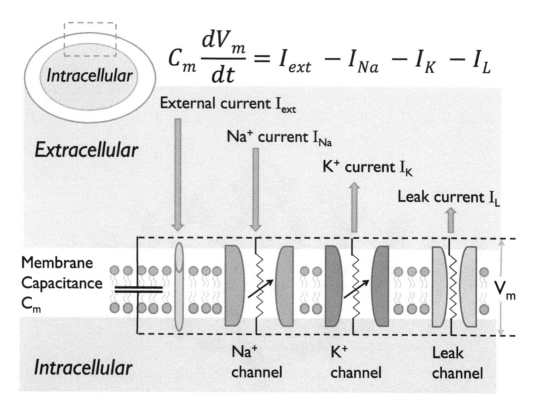

$$C_m \frac{dV_m}{dt} = I_{ext} - I_{Na} - I_K - I_L$$

FIGURE 2.2 Schematic for the equivalent electrical circuit of a cell membrane, (box with broken lines) that isolates the intracellular space from the extracellular space. The lipid bilayer is represented by a capacitor that maintains the potential V_m across the cell membrane and the ion-channels for the inward Na^+ and outward K^+ currents act as rheostats with their conductances varying with voltage. I_{ext} is the externally applied current that depolarizes the cell membrane, resulting in an action potential.

to measure the voltage across the membrane. So while scientists knew that the cellular membrane served to maintain ionic concentration differences and that it was semi-permeable to potassium ions, the role of sodium was unknown. With the introduction of the *space clamp* and *voltage clamp* techniques in the 1940s, it became possible to measure and characterize the cellular membrane potential. In a series of five papers, Alan Hodgkin and Andrew Huxley along with Bernard Katz, described the mechanism of the initiation of action potential in a *squid giant axon* and suggested a theoretical model caricaturing the electrical profile of the cell through an *RC circuit* [Hodgkin and Huxley 1952a,b,c,d; Hodgkin et al. 1952]. For a more detailed historical perspective of their Nobel prize winning work, refer to [Nelson and Rinzel 1998].

In the Hodgkin–Huxley formalism, the biological components of the cell are represented as electrical elements as shown in Fig. 2.2. The lipid bilayer acts as a capacitance C_m that allows charge to build up across the cell membrane, while the different ion channels are represented by the voltage dependent conductances g_i. The transmembrane voltage is V_m. (In this book we often use V interchangeably with V_m.) In such a circuit, the net current is the sum of the components flowing through the capacitive membrane (I_m) and the conducting ion channels (I_i). The ionic term consist of three components, one for each kind of current, namely sodium (I_{Na}), potassium (I_K), and leakage (I_L). Sodium current contributes to the depolarization, resulting in a rapid rise of the membrane potential, while the repolarization is because of the slower flow of potassium ions. One can write the equation of the equivalent circuit as:

$$C_m \frac{dV}{dt} = -I_{ion} + I_{ext}, \tag{2.2}$$

where the term on the left-hand side of the equation represents the current that charges the lipid bilayer membrane, I_{ion} is the ionic currents, and I_{ext} is the externally applied current that depolarizes the cell membrane. In general, I_{ion} is the sum of the various ionic currents, i.e.,

$$I_{ion} = \Sigma I_i, \tag{2.3}$$

with each of the ionic currents given by

$$I_i = g_i(V - E_i). \tag{2.4}$$

Here, E_i is the equilibrium potential or the membrane reversal potential for the particular ionic current. The equilibrium potential corresponds to that value for which there is no net flow of charges across the membrane. An important assumption (which was later proved to be true) in this model is that the conductances of g_i are voltage gated. That is, the activation and deactivation of these channels depends on the membrane voltage. Each of these conductances is assumed to be a result of the activity of many gates that can be either in an ON state or OFF state. Correspondingly, when all the gates of an ion-channel are in the ON state, the channel is considered to be open and if any of the gates are in the OFF state, the ion-channel is closed. In the Hodgkin–Huxley model, the probability that a given gate is in the ON or OFF state is a function of the voltage. The transition between the ON and OFF states is modelled as

$$\frac{dp_i}{dt} = \alpha_i(V)(1 - p_i) - \beta_i(V)p_i, \tag{2.5}$$

where p_i is the probability of the transition of gate i and α_i and β_i are the voltage dependent rate constants for the transition from OFF to ON and ON to OFF states, respectively.

While g_i is the conductance of the individual gates in an ion channel, the macroscopic conductance of the ion channel G_{ic} itself is basically the product of the individual probabilities of a gate being in an ON or OFF state. That is:

$$G_{ic} = \bar{g}_{ic} \prod p_i, \tag{2.6}$$

where $\bar{g}_{ic}$ is the maximum conductance of the ion channel when all its gates are in the ON state. For example, the conductance of the sodium channel is found to be dependent on the gates m and h, and is given by $G_{Na} = \bar{g}_{Na} p_m^3 p_h$.

Before we write the complete equations describing the Hodgkin–Huxley framework, we will introduce a small change in terminology. As in the original paper, we will name the probabilities of the individual gates p_i as m, n, h, etc. Thus the conductance of the sodium channel in the new notation is $G_{Na} = \bar{g}_{Na}m^3h$. Consolidating Eq. (2.2), Eq. (2.3), Eq. (2.4) and Eq. (2.6) we have,

$$I_{ion} = \bar{g}_{Na}m^3h(V - E_{Na}) + \bar{g}_K n^4(V - E_K) + \bar{g}_L(V - E_L) \qquad (2.7)$$

and the equations for each of the gates m, n, and h become,

$$\frac{dm}{dt} = \alpha_m(V)(1 - m) - \beta_m(V)m,$$
$$\frac{dh}{dt} = \alpha_h(V)(1 - h) - \beta_h(V)h, \qquad (2.8)$$
$$\frac{dn}{dt} = \alpha_n(V)(1 - n) - \beta_n(V)n.$$

The rate constants α_i and β_i need to be determined next. In the original experiment this was achieved by clamping the voltage to a fixed value and measuring the time course of the probability of opening or closing of gates. Knowing the asymptotic steady state corresponding to zero rate of change in the transition probability ($\frac{dp_i}{dt} = 0$),

$$p_{i,t\to\infty} = \frac{\alpha_i(V)}{\alpha_i(V) + \beta_i(V)}, \qquad (2.9)$$

and the time taken to reach this value, $\tau_i(V)$

$$\tau_{i,t\to\infty}(V) = \frac{1}{\alpha_i(V) + \beta_i(V)}, \qquad (2.10)$$

allows us to determine the time evolution of the opening and closing of gates. Hodgkin and Huxley then did numerical fitting to the experimental data to determine the form of the conductance for each ion-channel.

It turns out that this formalism to determine the electrical properties is sufficiently general and can be used to describe other kinds of biological cells including cardiac and uterine myocytes. While Hodgkin and Huxley considered only 3 kinds of ionic currents, later variations of the model began to incorporate the effect of the different types of potassium currents and calcium current too.

2.3   A phenomenological two-variable model for excitability

The system of equations representing the Hodgkin–Huxley formalism consists of 3 equations representing the gates m, h, and n and one for the voltage V. Analyzing such high dimensional systems is very difficult, but can be attempted if the number of dimensions is reduced. As already mentioned in Chapter 1, FitzHugh provided a method to understand the behavior of the Hodgkin–Huxley system by proposing a simple model with two variables that phenomenologically reproduces the key features of the former [FitzHugh 1960, 1961].

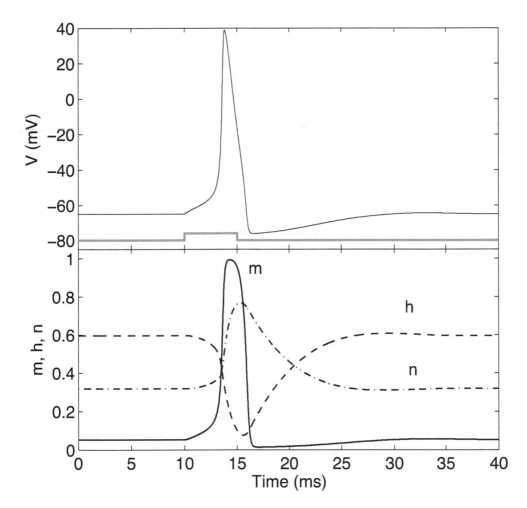

FIGURE 2.3 Dynamics of the variables on the application of an external current in the Hodgkin–Huxley system of equations. The time evolution of voltage V (top) and the dynamics of the different gates m, n, and h (bottom) as a response to a current of strength $4\mu A$ applied for a duration of 5 ms.

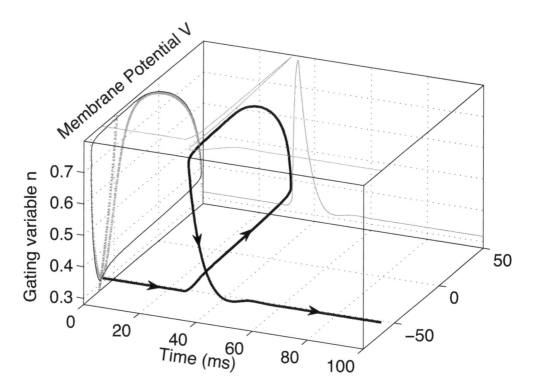

FIGURE 2.4 Phase portrait of the equations corresponding to the reduced model of the Hodgkin–Huxley equations. The nullclines $\frac{dn}{dt} = 0$ and $\frac{dV}{dt} = 0$ are plotted on the $V - n$ plane. On the application of a supra-threshold stimulus the system executes an action potential, the trajectories being marked on the phase-plane by the arrows.

Based on the observation that two of the variables m and V change much faster than n and h, FitzHugh classified them into two groups: fast (m and V) and slow (n and h). That is, the activation of sodium and change in membrane voltage are fast compared to the slow processes of inactivation of sodium and activation of potassium. The dynamics of variable m is so rapid that it reaches the asymptotic state almost immediately and hence FitzHugh replaced the instantaneous value of m with m_∞. Further, he eliminated one of the slow variables h based on the observation that during the course of an action potential $n + h \sim 0.8$. Thus, the original four-dimensional system was reduced to one with two variables V and n. Figure. (2.4) shows the dynamics in the phase plane for the variables V and n and their respective nullclines ($\frac{dn}{dt} = 0$ and $\frac{dV}{dt} = 0$). The arrows indicate the trajectory in the $V - n$ phase space. On the application of a supra-threshold stimulus, the cell depolarizes with a rapid increase in V. The dynamics of the gating variable n determines the slow recovery process that follows the depolarization and the cell repolarizes back to its resting state. This representation of a high dimensional complex system by using just two variables is extremely useful in practice and led to the development of the FitzHugh–Nagumo set of coupled differential equations that we will discuss in detail in the next section.

2.3.1 Modeling action potential with two variables

The FitzHugh–Nagumo (FHN) model qualitatively describes the features of the biological action potential using two variables u and v. The variable u (corresponding to voltage V in Hodgkin–Huxley) is the excitation variable representing the rapid depolarization of the cell on the exceeding of a threshold. On the other hand the recovery of the cell to a resting state is captured by the variable v. Since the process of excitation is much faster than that of recovery, there has to be a parameter ϵ which describes the two time scales involved. To be more precise, ϵ is the ratio of the time scale of excitation to that of recovery, which for excitable media is small. The generic form of such models is,

$$\dot{u} = F(u, v),$$
$$\dot{v} = G(u, v), \tag{2.11}$$

$F(u, v)$ captures the excitation and is given as $u(1 - u)(u - a) - v$, while $G(u, v) = \epsilon(ku - v)$. Just as in Fig. 2.4, the dynamics of the variables u and v can be studied via a phase-plane diagram. Figure 2.5 captures the effect of a supra-threshold stimulus applied to the variable u in the FitzHugh–Nagumo coupled ordinary differential equations. The nullclines of the system, obtained by setting $\dot{u} = 0$ and $\dot{v} = 0$, are described by the broken lines on the u-v plane. One can observe the u nullcline has three branches and resembles the letter N. The v nullcline on the other hand is a straight line and intersects the u nullcline on the left branch. This point of intersection is the stable resting state for this excitable system. The trajectory of the action potential on this u-v phase plane is shown and can be understood as follows. Upon the application of a supra-threshold stimulus the fast variable u increases rapidly and hits the right-most branch of the N-shaped curve. The trajectory then moves very slowly along this branch until it reaches the maximum value and then executes a quick jump to the left-most branch and crawls along it to slowly return to the original resting state.

When the perturbation applied to the u equation is increased, the v nullcline intersects the u nullcline on its middle branch. In this scenario, there is no stable fixed point in the system. The point of intersection of u and v nullclines corresponds to a stable periodic orbit, referred to in non-linear dynamics literature as a "limit cycle." Instead of returning to the resting state after an excitation, the system jumps periodically back and forth between the left and the right branches of the N-shaped u nullcline. Note that this change in the characteristics of the dynamics, from excitable to oscillatory, can also be obtained by shifting the v nullcline to the right instead of applying an external perturbation to the u equation. The equation for $G(u, v)$ is modified to $G(u, v) = \epsilon(ku - v - b)$, where b corresponds to the shift of the v nullcline toward the right. Figure 2.6 shows the onset of oscillations, marked by the periodic trajectory, when the v nullcline is shifted to the right.

Another dynamical behavior observed in the FHN model is the phenomenon of bistability. As the name suggests, this state corresponds to the simultaneous existence of two stable attractors. In the FHN model, decreasing the slope k in $G(u, v)$ can result in the intersection of the v-nullcline and u-nullcline at three points.

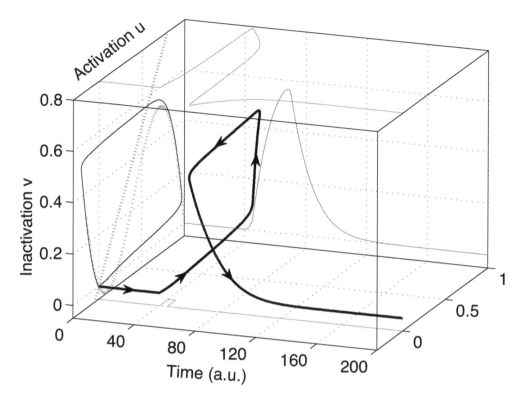

FIGURE 2.5 Phase portrait of the FitzHugh–Nagumo equations corresponding to an excitable system having one stable resting state. On the application of a supra-threshold stimulus the system executes an action potential, the trajectories being marked on the phase-plane by the arrows.

While the middle intersection corresponds to an unstable fixed point, the upper and lower are the stable equilibrium points. Depending on the initial condition, the dynamics of the system will go to one or the other attractor.

2.4 "Daughters" of FitzHugh–Nagumo

The FHN equations have spawned several models that have been used by scientists to describe the various excitable systems in the context of physical, biological, and chemical systems. In the following sections we will describe some of these "daughter" models.

2.4.1 Piecewise linear kinetics

A model where the cubic non-linearity of FHN is replaced by a piecewise linear function ("Puschino" kinetics) was developed by Alexander V. Panfilov to describe the nature of refractory periods in the excitable cardiac cell in a more realistic manner [Panfilov 1998; Panfilov and Hogeweg 1993]. As mentioned earlier, the initiation of action potential is encoded in the non-linear function F, which for the Panfilov

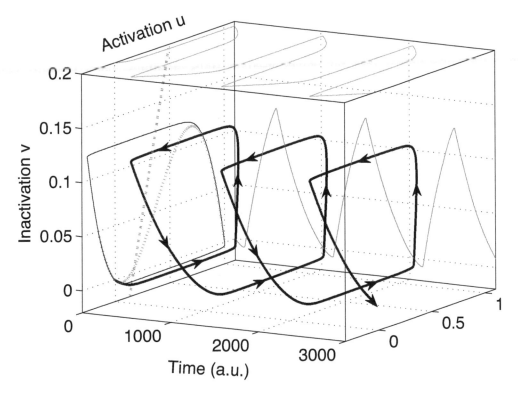

FIGURE 2.6 Phase portrait of the FitzHugh–Nagumo equations corresponding to an oscillatory system having a stable limit cycle. The periodic evolution of the variables u and v are marked, with the arrows indicating the trajectories on the phase-plane. The system is made oscillatory by setting $G(u, v) = \epsilon(ku - v - b)$.

model is described as:

$$F = C_1 u, \quad \text{for} \ \ u < e_1,$$
$$F = -C_2 u + a, \quad \text{for} \ \ u_1 \leq u \leq u_2, \quad \text{and,} \qquad (2.12)$$
$$F = C_3(u - 1), \quad \text{for} \ \ u > u_2.$$

The physiologically appropriate parameter values used in Refs. [Panfilov 1998; Panfilov and Hogeweg 1993] are $u_1 = 0.0026$, $u_2 = 0.837$, $C_1 = 20$, $C_2 = 3$, $C_3 = 15$, $a = 0.06$ and $k = 3$. The function $\epsilon(u, v)$ determines the dynamics of the recovery variable:

$$\epsilon(u, v) = \epsilon_1 \ \text{for} \ \ u < u_2,$$
$$\epsilon(u, v) = \epsilon_2 \ \text{for} \ \ u > u_2, \quad \text{and,} \qquad (2.13)$$
$$\epsilon(u, v) = \epsilon_3 \ \text{for} \ \ u < u_1 \ \ \text{and} \ \ v < v_1.$$

The corresponding parameter values used in Refs. [Panfilov 1998; Panfilov and Hogeweg 1993] are $v_1 = 1.8$, $\epsilon_1 = 0.01$, $\epsilon_2 = 1.0$, and $\epsilon_3 = 0.3$.

The Panfilov model captures several essential features of the spatiotemporal evolution of V in cardiac tissue such as breakup of spiral waves into spatiotemporal chaos [Pandit et al. 2002; Panfilov 1998; Panfilov and Hogeweg 1993]. The model describes the existence of two kinds of refractory periods, absolute and relative. Absolute refractory period corresponds to the time during which, irrespective of the strength of the external signal, the system remains at the resting state. On the other hand, during the relative refractory period, a sufficiently strong suprathreshold stimulus can still generate an action potential.

2.4.2   Models with restitution and dispersion

Important non-linear behavior observed in cardiac cells includes, the phenomena of *restitution* and *dispersion*. Restitution is the non-linear response in the duration of action potential observed during the rapid stimulation (pacing) of cardiac cells. In the cardiac tissue, the conduction velocity of the wave of excitation displays a non-linear relationship with respect to the resting time between two successive action potentials. This relationship is called the dispersion of the conduction velocity. We will discuss more about these non-linear aspects of excitation dynamics in the later chapters. Karma model [Karma 1994] incorporates both the restitution and dispersion effects and has the form:

$$F = \tau_u^{-1}[-u + (u^* - v^M)\{1 - \tanh(u - u_h)\}\frac{u^2}{2}],$$
$$G = \tau_v^{-1}[\frac{1 - v(1 - e^{-Re})}{1 - e^{-Re}} \ \Theta(u - u_n) - v\{1 - \Theta(u - u_n)\}], \qquad (2.14)$$

where u is a dimensionless representation of the membrane potential and v is the effective slow ionic current channel gating variable. $\Theta(x) = 0$ for $x \leq 0$, $\Theta(x) = 1$ otherwise, is the Heaviside step function, and the parameters Re and M control the restitution and dispersion effects, respectively. Increasing Re makes the restitution curve steeper and makes alternans more likely, while increasing M weakens dispersion.

Another model that incorporates details of the restitution curve is the Aliev–Panfilov model [Aliev and Panfilov 1996]. It has the form

$$F = -K(u - a)(u - 1) - uv,$$
$$G = (\epsilon + \frac{\mu_1 v}{\mu_2 u})(-v - ku(u - b - 1)) \tag{2.15}$$

The parameter values of K, a, ϵ, μ_1, μ_2, and b are evaluated based on real experiments.

2.4.3 Barkley model

The Barkley model is a variation of the FHN that allows for rapid computation by exploiting the simplicity of the u nullclines and minimizing the number of floating point operations. The basic form of the model is given by:

$$F = \frac{1}{\epsilon}u(1 - u)[u - (v + \frac{b}{a})]$$
$$G = u - v \tag{2.16}$$

Here ϵ, a, and b are parameters, with a determining the APD and the ratio b/a determining the excitation threshold.

2.5 Cellular automata

So far we have looked at models that describe the excitable dynamics in terms of differential equations. While this is by far the most popular approach to modeling excitable media, there exists another class of discrete state models, namely *cellular automata* that are simpler and chronologically older. Cellular automata (CA) provide a simplified description of the medium, where each excitable unit can only be in a finite number of states. The updating dynamics for every excitable unit depends on its current state and the state of its neighbors. The earliest description of a biological excitable medium, the cardiac tissue, using a cellular automata was provided by Wiener and Rosenblueth in 1946 [Wiener and Rosenblueth 1946]. They considered three dynamical states, namely excited, refractory, and excitable, and used them to formulate a mathematical description of reentrant spiral waves. Greenberg et al., came up with a simple deterministic CA model that had 3 dynamical states [Greenberg et al. 1978]. The update rule for a two-dimensional lattice in this system was that a cell jumps from the resting to the excited state only when at least one of its neighbors is excited. In the successive time steps, the system shifts from the excited to a recovery state and then returns back to the resting state.

2.6 Difference equation description of excitability

A set of two coupled difference equations (i.e., a two-dimensional map) has been proposed by Dante R. Chialvo to describe the dynamics of excitable systems evolving in discrete time intervals [Chialvo 1995]. The time-evolution of the pair of variables x_t, y_t, where x is the activation variable, y is the recovery variable, and t corresponds to a time step is governed by the equations:

$$
\begin{aligned}
x_{t+1} &= f(x_t, y_t) = x_t^2 e^{(y_t - x_t)} + k, \\
y_{t+1} &= g(x_t, y_t) = a y_t - b x_t + c,
\end{aligned}
$$

The model has four parameters: k represents an external stimulus applied to the system, $a(< 1)$ corresponds to a time-constant for the recovery dynamics, $b(< 1)$ determines the activation-dependence of the recovery process and c is an offset. In the absence of any external stimulation, i.e., $k = 0$, the point $x^* = 0, y^* = c/(1 - a)$ is a stable equilibrium of the dynamics. When this system is excited with a supra-threshold stimulation, the fast variable x shows an abrupt increase. This triggers changes in the slow variable y, such that the state of the cell is gradually brought down to that of the resting state. Once excited, the cell remains impervious to stimulation up to a refractory period, the duration of which is governed by the parameter a. Typical parameter values used in several studies are $a = 0.89$, $b = 0.6$, and $c = 0.28$ while k is varied to observe different behavior.

2.7   Modeling natural systems using excitable media models

2.7.1   Deposition of CO on platinum crystal

When a mixture of carbon-monoxide and oxygen, both at extremely low partial pressures, is heated to $\simeq$ 400 K in a chamber containing a single-crystal wafer of platinum [surface Pt (110)], these gases adsorb onto the crystal surface and react to form carbon dioxide, CO_2. The concentration patterns of the adsorbed gases, imaged by Photo Electron Emission Microscopy (PEEM), show target patterns, traveling pulses, states with steadily rotating spiral waves, and turbulent states with spiral creation and annihilation [Jakubith et al. 1990]. In this temperature regime the system displays *excitability*. There is a threshold concentration of oxygen below which the CO covered platinum surface is stable against sub-threshold increases in the local oxygen concentration. A supra-threshold increase triggers a large transient in the local oxygen concentration and may lead to traveling waves and rotating spirals.

We use the Bär model (Ref. [Hildebrand et al. 1995]) for the oxidation of CO on Pt(110) which consists of the following PDEs in two spatial dimensions $\mathbf{x}$:

$$
\begin{aligned}
\frac{\partial u}{\partial t} &= \nabla^2 u - \frac{1}{\varepsilon} u (u - 1)[u - (v + b)/a], \\
\frac{\partial v}{\partial t} &= F(u) - v.
\end{aligned}
\tag{2.17}
$$

where the fields u and v are related to CO and O coverages [Hildebrand et al. 1995], a, b, and ε are control parameters related to rate constants, etc., for the chemical reactions involved, t denotes time, and $F(u) = 0$ if $u < \frac{1}{3}$, $F(u) = 1 - 6.75u(u - 1)^2$ if $\frac{1}{3} \le u < 1$, and $F(u) = 1$ if $u \ge 1$. The numerical studies of Ref. [Hildebrand et al. 1995] have yielded a stability diagram for the statistical steady states of Eq. (2.17) in the $b - \varepsilon$ plane with $a = 0.84$. As ε is increased from 0, say with $b = 0.07$, a transition occurs from a state S, with rigidly rotating spirals, to another state M with meandering spirals; on further increasing ε, M evolves into states $T1$ and $T2$ that exhibit spiral turbulence [Hildebrand et al. 1995].

2.7.2 Oscillating chemical reactions

The Brusselator, proposed in 1967 by Ilya Prigogine's group in Brussels, was the first mathematical model to explain the mechanism of oscillations in chemical concentrations observed in the Belousov–Zhabotinsky reaction [Prigogine and Lefever 1968]. The chemical reaction scheme underlying this model is as follows:

$$
\begin{aligned}
A &\longrightarrow X \\
B + X &\longrightarrow Y + D \\
2X + Y &\longrightarrow 3X \\
X &\longrightarrow E
\end{aligned}
\tag{2.18}
$$

and the rate equations given by,

$$
\begin{aligned}
F &= a - (b+1)u + u^2 v, \\
G &= bu - u^2 v,
\end{aligned}
\tag{2.19}
$$

where u and v are the concentration of the chemical species. The system shows relaxation oscillations for $b > 1 + a^2$.

The Oregonator is another simple model that qualitatively describes the Belousov–Zhabotinsky chemical reaction. A mechanism for the complex chemical steps involved in the Belousov–Zhabotinsky reaction was formulated by Field, Körös, and Noyes. It is this FKN mechanism (named after the first letters of their surnames) that has served as the basis for the development of Oregonator. The initial model captured the five essential steps of the FKN mechanism, namely

$$
\begin{aligned}
A + Y &\longrightarrow X + P \\
X + Y &\longrightarrow 2P \\
A + X &\longrightarrow 2X + 2Z \\
2X &\longrightarrow A + P \\
Z &\longrightarrow fY
\end{aligned}
\tag{2.20}
$$

in terms of 3 coupled differential equations [Field and Noyes 1974], X representing the activator $HBrO_2$, Y the inhibitor Br^- and the oxidized catalyst Z. A and P represent the reactant and product species. The three-variable Oregonator was simplified to a two-variable model by assuming that Br^- varies quasistatically with respect to both $HBrO_2$ and the oxidized catalyst. The resulting two variable model has the form:

$$F = \frac{1}{\epsilon}[u(1-u) - fv\frac{u-q}{u+q}],$$
$$G = u - v,$$

(2.21)

Here u and v correspond to the concentration of the activator ($HBrO_2$) and the oxidized form of the catalyst, f is the stoichiometric factor, and ϵ and q are parameters.

2.7.3 Electrical activity in the heart

Now that we have a set of equations for describing excitable media, the natural question is how do we connect this to the variables that we normally associate with a cardiac cell, i.e., transmembrane potential, ionic currents, etc. It was shown by FitzHugh that the two-variable model is the simplest of a family of reaction-diffusion models of the form:

$$\frac{dV}{dt} = \frac{-I_{ion}}{C_m},$$

(2.22)

where V is the transmembrane potential, C_m is the capacitance across the cell membrane ($\simeq 1\mu F/cm^2$), and I_{ion} is the total ionic current across the membrane. Comparing with Eq. (2.22), we note that the x variable of the FitzHugh–Nagumo model can be identified with the voltage V and y corresponds to the slow repolarizing component of the total ionic current I_{ion}. Different models of cardiac cells differ in the number and details of the ionic currents. The older models invariably described the ion-channel dynamics as deterministic processes. But more recently Markov models have been used to capture the stochastic aspects of the ion-channel dynamics [Greenstein and Winslow 2002]. We shall discuss a few detailed models of cardiac electrical activity in the following sections.

2.8 Physiologically realistic models of cardiac excitation

In this section we shall discuss two single cell models, one describing the electrical activity of a myocardial cell in the guinea pig ventricle and the other, a human ventricular myocyte.

2.8.1 Luo–Rudy I model

Ching-Hsing Luo and Yoram Rudy in 1991 introduced a model for a myocyte in the guinea pig left ventricle [Luo and Rudy 1991]. This model is based on the Hodgkin–Huxley formalism that we discussed earlier in this chapter. The Luo–Rudy I (LRI) model was built on the basis of single cell and single channel data and captures many essential features of real ventricular myocytes. It consists of eight variables and has a large number of parameters that have to be fitted with experimental measurements on cardiac cells. The specific functional form of I_{ion} is the sum of six distinct currents,

$$I_{ion} = I_{Na} + I_{si} + I_K + I_{K1} + I_{Kp} + I_b, \qquad (2.23)$$

The different terms represented in the above equation are: the fast outward sodium (I_{Na}), slow inward calcium (I_{si}), plateau potassium (I_{Kp}), time dependent (I_K), time independent potassium (I_{K1}), and background (I_b) currents. Each of these currents is in turn determined by time-dependent ion-channel gating variables ξ. The time-evolution of these gates is the same as was described for the Hodgkin–Huxley formalism, namely

$$\frac{d\xi}{dt} = \frac{\xi_\infty - \xi}{\tau_\xi}, \qquad (2.24)$$

The parameters in these equations are the steady-state values of ξ, $\xi_\infty = \alpha_\xi/(\alpha_\xi + \beta_\xi)$, and the time constants, $\tau_\xi = 1/(\alpha_\xi + \beta_\xi)$, which are governed by the voltage-dependent rate constants for the opening and closing of the channels, α_ξ and β_ξ, themselves complicated functions of V. The currents in the LRI model are given by:

$$
\begin{aligned}
I_{Na} &= G_{Na}m^3hj(V - E_{Na}), \\
I_{si} &= G_{si}df(V - E_{si}), \\
I_K &= G_K x x_i(V - E_K), \\
I_{K_1} &= G_{K_1}K_{1\infty}(V - E_{K_1}), \\
I_{Kp} &= G_{Kp}K_p(V - E_{Kp}), \\
I_b &= 0.03921(V + 59.87),
\end{aligned}
$$

and $K_{1\infty}$ is the steady-state value of the gating variable K_1. All current densities are in units of $\mu A/cm^2$, voltages are in mV, and G_ξ and E_ξ are, respectively, the ion-channel conductance and reversal potential for the channel ξ. The ionic currents are determined by the time-dependent ion-channel gating variables h, j, m, d, f, x, x_i, K_p, and K_1 generically denoted by ξ, which follow ordinary differential equations of the type

$$\frac{d\xi}{dt} = \frac{\xi_\infty - \xi}{\tau_\xi},$$

where $\xi_\infty = \alpha_\xi/(\alpha_\xi + \beta_\xi)$ is the steady-state value of ξ and $\tau_\xi = \frac{1}{\alpha_\xi + \beta_\xi}$ is its time constant. The voltage-dependent rate constants, α_ξ and β_ξ, are given by the following empirical equations:

$$
\begin{aligned}
\alpha_h &= 0, \text{ if } V \geq -40 \text{ mV}, \\
&= 0.135 \, \exp\left[-0.147 \, (V + 80)\right], \text{ otherwise}; \\
\beta_h &= \frac{1}{0.13 \, (1 + \exp\left[-0.09(V + 10.66)\right])}, \text{ if } V \geq -40 \text{ mV}, \\
&= 3.56 \, \exp\left[0.079 \, V\right] + 3.1 \times 10^5 \, \exp\left[0.35 \, V\right], \text{ otherwise},
\end{aligned}
$$

$$\alpha_j \;=\; 0, \text{ if } V \geq -40 \text{ mV},$$

$$=\; [\frac{(\exp[0.2444\ V] + 2.732 \times 10^{-10}\ \exp[-0.04391\ V])}{-7.865 \times 10^{-6}\{1 + \exp[0.311\ (V + 79.23)]\}}]$$
$$\times (V + 37.78), \text{ otherwise;}$$

$$\beta_j \;=\; \frac{0.3\ \exp[-2.535 \times 10^{-7}\ V]}{1 + \exp[-0.1\ (V + 32)]}, \text{ if } V \geq -40 \text{ mV},$$

$$=\; \frac{0.1212\ \exp[-0.01052\ V]}{1 + \exp[-0.1378\ (V + 40.14)]}, \text{ otherwise;}$$

$$\alpha_m \;=\; \frac{0.32\ (V + 47.13)}{1 - \exp[-0.1\ (V + 47.13)]},$$

$$\beta_m \;=\; 0.08\ \exp[-0.0909\ V],$$

$$\alpha_d \;=\; \frac{0.095\ \exp[-0.01\ (V - 5)]}{1 + \exp[-0.072\ (V - 5)]},$$

$$\beta_d \;=\; \frac{0.07\ \exp[-0.017\ (V + 44)]}{1 + \exp[0.05\ (V + 44)]},$$

$$\alpha_f \;=\; \frac{0.012\ \exp[-0.008\ (V + 28)]}{1 + \exp[0.15\ (V + 28)]},$$

$$\beta_f \;=\; \frac{0.0065\ \exp[-0.02\ (V + 30)]}{1 + \exp[-0.2\ (V + 30)]}, ;$$

$$\alpha_x \;=\; \frac{0.0005\ \exp[0.083\ (V + 50)]}{1 + \exp[0.057\ (V + 50)]},$$

$$\beta_x \;=\; \frac{0.0013\ \exp[-0.06\ (V + 20)]}{1 + \exp[-0.04\ (V + 20)]},$$

$$\alpha_{K1} \;=\; \frac{1.02}{1 + \exp[0.2385\ (V - E_{K1} - 59.215)]},$$

$$\beta_{K1} \;=\; \frac{[0.49124\ \exp[0.08032\ (V - E_{K1} + 5.476)]}{1 + \exp[-0.5143\ (V - E_{K1} + 4.753)]}$$
$$+ \exp[0.06175\ (V - E_{K1} - 594.31]].$$

The gating variables x_i and K_p are given by

$$x_i \;=\; \frac{2.837\ \exp 0.04(V + 77) - 1}{(V + 77) \exp 0.04\ (V + 35)}, \text{ if } V > -100 \text{mV},$$

$$=\; 1, \text{ otherwise,} \qquad (2.25)$$

$$K_p \;=\; \frac{1}{1 + \exp[0.1672\ (7.488 - V)]}. \qquad (2.26)$$

The values of the channel conductances G_{Na}, G_{si}, G_K, G_{K_1}, and G_{Kp} are 23, 0.07, 0.705, 0.6047, and 0.0183 mS/cm^2, respectively. The reversal potentials are $E_{Na} = 54.4$ mV, $E_K = -77$ mV, $E_{K1} = E_{Kp} = -87.26$ mV, $E_b = -59.87$ mV, and $E_{si} = 7.7 - 13.0287 \ln Ca$, where Ca is the calcium ionic concentration satisfying

$$\frac{dCa}{dt} = -10^{-4} I_{si} + 0.07(10^{-4} - Ca).$$

The times t and τ_ξ are in ms; the rate constants α_ξ and β_ξ are in ms^{-1}.

In the first stage of the development, the model had no explicit variable to capture the complicated intracellular Ca^{2+} dynamics. The LR dynamic model that the duo introduced in 1994 had a more detailed description of the various intracellular Ca^{2+} processes [Luo and Rudy 1994]. However in this book we use only the LR I model proposed in 1991. The LR I model reproduces several features that are observed in the real cardiac cells including alternans, Wenckebach periodicities, chaotic activity reminiscent of fibrillation, etc.

2.8.2   TNNP model

The model developed by ten Tusscher et al. (referred to as TNNP) describes the action potential of human ventricular cell [ten Tusscher et al. 2004]. The model incorporates most of the experimentally determined currents including the fast sodium, L-type calcium, transient outward, and the rectifier currents and also captures the basic Ca^{2+} dynamics such as transients and inactivation of calcium current. Thus the I_{ion}, which is the sum of the transmembrane currents for this model is given by:

$$I_{ion} = I_{Na} + I_{K1} + I_{to} + I_{Kr} + I_{Ks} + I_{CaL} + I_{NaCa} + I_{NaK} + I_{pCa} + I_{pK} + I_{bCa} + I_{bNa} \tag{2.27}$$

where the currents not captured in the LRI model include the sodium-calcium exchanger current (I_{NaCa}), Na^+/K^+ pump current(I_{NaK}), plateau currents I_{pCa} and I_{pK} and the background currents I_{bCa} and I_{bNa}. TNNP also captures the differences in the action potentials across the different type of cells such as endocardial, epicardial, and M cells. Further it reproduces the experimentally observed data for important electrocardiological features such as the APD restitution and conduction velocity dispersion. These are some fundamental characteristics in the context of the onset of cardiac arrhythmias and we will discuss them in more detail in Chapter 3.

2.9   Spatial propagation

So far we have looked at models that describe the properties of a single excitable unit. This would correspond to a single cell in biological systems like the heart and a well-mixed homogeneous medium in the case of chemical oscillators like the Belousov–Zhabotinsky systems. But real excitable systems are typically spatially extended and often display dynamics that cannot be observed in single cells. In fact the most interesting properties of excitable media, such as pattern formation and spatiotemporal chaos, are manifested only in spatially extended systems.

As an example of a higher dimensional spatial system, we will consider the cardiac tissue. The heart cells function as a *syncytium* that are electrically synchronized through specialized gap junctions [Clayton et al. 2011]. Although there exists a delay in the spread of depolarization to neighboring cells via the gap junctions, for most practical purposes the discrete nature of cardiac myocytes can be ignored and the propagation of excitation assumed to be continuous. Mathematically, one can move from a discrete description of a single cardiac cell to a spatially extended system by approximating the spread of depolarization via gap junctions across the tissue as a diffusive process. In such a formalism, the individual cells communicate with each other purely through diffusion. The ordinary differential equations describing the electrical activity of the single cell are then modified to obtain a partial differential equation of the form

$$\frac{\partial V}{\partial t} = D\,\frac{\partial^2 V}{\partial x^2} - \frac{I_{ion}}{C_m}, \tag{2.28}$$

where D represents the diffusion constant and captures the spatial propagation of action potentials across a medium. For an isotropic tissue, with properties being the same along all directions, D is a scalar. The value of D is determined as,

$$D = \frac{G_i}{S_v C_m}, \tag{2.29}$$

G_i is the bulk intracellular conductivity and S_v is the surface to volume ratio of cells. The real heart is an anisotropic medium, with the propagation of the action potential along the direction of muscle fibers being two to three times faster than that along the orthogonal direction [Hooks et al. 2002]. For an anisotropic tissue, D is a tensor.

To see why such an equation can describe the passage of excitation across a piece of cardiac tissue, consider a one-dimensional chain of cells, with nearest neighbors connected to each other via gap junctions. Let the transmembrane potential of the n-th cell be V_n and the current from the n-th to the $(n-1)$-th cell be I_n. Then the net current that passes through gap junctions of the n-th cell is:

$$I_{junction} = I_n - I_{n+1} = g_{gap}(V_n - V_{n-1}) - g_{gap}(V_{n+1} - V_n),$$

where g_{gap} is the gap-junction conductance. Using the continuum approximation, we can write

$$I_{junction} = -g_{gap}\frac{\partial^2 V}{\partial x^2},$$

so that

$$C_m\frac{\partial V}{\partial t} = -I_{ion} - I_{junction} = g_{gap}\frac{\partial^2 V}{\partial x^2} - I_{ion},$$

that is essentially identical to the non-linear diffusion equation form given above.

2.10   Compartmental models

So far in our approach to modeling, we have been assuming that the cardiac cell consists of a single domain. The extracellular space has been assumed to be more conducting than the intracellular space and so its contribution to the transmembrane potential has been ignored. The *bidomain* approach on the other hand explicitly considers the cardiac tissue to be comprised of two sections, an intracellular and an extracellular space [Henriquez 1992; Keener and Bogar 1998]. The transmembrane potential in bidomain is explicitly calculated as the difference between the intracellular and the extracellular potentials.

2.10.1   Comparing bidomain and monodomain

While bidomain models certainly provide a more accurate description of the cardiac tissue, it is computationally much more expensive and difficult to solve than the monodomain equations. In terms of computation time, the latter model can be up to 10 times faster than the bidomain [Clayton and Panfilov 2008]. Further, monodomain models reproduce many experimentally observed features, especially with respect to propagation of waves of electrical activity. In fact it has been shown that there is no major difference between the monodomain and bidomain models, as long as large currents are not applied [Potse et al. 2006]. But for situations involving the application of large currents, such as for modeling conventional defibrillation, it is vital to use bidomain models as only these can account for experimentally observed patterns of polarization and depolarization produced due to unequal anisotropy of extracellular and intracellular spaces [Trayanova 2006].

2.10.2   Fibroblasts and other compartments

At this stage one might ask the question, "How about the other components of the extracellular matrix, like fibroblasts. Shouldn't their properties be accounted for in our models?" Fibroblasts are connective cells forming the bulk of the heart and have been recognized to have important roles in the growth and regeneration of cardiac myocytes, especially after incidents of myocardial infarction (commonly referred to as heart attack). Fibroblasts influence the electro-chemical properties of the surrounding cells via coupling [Chilton et al. 2007] and their effects need to be incorporated for developing an accurate description of the electrical activity of cardiac myocytes. Various modeling studies have looked at the coupling between myocytes and fibroblasts [Sachse et al. 2008] and their effect on propagation of electrical activity in one and two dimensions [Jacquemet and Henriquez 2008; Nayak et al. 2013; Xie et al. 2009]. One simple way to incorporate the effect of the fibroblast is to model it as a passive cell, which does not display an action potential, connected to a cardiac myocyte. The equation for the membrane potential for the fibroblast, V_f is given as:

$$\frac{\partial V_f}{\partial t} = \frac{-I_{ion_f}}{C_f},$$

(2.30)

where C_f is the capacitance of the fibroblast and I_{ion_f} the sum of all ionic currents for the fibroblast.

2.11 Numerically solving excitable media models

Solutions to many excitable media models typically involve solving differential equations, either ordinary (ODE) or partial (PDE). Finding analytical solutions to equations with the generic form as in Eq. (2.28) is often not possible and one is forced to rely on numerical solutions. While the models have to reproduce the experimentally or naturally observed phenomenon, their utility extends far beyond that. The most important contribution of any computational study would be predictions and suggestions that would aid in the exploration of as yet unobserved phenomenon in experiments. Sometimes computational studies serve to complement experiments. For example, sophisticated cardiac models are crucial in scenarios where experimental approaches would have severe limitations. An obvious example is the study of electrical activity in the bulk of the heart tissue. A typical cardiological experiment involves the optical mapping of the electrical activity on the surface of the heart. But the heart is a three- dimensional organ and understanding the electrical activity in its bulk is just as crucial. Numerical solutions come to the rescue in this situation allowing us to predict the electrical activity in the bulk of the heart. The caveat here is that the validity of our solution would depend on the extent of realism in the details incorporated in the model. Similarly, modeling fibroblasts as passive units that do not show a depolarization allows us to bypass the fact that there is a lot we currently do not know about the structural and functional features of these connective cells. In this regard, sophisticated techniques have been developed to solve the partial differential equations even for complicated geometry such as the ventricular anatomy. In the following subsections we will give a small introduction to some of the important techniques that are used to numerically solve the ODEs and PDEs describing cardiac dynamics. Our goal is not to give a formal tutorial on computational methods but rather to provide the reader with a flavor for some of the numerical techniques used in this field. The interested reader is encouraged to refer to textbooks on numerical techniques for a more detailed treatment.

2.11.1 Basics of numerically solving an ordinary differential equation

Let us consider a simple system, namely the first-order ordinary differential equation with a single dependent variable,

$$\frac{dy}{dt} = f(y), \qquad (2.31)$$

where f is some function of the variable y. Solving this equation is basically finding the value of y at different times via numerical integration. The typical procedure is to approximate the given function $f(y)$ over some interval by using other functions that can be integrated easily, for example polynomials. Information required for doing this include the initial value of the function and the value of the function's derivatives at certain times, etc. The Euler method is the simplest and most often used technique for such an approximation. In this scheme the value of the function to be integrated ($f(y)$) is determined at a point by approximating it by a straight line passing through it. The slope of this line is given by the first derivative of the function at the point. Thus, given the value of y_0 at time $t = 0$, the equation

$$y_h = y_0 + hf(y_0), \tag{2.32}$$

determines y_h, its value at time $t = h$. By dividing the time interval over which we want to integrate into steps of h, the value of y at any step nh can be evaluated as:

$$y_{nh} = y_{(n-1)h} + hf(y_{(n-1)h}), \tag{2.33}$$

While this is a simple update rule which can be easily implemented (as shown in Code 1 in the Appendix), it has several limitations. For the non-monotonic functions often encountered in non-linear systems, the process of approximating the function by a straight line is not always a good idea. This is because the error in the accuracy of the Euler scheme goes as the order of h. So, to minimize the error, the time step has to be very small. The trouble with choosing very small values of h is that the computation time then becomes very large as the number of integration steps increases substantially. The Runge–Kutta scheme, an improvement over the Euler method, was proposed in the 1900s and is a higher order scheme, requiring more calculations but giving more accurate results. To determine the function $y_{(n+1)h}$ from y_{nh} we first calculate the four numbers, $k_1 = hf(y_{nh})$, $k_2 = hf(y_{nh} + k_1/2)$, $k_3 = hf(y_{nh} + k_2/2)$, $k_4 = hf(y_{nh} + k_3)$. The equation for the evolution is thus the weighted average, given by

$$y_{(n+1)h} = y_{nh} + \frac{1}{6}(k_1 + 2k_2 + 2k_3 + k_4), \tag{2.34}$$

and the error in this method is of the order of h^4, much lower than the error in the Euler scheme which usually goes as h.

2.11.2 Basics of numerically solving a partial differential equation

ODEs are not a sufficient tool to describe the phenomenon of pattern formation in excitable systems. Spatial evolution of these systems can be described only by using partial differential equations (PDE) which are computationally more challenging especially when the system under consideration has several variables and needs to be solved on more than one spatial dimension. In the section on high performance computing we will discuss some approaches to reducing the computational time involved in solving these higher dimensional partial differential equations. For the moment though we will briefly outline a standard technique to solve a PDE, namely the "finite difference scheme." Let us consider Eq. (2.28) on one-dimension,

$$\frac{\partial V}{\partial t} = D\ \frac{\partial^2 V}{\partial x^2} - \frac{I_{ion}}{C_m}, \tag{2.35}$$

The second term on the right-hand side are the ODEs that will be integrated using schemes like Euler and Runge–Kutta. To solve the spatial term $\frac{\partial^2 V}{\partial x^2}$, we first discretize the system on a grid or lattice. That is, we divide the domain into points separated by a spacing dx. We then approximate the spatial derivatives by a difference equation,

$$V(x, t + dt) = V(x, t) + [V(x - dx, t) - 2V(x, t) + V(x + dx, t)]\frac{dt}{dx^2}, \qquad (2.36)$$

with dt being the integration time step. Note that we have chosen the neighboring points $x + dx$ and $x - dx$ to evaluate the value of V at point x. This choice of the geometrical arrangement of points that contribute to the evaluation of a variable at a given position is called a stencil. In two-dimensional systems, a commonly used stencil is the five-point stencil, wherein the points $x + dx$, $x - dx$, $y + dy$, and $y - dy$ are the chosen neighbors that determine the value of the variable at point (x, y). The corresponding equation for the evaluation of $V(x, y)$ (choosing $dx = dy$) is

$$V(x, y, t + dt) = V(x, y, t) + ([V(x - dx, y, t) - 2V(x, y, t) + V(x + dx, y, t) +$$
$$V(x, y - dy, t) - 2V(x, y, t) + V(x, y + dy, t)])\frac{dt}{dx^2},$$
$$(2.37)$$

To obtain a more continuous and smoother spatial description in two dimensions, higher order stencils, such as the 9-point stencil, can be used. This approach of discretizing on uniform grid can be extended to a non-uniform grid too. While there are other more sophisticated methods like the finite element method that allows modeling of complicated surfaces, the biggest advantage of the finite difference scheme lies in its simple and straightforward implementation.

2.12   Role of high performance computation

We have seen that solutions of spatially extended excitable media involve solving partial differential equations, often for very large systems. Typical heart models consist of several equations. For example, the equation for the time variation of the transmembrane potential (Eq. 2.28) of the Luo–Rudy I model for a single cell would involve the integration of 8 differential equations at every time step. For a two-dimensional medium we typically simulate a system of size 400×400. To observe interesting dynamics, one would have to run the simulations for at least 10^5 to 10^6 iterations. Thus a typical simulation even for this relatively simple system could run on a single machine for hours. More realistic models of the heart would have an even larger number of equations. While scientists have developed many smart ways to reduce the computation time, such as by using look-up tables wherever pre-computation of values is possible, it turns out that this is still not sufficient. And often one might have to consider several different initial conditions.

One possible way to decrease computational time is to perform a parallel implementation of the program. The term "parallel computing" refers to the process of running the program on multiple processors, as opposed to serial computing where the code runs only on a single processor. Broadly, there are two ways to parallelize the numerical program. The easier (and in some sense dumber) way to parallelize

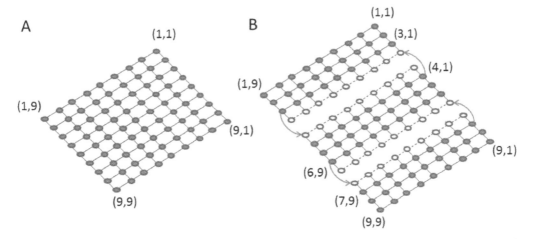

FIGURE 2.7 Parallelization via 1-D domain decomposition. (A) A domain of size 9×9 points (gray) is (B) divided among 3 processors, each of which has to now perform computations for 3×9 points. The boundary points on each node send and receive data from their neighbor. The data received from the adjacent processor is stored in the ghost points (open).

is to run the code separately on multiple nodes of a computational cluster or on parallel machines with shared memory. While this is a very useful technique to explore the parameter space of a model in detail or run several initial conditions simultaneously, it does not serve to reduce the computational time for an individual simulation.

The other kind of parallelism is based on the technique of Message Passing Interface (MPI), an application interface that allows communication between programs running on different machines. Typically this involves the usage of High Performance Computation (HPC) machines with distributed memory where each processor retains a copy of the section of the variables. The original program is modified to include instructions that enable the division of the simulation domain into smaller components, in a process called "domain decomposition." The different domains can then communicate with each other when necessary. As a simple example let us consider the domain shown in Fig. 2.7. Using MPI the original lattice of size 9×9 is divided into 3 domains with each domain of size 3×9 allotted to one processor. For a system of size $L \times L$, each processor would have $L \times (L/N)$ grid points at which it would have to do the computation. Since calculating the Laplacian term would require the knowledge of the state of the neighbors, the grid points at the boundary of each domain would have to receive data from the neighboring processor. Similarly, data of its own boundary points would have to be sent to its own neighbor. To enable this, we need to allocate "ghost points" to every processor. In Figure. (2.7), the open circles correspond to the ghost points which store data received from the neighbors. In order to transfer the data from one processor to another during every iteration of the program, a sequence of "Send" and "Receive" commands have to be executed. An efficient domain decomposition would involve choosing the right num-

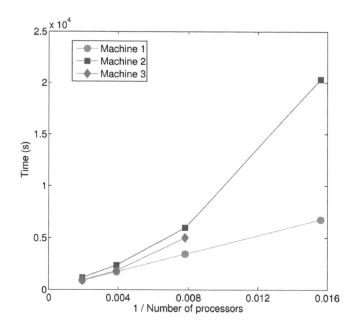

FIGURE 2.8 Computational time for running 5000 iterations of the LRI model as a function of the number of processors used to parallelize the domain. The system is a 3-dimensional cube of size $512 \times 512 \times 512$. The scaling results are shown for 3 machines with different architecture.

ber of processors that would be able to balance the time for computation at each node and communication between different nodes. We will consider one example, that will make this point clearer. Figure (2.8) compares the scaling of computation time as a function of the number of processors used. The system numerically solved is 50 ms of LRI dynamics on a $512 \times 512 \times 512$ domain. The run times are compared for computational clusters with different architectures. While there is considerable variation in the computation times across different architectures for a small number of processors, the performance of the machines becomes comparable as the number of nodes increases. For the machines indicated by the square and diamond, the load between the computation and communication time is not balanced when the number of processors is small. More time is spent on computation at each node and less on communication between nodes. But as the number of processors increases, this imbalance begins to be rectified. For a linear scaling in performance, the communication and computation time have to be balanced, as is seen in the case of the machine indicated by the circle.

3

Dynamics of patterns in excitable media

3.1 Introduction

In this chapter we will discuss the theory of pattern formation in excitable media, specifically spirals and scrolls. Spirals are self-organized two-dimensional waves of excitation whose front can be described as an involute of a circle. In three dimensions, the corresponding waveform is called a scroll, which can be thought of as contiguous spirals stacked on top of each other. The two-dimensional cross-section of the scroll would thus be a spiral wave. In this chapter, we will describe the origin and dynamics of these patterns of excitation in homogeneous excitable media. Here the word homogeneous implies that the properties of the medium such as its excitability, diffusion coefficient, ionic currents, and conductances, etc., are uniform and do not vary across the medium.

3.2 What is a spiral wave? The phase singularity at the spiral core

The earliest understanding of the formation of the spiral was an extension of the idea of a one-dimensional reentrant wave, which is the rotation of a single pulse of activity around a circular region [Mines 1913; Wiener and Rosenblueth 1946; Winfree 2000]. Based on a series of path-breaking experiments carried out on animal hearts, George R. Mines proposed the idea that if the wavefront of electrical activity in an excitable tissue can propagate in a complete circuit it can result in persistent periodic activation of the system. Figure 3.1 adopted from a pioneering article published more than a hundred years ago by Mines illustrates the principle underlying such a one-dimensional reentrant circuit in excitable media.

In two and three dimensions, reentry is related to the idea of a spiral or a scroll wave rotating around an inexcitable region or obstacle [Wiener and Rosenblueth 1946; Winfree 2000]. Here, the spiral wave of excitation is anchored or pinned to

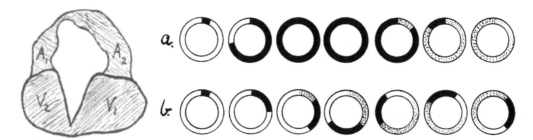

FIGURE 3.1   (left) Ring of cardiac excitable tissue created by G. R. Mines from a tortoise heart. After stimulating the system several times, a cycle of propagating activity resulting in successive contractions of V_1, V_2, A_1, A_2, respectively was observed. The cycle occurred over and over again without any additional stimulation indicating that a self-sustaining wave of excitation was circulating around the system. (right) Schematic representation of the spatiotemporal evolution of activity in the ring following a stimulation that results in a unidirectionally propagating wave. The top panel (a) shows that if the conduction speed is rapid so that the wave returns to its initial position while the latter is still in the refractory state, the excitation dies out. The bottom panel (b) indicates that for slow conduction speed, when the initial starting position of the wave recovers by the time the wave comes back after completing one circuit, reentrant activity will become persistent. The region in absolute refractory state (i.e., which cannot be excited by any supra-threshold stimulus) is indicated in black, while those in relative refractory state are indicated by dots. This figure is adapted from G. R. Mines, On dynamic equilibrium in the heart, *J. Physiology (London)* 46 (1913) 349-383.

the obstacle and repeatedly excites the medium as it moves around it. The refractory period of the medium determines the minimum critical size of the obstacle that would support such repeated excitations. The obstacle has to have a circumference larger than this critical size, as otherwise the excitation front would hit the recovering medium behind the previous wave. This phenomenon is called *anatomical reentry*. In the 1960s Russian scientists working at Puschino led by Valentin Krinsky showed that the presence of an inexcitable obstacle is not a necessary condition for two-dimensional reentrant waves of excitation and even a completely homogeneous medium can sustain such activity. In this case, the dynamics of the reentrant "free spiral" is determined purely by the refractory properties of the medium and this phenomenon is known as *functional reentry*. The region determined by the refractory properties, around which the spiral rotates is referred to as the *core*.

3.3 The genesis of a spiral wave

We explained in Chapter 1 that an excitable system is characterized by the presence of one stable resting state, a threshold and a refractory period. In spatially extended systems (i.e one-dimension and above), these properties can result in the observation of distinct features. Specifically, the phenomenon of patterns of excitation is a characteristic feature of spatial extended excitable media. Spatial patterns in an excitable system arise from the distinct property of mutual annihilation on collision of interacting excitation waves. This property is a consequence of the refractory period of the excitable medium. For a traveling wave of excitation, this corresponds to the region immediately behind the wavefront. This refractory region (also referred to as the waveback) consists of cells that are in the recovery phase and cannot be stimulated by another excitation front. As a result, two fronts on collision cannot penetrate into the recovering waveback of the other and hence mutually annihilate.

These spatial patterns are referred to variously as *reentrant excitations* (in one-dimension), *vortices* or *spiral waves* (in two-dimensions), and *scroll waves* (in three-dimensions). When an excitation wave propagates across partially recovered tissue or encounters an inexcitable obstacle [Jalife et al. 1999], it can form a broken wavefront. This broken wavefront has two velocity components, one along the direction of motion of the plane wavefront (longitudinal) and the other in the direction perpendicular to the motion of the plane wavefront (transverse). The asymmetry in the magnitude of the longitudinal and transverse velocities results in the gradual curling around of the free end of the broken wavefront to form a spiral wave as in Fig. (3.2). Figure 3.3 depicts another technique often used in experiments, termed as S1S2 protocol, that is used to generate spiral waves. This method exploits the condition that there exists a transient heterogeneity in the medium in the wake of an excitation wavefront S1. The second stimulus S2, if applied prematurely, i.e., before the complete recovery of the medium, can result in what is called a "figure of eight," with two spiral hooks.

Once formed, these waves evolve to become self-sustained sources of high-frequency excitation in the medium, and usually can only be terminated through ex-

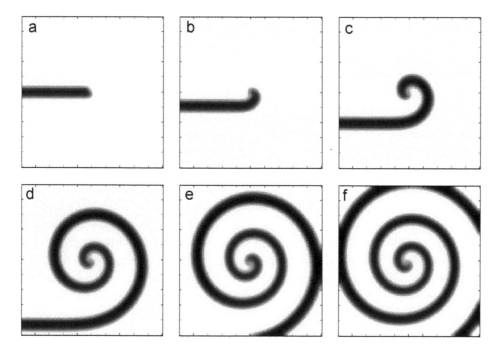

FIGURE 3.2 The formation of a spiral wave from a broken wavefront in the Bär model. The dark region corresponds to the excited wavefront. A broken wavefront is produced by breaking a plane wave (a), which develops velocities in both longitudinal and transverse directions and in the process begins to curl (b)-(c) and forms a fully developed spiral with multiple turns (d)-(f).

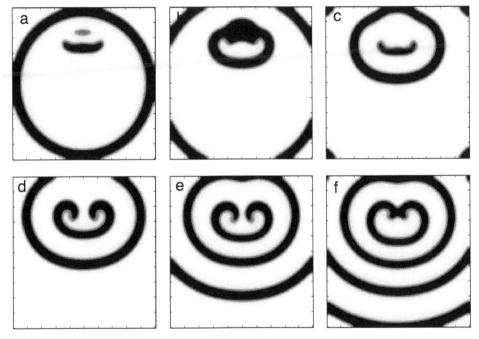

FIGURE 3.3 T1T2 protocol to generate a spiral wave in the Bär model. The first stimulus is applied at time T1 at the center of the domain (not shown in the figure). This is followed by a premature second stimulus given at time T2 over a different set of points that is not fully recovered (a). The wave generated by T2 initially propagates only in one direction (b). This results in the formation of a two-hook structure (c), that evolves into two spirals (d-e).

ternal intervention. In general, spiral waves are associated with periodic or quasiperiodic patterns of temporal activity. Non-linear properties of excitable media, such as the dependence of the action potential duration as well as the conduction velocity on the time difference between two successive excitations can result in more complex chaotic or non-chaotic spatiotemporal rhythms. We will discuss some of these complex rhythms in Chapter 10.

3.3.1 Geometry of the spiral

A simple way to understand the dynamics of spiral waves is to study it as a motion of two curves, one representing the excitation front and the other the recovering waveback. While for a plane wave the front and the back do not touch, this is not true in the case of a spiral. The spiral wave is unique in that, there exists a point where the excitation front and recovering back actually join. This point is a phase singularity, referred to as the spiral tip. Unlike in the case of the plane or circular wavefront, the curvature of a spiral wave is not a constant throughout. Instead, it varies continuously along the wavefront, becoming maximum at the tip of the spiral. The curvature at the spiral tip exceeds the maximum critical value that can result in the propagation of the wave along the transverse direction. This creates a zone of zero conduction, as the tip does not have enough strength to invade into this core region. Not surprisingly, this unexcited region is called the spiral core.

The formation of the spiral core is a consequence of the source-sink mismatch created due to the maximum curvature of the wavefront at the tip.[†] But what does curvature have to do with diffusion of excitation? The strength of the diffusive current depends on the number of cells in the resting state (sink) that are potentially excitable relative to the number of already excited cells (source). For example, each excited unit in a plane wavefront (curvature = 0) has to pass current only to one cell in front of it. On the other hand, if the wavefront is convex (curvature > 0), each unit has to excite more than one cell. Thus current load on the convex wavefront is larger than that in the case of a plane wave. This means that the convex wavefront propagates more slowly than the plane wave. When the curvature of the convex front becomes very large, this causes a source-sink mismatch creating a region of zero propagation. The spiral core is such a region of zero propagation which is potentially excitable but not actually excited. One might ask the question: What happens when the curvature becomes negative? Such waves would have a concave shape and their velocity of propagation would be higher than that of the plane wave.

[†]Historically the idea of the source-sink relationship determining the dynamics of the spiral wave was inspired from the concept of a liminal length that was introduced by Rushton [Rushton 1937]. Applying Kirchoff's law to a one-dimensional network of nerve cells (constituting a nerve fiber) that were represented as a series of connected condensers and resistors, Rushton established that there exists a minimum length of the fiber that has to be excited simultaneously for the successful propagation of a wave of excitation. No propagating wave will be set up, in case this minimal length of cells is not simultaneously stimulated. The existence of this length was later shown in cardiac Purkinje fibers by Noble [Noble 1972].

Experimentally the relationship between curvature and wavefront propagation has been explored in both chemical and biological systems. Cabo et al. demonstrated the dependence between curvature and dynamics of wave propagation in an isolated sheep ventricular epicardial muscle sheet [Cabo et al. 1994]. Using the similarity between waves diffracted through a narrow isthmus and those produced via a point stimulation, they established the relation between curvature and velocity of propagation. Wavefronts with negative curvature have been observed in both chemical and biological systems [Foerster et al. 1988, 1990]. Let us now discuss the spiral core in some more detail.

3.3.2 Spiral core

Th spiral core can be defined as the trajectory traced by the "spiral tip," which is the intersection of the excitation wavefront and the recovery waveback. Mathematically, this is a singularity in phase, i.e., the phase at that point is undefined. Several techniques have been developed to determine the location of the tip and use it to track the spiral core. One simple technique exploits the fact that the point at the tip sees all possible phases of excitation. For example, these phases could be broadly classified as refractory, excited, excitable, etc. The tip is determined by identifying those cells that have neighbors in each of these possible phases of excitation [Biktashev et al. 1994]. One disadvantage of this method is that the accuracy of the computation of the core depends on the spatial resolution. Another way to determine the tip is to find intersection between the contours of two variables. For example in the case of the Barkley or FHN model this would mean finding the intersection of the contours of the fast and slow variables [Barkley et al. 1990]. This method has the advantage that it is not very sensitive to spatial resolution and can be used to determine the spiral core in FHN-like models where a clear slow variable exists. Exploiting the fact the velocity at the spiral tip is zero (because of maximum curvature), Fenton and Karma [Fenton and Karma 1998] used an arbitrary isopotential line separating the excitation front from the recovering back and determined the points having zero normal velocity. These points together form the spiral core. This method has been used quite extensively as it is robust with respect to spatial resolution and can be used to determine the spiral tip even in models where the slow dynamics is not governed by one single variable.

3.4 Excitability of the medium and the nature of the spiral wave

A spiral wave can be characterized by how closely the turns of the spiral are packed, i.e., its wavelength and its period of rotation. These properties depend on the excitability of the underlying medium. Excitability as defined earlier, determines the ease with which an excitation can be elicited by a supra-threshold stimulus. The dynamics of the spiral wave varies with the excitability of the underlying medium. Spiral dynamics depend crucially on the curvature of the spiral tip. This, in turn,

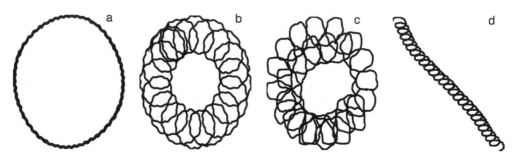

FIGURE 3.4 A "bestiary" of trajectories traced by the phase singularity on the tip of the spiral wave produced using the Barkley model. As parameter a is varied, $(b = 0.05)$, the trajectories of the spiral tip take on various forms such as (a) large circular core corresponding to low excitability, (b) inward petals, (c) outward petals due to a quasiperiodic transition, and (d) linear drift like behavior.

is determined by the local excitability of the medium. Thus, in a spatially extended system, the effect of the excitability of the medium would manifest in terms of the size and geometry of the trajectory of the spiral tip (spiral core). For a weakly excitable system, the spiral core would be very large and circular (Fig. (3.4)(a)). As the excitability increases, spirals undergo a supercritical Hopf bifurcation that results in the introduction of an additional frequency. This second frequency leads to a quasiperiodic motion or meandering of the spiral core. The refractory tail of the spiral causes the periodic alteration of the excitability of the medium which subsequently results in the production of complicated patterns. Depending on the ratio of the two frequencies, the pattern could change from an epicycloid (Fig. (3.4)(b)) to a hypocycloid (Fig. (3.4)(c)). The meandering motion could also take on a linear translational nature where the spiral core drifts along a particular direction (Fig. (3.4)(d)). A further increase in excitability leads to a transition to a linear core, where the radius of rotation due to the larger frequency is actually smaller than the minimum spiral wavelength. In terms of morphology, the spirals in a low-excitability domain are thin and sparse, i.e., its wavelength is large. The thickness and number of turns of a spiral increases with excitability. At the other end of the spectrum is the sub-excitable medium. In this limit, a broken wavefront does not become a spiral. This is because the transverse velocity of the broken wavefront for this medium is negative. This means that while an unbroken plane wavefront can propagate in the sub-excitable medium, a broken wave will retract and shrink with time. Thus, a sub-excitable medium will not support the formation of a spiral.

3.4.1 An example

In both abstract and realistic models of excitable media, the dynamics of a spiral changes from periodic rotation to meander to spiral breakup as one or more parameters are varied. In the process, spirals can display behavior, ranging from rigid rotation to meander, drift, and finally break up into a spatiotemporally chaotic state, manifested via the formation of multiple spiral fragments. We will discuss the mechanisms of spiral drift and breakup in Chapters 6 and 5, respectively. The time series of the excitation variable or voltage for a spiral wave can be used to reveal the underlying dynamics of the system. For a spiral wave doing a simple rotation, the time series is regular and periodic. On the other hand for the meandering spiral, the time series would indicate the presence of more than one frequency. A fully chaotic state with multiple spirals would show a highly aperiodic time-series, with a broad spectrum of frequencies present. We will discuss the quantification of this chaotic state in Chapter 12. Figure 2 in Ref. [ten Tusscher and Panfilov 2003a] is an example of the change in the dynamics of the spiral core as one parameter, the conductance of the slow inward calcium current, is varied in the LR I model. As G_{si} is increased the spiral core changes from circular to hypocycloidal with the size of the meandering core becoming larger with $G_s i$. Increase in $G_s i$ beyond 0.045 results in spiral breakup and transition to chaos (not shown in the figure).

While we have so far discussed the qualitative aspects of the dynamics of spiral waves, there exists a mathematical formulation for the propagation of waves of excitation, based on a kinematical approach, referred to as the *Eikonal equation*. We will discuss these equations in the next section.

3.5   Kinematical approach to spiral wave dynamics

To understand the formation of the spiral and the entire spectrum of dynamics that they exhibit, it is crucial to understand the role of the free end of the broken wavefront which evolves to form the spiral tip. It turns out that this analysis can be done using a kinematical approach. Here, the distinction between the front and back is ignored and the dynamics of the wave is described using a single curve. It is immediately obvious that this approximation is only valid for the low-excitability limit, where the thickness of the wave is small. Immediately after the formation of the broken wave the curve is flat and the end point grows with a fixed velocity. But with time the tip starts increasing in curvature and the velocity starts changing. Due to the dependence of wave speed on curvature, because of the source-sink mismatch, the velocity at the free end of the curve will necessarily be smaller than the normal velocity. Based on this idea, Mikhailov et al. [Mikhailov et al. 1994] came up with the eikonal equation for the motion of curves with free ends. In the eikonal equation, the curves with free ends have two velocities V_N and V_T, corresponding to the normal and transverse propagation, respectively, given by:

$$V_N = V_0 - D\kappa,$$
$$V_T = V_{T_0} - \gamma\kappa_0, \qquad\qquad (3.1)$$

where D and γ are positive coefficients and V_0 and V_{T_0} are the velocity of propagation for a wavefront with zero curvature, i.e., a plane wave, in normal and transverse directions, respectively. Curvature for the normal and transverse direction are given by κ and κ_0, respectively. Equation (3.1) suggests that the rate of growth of the tip V_{T_0}, decreases with curvature k_0 and changes sign at a critical value $k_c = G_0/\gamma$. On increasing curvature beyond k_c, the curve retracts. Thus, the tip of the free end of a spiral in the kinematic formulation is characterized by this curvature. It has been shown by Mikhailov [Mikhailov et al. 1994] and others [Hakim and Karma 1999] that by using the above equations, both simple rotation and complicated spiral dynamics like drift can be understood, at least in the limit of weak excitability. It was later shown that APD also has a strong dependency on the curvature of the medium [Comtois and Vinet 1999].

$$APD = APD_0 + \beta\kappa, \tag{3.2}$$

The above is a relation analogous to Eq. (3.1) and holds for spirals with small curvatures [Qu et al. 2000d].

Note that Eq. (3.1) is valid only for a purely isotropic medium and needs to be modified for anisotropic media such as cardiac tissue. Cells in the cardiac tissue are coupled to each other via gap junctions and are organized into fibers. These, in turn, are arranged into even larger structures. Propagation of a wave of action potential is two to three times faster along the fibers than orthogonal to it. Further, the propagation is at least two times slower in the direction orthogonal to the larger structures than in the direction perpendicular to the fibers [Hooks et al. 2002]. Further, the orientation of the fibers and these structures varies through the heart [Nielsen et al. 1991]. Thus to capture the effect of anisotropy, it is vital to incorporate the dependence of direction in the eikonal equation.

3.6 Scroll waves and filament dynamics

Scroll wave is a three-dimensional analog of the spiral. A simple way to visualize scrolls is to think of them as a set of contiguous rotating spirals whose phase singularities (spiral tips) describe a continuous line called *filament* along the rotation axis perpendicular to the plane of the spirals [Winfree 1987]. The filament is thus a higher dimensional generalization of the core and is just as useful (if not more) in tracking the dynamics of the spatial patterns. Filaments can be straight, curved, twisted and can even form a closed loop called a scroll ring.

Winfree first showed scrolls in chemical systems in 1973 [Winfree 1973] and further proposed that they might also exist in heart tissue, a suggestion verified experimentally in the eighties [Chen et al. 1988; Frazier et al. 1989]. Scroll waves have also been experimentally observed in many other natural systems including chemical waves in the Belousov–Zhabotinsky reaction [Fast and Pertsov 1990; Welsh et al. 1983; Winfree 1973], and patterns of aggregation during *Dictyostelium* morphogenesis [Steinbock et al. 1993; Weijer 1999].

3.6.1 Positive and negative tension

A unique characteristic of the untwisted scroll ring is the *filament tension*, which is the property of the change in the radius of a scroll ring, which tends to either shrink or grow in size. Under normal or high excitability regimes the scroll filaments tend to contract and disappear, while lowering the excitability makes them grow.

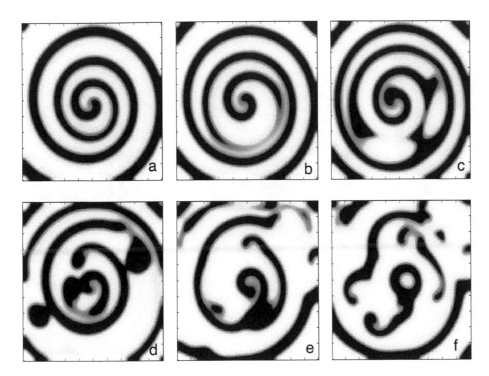

FIGURE 3.5 Spiral breakup resulting in spatiotemporal chaos in the Bär model. The breakup starts away from the core in (b), with a wavefront colliding with an inexcitable region, in the process giving rise to new spiral filaments which interact among each other to form a spatiotemporally chaotic state.

The equation for the evolution of the radius of the scroll ring was introduced by Biktashev et al. [Biktashev et al. 1994]. The local radial velocity V_r, which is the rate of change of the radius r of the filament, was determined to be,

$$V_r = -\gamma/r, \qquad (3.3)$$

where γ is the filament tension. The sign of γ determines the stability of the filament. A positive value of γ (positive tension) would cause the filament to shrink and either disappear or stabilize to a straight filament between the boundaries. On the other hand, scroll rings with negative tension ($\gamma < 0$) will expand and can collide with the boundaries of the medium [Alonso et al. 2003]. This collision with the boundaries can result in the creation of newer filaments.

3.7 Onset of spatiotemporal chaos

Under certain conditions spiral or scroll waves become unstable, eventually breaking up into multiple wavelets, leading to a spatiotemporally chaotic state as shown in Fig. 3.5. The earliest numerical studies by Wiener–Rosenblueth [Wiener and Rosenblueth 1946] and Moe [Moe et al. 1964] explored this possibility. Increased computational power of the 1990s showed the possibility of spiral breakup in many

T = 3850 ms   T = 5720 ms   T = 8690 ms

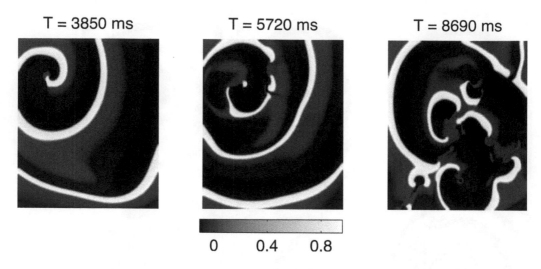

0   0.4   0.8

FIGURE 3.6   Onset of spatiotemporal chaos in the Panfilov model in a 2-D simulation domain with linear dimension $L = 256$. The initial condition is a broken plane wave that is allowed to curl around into a spiral wave (left). Meandering of the spiral focus causes wavebreaks to occur (center) that eventually results in spiral turbulence, with multiple independent sources of high-frequency excitation (right). The colorbar indicates the value of the activation variable e at each spatial point.

kinds of models [Ito and Glass 1991; Panfilov and Holden 1990, 1991; Winfree 1989]. Various mechanisms of spiral and scroll breakup have been identified, including the interaction of high frequency excitation waves with inexcitable obstacles [Panfilov and Keener 1993], quasiperiodic meandering of the spiral [Chen et al. 1997; Qu et al. 2000c], steep APD restitution slope [Karma 1993, 1994], negative filament tension [Alonso et al. 2003; Biktashev et al. 1994], rotational anisotropy [Fenton and Karma 1998; Qu et al. 2000b; Rappel 2001], etc. We will briefly describe a few of these modes of spiral breakup. For a detailed discussion of the multiple scenarios of spiral wave breakup, see Refs. [Bär et al. 1999; Fenton et al. 2002].

3.7.1   Meandering induced breakup

In a system where meandering is sufficiently high, part of the spiral wave can collide with another part of itself and break up spontaneously, resulting in the creation of multiple smaller spirals (Fig. 3.6). The process continues until the spatial extent of the system is spanned by several coexisting spiral waves that activate different regions without any degree of coherence. This state of *spiral turbulence* marks the onset of spatiotemporal chaos, as indicated by the Lyapunov spectrum and Kaplan–Yorke dimension [Pandit et al. 2002]. This mechanism was experimentally verified using a Belousov–Zhabotinsky system by Ouyang et al. They showed that as a parameter is varied, the spiral core changes from a circular tip to a flower with either inward or outward petals via a supercritical Hopf bifurcation. The Doppler effect stemming from increased meandering can result in the interaction of two wavefronts, resulting in instability and eventual breakup. The resultant breakup occurs close to the original core and subsequent interactions can create many spiral fragments [Zhou and Ouyang 2000, 2001].

3.7.2 Steep APD restitution curve

The restitution curve (Fig. 3.7) captures the non-linear relationship between the action potential duration (APD) and the diastolic interval (DI). A similar non-linear relationship exists between the conduction velocity (CV) and DI, referred to as the dispersion curve. Originally the "restitution hypothesis" suggested that spiral breakup occurs when the maximum slope of the curve exceeds 1 [Karma 1993, 1994]. When this condition is satisfied, complex oscillations in cycle length (CL) and APD set in via a period-doubling bifurcation. In cardiac literature, such oscillations are called *alternans*. When the magnitude of these oscillations becomes large, they would lead to a conduction block in the wavefront. At the point of the block, the wave would not be able to propagate locally, giving rise to a breakup. However, later studies have shown that the slope of the restitution curve is not a sufficient predictive condition of such a breakup scenario. Other factors such as the spatiotemporal heterogeneity of APD or CV, cardiac memory, etc., [Echebarria and Karma 2002a; Fox et al. 2002a; Qu et al. 1999] also determine whether a spiral breakup occurs or not. We will discuss the phenomenon of alternans and its control in detail in Chapter 10.

3.7.3 Long wavelength instability

Another kind of spiral breakup scenario involves long-wavelength instability, which is a result of spatial modulation of wavefronts far from the core. Under certain conditions, such as in a spatially heterogeneous medium, the distance between two successive waves away from the core can spatially vary, depending on the local recovery properties of the medium. This can sometimes result in a front getting too close to the recovering tail of the wave ahead of it. Collision with the recovering tail may cause a conduction block, which can lead to the formation of more fragments far from the original spiral core. Such a situation was experimentally verified in the Belousov–Zhabotinsky chemical system [Zhou and Ouyang 2001].

3.8 Mother rotor and multiple wavelets

The scenarios explained above are general and can be observed in many naturally occurring excitable systems. In the context of spatial patterns in the heart, these breakup mechanisms can be broadly classified either under mother rotor or multiple wavelet hypothesis.

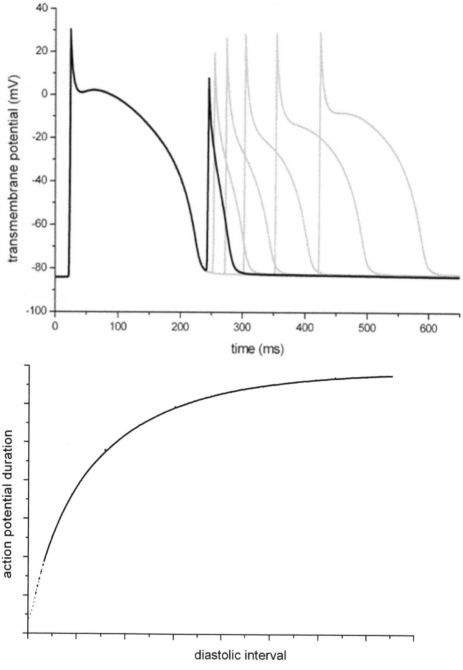

FIGURE 3.7   (top) Non-linear response of the action potential duration to a second stimulation immediately following an excitation of a cell in the resting state. Stimulating the cell within a short time of the initial activity produces an action potential having a markedly reduced duration as the cell has not yet completely recovered. Increasing the interval between the successive stimulation results in longer action potential durations, which eventually converges to a value identical to the first action potential duration. (bottom) This non-linear response of an excitable cell is characterized by the restitution curve. This is the functional relation between the interval between two successive stimulations of a cell (diastolic interval), and the duration of the action potential resulting from the second stimulation.

3.8.1 Multiple wavelet hypothesis

In the 1960s Moe used a model of cellular automata to show that spiral breakup can happen as a result of heterogeneity in the distribution of refractory periods in the excitable medium [Moe et al. 1964]. According to this hypothesis, the heterogeneities facilitate the formation of multiple spiral fragments that continuously interact with each other, resulting in a self-sustained turbulent state of excitation. Moe's work was later developed by Krinsky, who derived conditions for the extent of tissue heterogeneity in order to have such a self-sustaining activity [Panfilov and Pertsov 2001]. The *multiple-wavelet* mechanism put forth by Moe has been experimentally verified in the hearts of different species including dogs [Kirchhof et al. 1993], rabbits [Wu et al. 2004], and humans [Nash et al. 2006].

3.8.2 Mother rotor hypothesis

Advancement in experimental techniques has allowed the quantification of patterns of excitation during fibrillation [Gray et al. 1995a; Witkowski et al. 1998]. These studies have shown that the number of wavelets observed in experiments can be significantly lower than that predicted by Moe's hypothesis. Further, the lifespan of many of these fragments was found to be smaller than the period of a single reentrant rotation. These observations led scientists to propose an alternate mechanism for the origin and maintenance of fibrillation in the heart via the formation of a mother rotor. According to this hypothesis, fibrillation is maintained by one or few stable but high frequency sources of excitation. The rapid waves generated by these "mother rotors" interact with pre-existing heterogeneites, such as gradients or partially or fully non-conducting regions in the medium. This interaction can result in the breakup and formation of multiple fragments, as the waves are able to conduct into some regions but are blocked elsewhere [Jalife 2000]. We will discuss the origin of mother rotors and their connection with heterogeneities like spatial gradients in excitable media in Chapter 6.

4

Excitable media across scales: A gallery of examples

In this chapter we will introduce several examples of systems that span an entire spectrum of length scales, ranging from waves of calcium concentration seen at the intracellular scale, all the way to large-scale patterns of plankton growth in oceans that can span hundreds of kilometers. Our examples are spread across a wide range of scientific disciplines, including the physical (patterns in surface reactions), chemical (Belousov–Zhabotinsky reaction), physiological (uterine contractions, spreading depressions in the brain), biochemical (intracellular calcium waves), biological (cAMP waves in *Dictyostelium* colonies), ecological (phytoplankton growth), and pathological (spatial spread of infectious diseases) systems. The common feature linking these diverse systems is that their dynamics can be modeled using excitable media models, with many of them displaying spatial patterns like spiral waves and scroll waves. We will briefly discuss each of these examples with the aim of giving a flavor of the diversity observed in naturally occurring excitable systems.

4.1 Intracellular: Calcium waves

It is significant that we start our exploration of the various examples of excitable systems with the waves of calcium in the intracellular medium of the cell. And not just because they probably constitute the lowest level at which patterns of excitations are observed. Calcium ions (Ca^{2+}) are absolutely vital for the living cell as they are involved in the regulation of many physiological events. Calcium is crucial for the muscular contraction in vital organs like heart and uterus [Bers 2002; Shmygol et al. 2007]. It was in the 1880s that Sidney Ringer serendipitously discovered the effect of calcium in biological cells [Carafoli 2002]. With the goal of identifying the role of various components in the blood on cardiac contraction, he suspended the extracted frog heart in a saline medium. He observed that when the hard water from the tap, that was used in the medium, was replaced with distilled water, the contractions grew weaker. In order to continuously maintain contractions he had to add calcium salts to water and in the process identified a key player in the contraction process of the heart.

Calcium also acts as a universal intracellular messenger carrying information for the regulation of several cellular processes such as fertilization, secretion, etc. [Carafoli et al. 2001]. On the downside, excess cellular Ca^{2+} can result in the cell killing itself, a phenomenon known as *apoptosis* [Keener and Sneyd 1998]. Thus, from the point of view of the cell, it is absolutely important that the concentration of free Ca^{2+} is carefully controlled. In fact, in evolutionary terms, the choice of calcium as the intracellular messenger is primarily motivated by its ability to easily form complexes, thereby limiting the free Ca^{2+} ions in the cellular medium [Carafoli et al. 2001]. Oscillations in the concentration of Ca^{2+} are associated with a wide variety of regulatory processes in the cell including the activation of oocytes during fertilization, cell migration, and development of muscles [Berridge et al. 2003]. Period-2 in the peak calcium concentrations during *calcium cycling* (which is the periodic movement of Ca^{2+} ions within and out of the cellular calcium storage sacs like endoplasmic reticulum (ER) and mitochondria) is known to trigger irregularity in the rhythmic actions of the heart [Weiss et al. 2006].

An interesting feature of intracellular calcium concentrations is their complex spatiotemporal dynamics. These complex dynamics arise out of the non-linearities in the calcium cycling process. The oscillations do not necessarily occur uniformly and often form spatial patterns like spirals. A classic case of spiral waves of intracellular Ca^{2+} was observed in *Xenopus* oocytes [Lechleiter et al. 1991](Fig. 4.1). These cells are large enough for the existence of a spiral with a complete turn, which is often not the case in many other kinds of cells that are typically an order of magnitude smaller [Keener and Sneyd 1998]. From a modeling point of view, the activator in these systems is the Ca^{2+} in the cellular medium and the positive feedback which leads to further release of calcium from the cellular stores such as ER. The inhibitor is the number of Ca^{2+} binding sites on ER, which closes the channel and inhibits the calcium dynamics at a much slower rate than the activation. Several models have been developed, each capturing one or more aspects of experimentally observed calcium dynamics [Atri et al. 1993; de Young and Keizer 1992; Othmer and Tang 1993].

4.2 Cellular aggregation: cAMP waves in slime mold

Next we will consider an example of biological excitable media at the level of a cell. There exists a species of amoeba *Dictyostelium discoideum* that normally grows and divides as individuals, but under conditions of extreme nutrient scarcity aggregates to form an excitable medium. These aggregates or *slugs* form colonies that slowly grow into tower like structures. The towers typically have a stalk and a head that acts as the repository of the spores. When external nutrient concentration turns favorable, the spores germinate and form new amoeba. Under stressful situations, these amoeba begin communicating with other cells, even across large distances, to begin the process of aggregation. Some individual amoeba cells may initiate

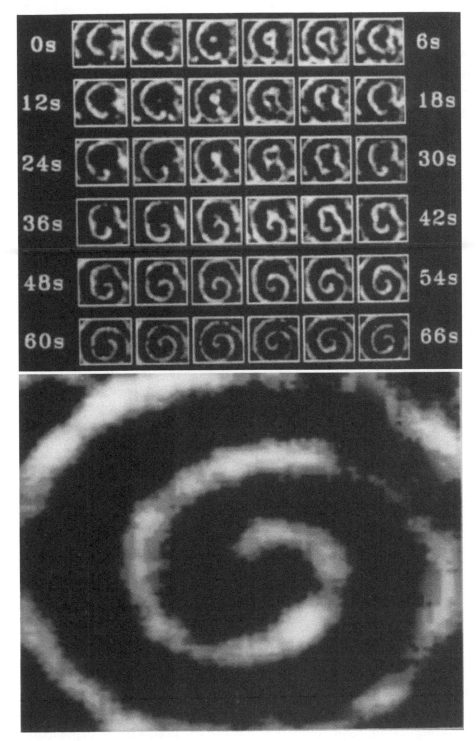

FIGURE 4.1 (top) The time-evolution of the spatial pattern of intracellular Ca^{2+} release in *Xenopus* oocyte. A magnified view of the last frame, corresponding to 66 s, exhibiting a fully formed spiral wave having wavelength $\sim$ 150 μm and period $\sim$ 6 s is shown in (bottom). This figure is used with permission from J. D. Lechleiter, Molecular mechanisms of intracellular calcium excitability in *X. laevis* oocytes. *Cell*, (1992) 69(2), 283–294.

communication by releasing a pulse of $3'5'$-cyclic adenosine monophosphate (cAMP) molecules. Other amoeba pick up this signal and respond by releasing more cAMP molecules. At the level of the whole colony, these pulses organize into large spiral waves of cAMP concentration (see Fig. 4.2 and Fig. 4.3). The other cells in the colony detect these cAMP waves and start moving toward the source of the signal (the center of the spiral) in order to aggregate and form slugs.

Levine et al., built a mathematical model to understand the mechanism used by the *Dictyostelium* to communicate across long distances [Levine et al. 1996]. If the concentration of cAMP that a cell detects in its environment is above a certain threshold, the cell gets excited and releases a pulse of cAMP for a fixed period of time and then returns to the unexcited state. Such repeated excitation and relaxation produces small circular or spiral patterns. How do random distributions of these pulses self organize into large spirals? It turns out that the excitability (the ease of creating an excitation) of this system is not a constant, but is actually dependent on the local concentration of cAMP. This positive feedback from local cAMP concentration to the cell's excitability seems to be a necessary condition for the creation of large spirals from small patterns [Levine et al. 1996]. With some modification, this model was shown to reproduce many of the spiral patterns observed in the wild type [Sawai et al. 2005].

An interesting feature observed in *Dictyostelium* populations is that the cell population density determines whether the large-scale pattern observed is a target wave or spiral. It has been observed in experiments that low densities favor target patterns, whereas populations of higher density prefer to form a spiral wave [Lee et al. 1996]. Once formed, the spirals entrain the pacemaker cells that were producing the concentric circular patterns.

4.3   Organ: Waves of spreading depression in the brain and the retina

Spreading depression (SD) is a term used to refer to a diffusive wave of depolarization typically propagating at the speed of a few millimeters/minute observed in the retina and the cortex of the brain. SD is the result of a temporary breakdown of the transmembrane potential as there occurs a large-scale movement of ions across the membrane. The process typically lasts for a couple of minutes before recovery to the original electrophysiological state is completed [Martins-Ferreira et al. 2000; Somjen 2001]. SDs have been associated with migraine and stroke, both of which are highly debilitating disorders.

Migraine is characterized by the pulsating headache typically on one side of the head lasting sometimes for days. While the association between the depolarizing wave and the headache is still debated [Moskowitz 2007], in humans at least it is known that SD in the cerebral cortex causes *migraine aura* [Lauritzen 1994]. Migraine aura is often perceived as a flickering sensation, lasting for several minutes, that can drastically reduce the field of vision. A type of SD called *periinfarct depolarization* that typically occurs around an infarct (a region of cells that have died

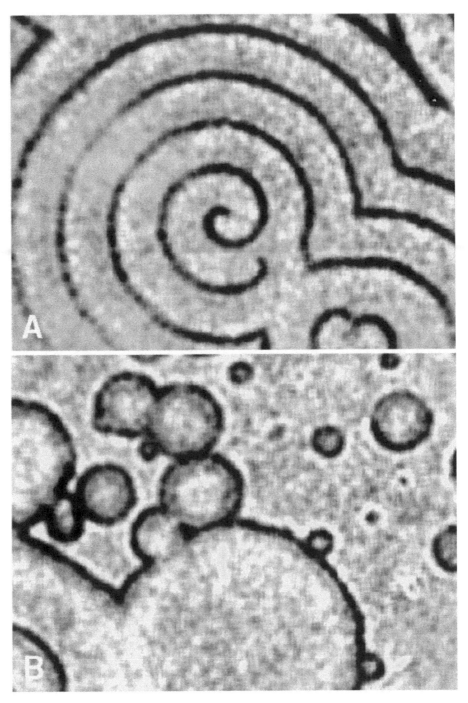

FIGURE 4.2 Pattern formation in starving populations of the amoebae *Dictyostelium discoideum* showing (A) a spiral wave state that develops following wave breaks after the initial appearance of target waves around a few pacemaker cells. These pacemakers are eventually entrained by the spiral wave which has a higher frequency of activation. (B) A homogeneous application of cAMP that causes signaling to cease for approximately two wave periods, resets the target wave pattern by extinguishing the spiral wave, and allows the pacemakers to regain control. This figure is used with permission from K. J. Lee, R. E. Goldstein and E. C. Cox, Resetting wave forms in *Dictyostelium* territories, *Phys. Rev. Lett.* 87 (2001) 068101.

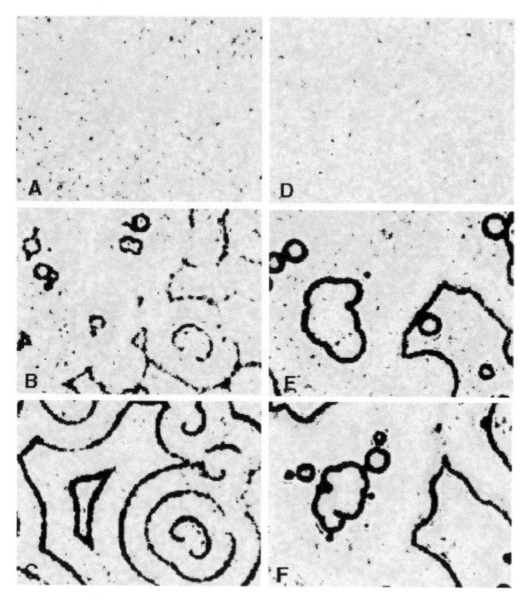

FIGURE 4.3 The time-evolution of spatial pattern in a monolayer of *Dictyostelium discoideum* following the deprivation of nutrition. (A) After 281 minutes, a homogeneous state is still observed. (B) Target waves that form around pacemaker cells and spiral waves emerging from broken wave segments are seen to coexist at 352 minutes. (C) After 408 minutes, fully developed spiral waves are observed. (D) Global application of cAMP at 413 minutes results in a homogeneous state within 4 minutes. Target waves from pacemaker cells are reestablished as observed (E) 16 minutes and (F) 46 minutes after the perturbation. This figure is used with permission from K. J. Lee, R. E. Goldstein and E. C. Cox Resetting wave forms in *Dictyostelium* territories, *Phys. Rev. Lett.* 87 (2001) 068101.

due to lack of oxygen) is associated with stroke [Hossmann 1995]. Stroke happens when blood flow to the brain is disrupted, due to either a block in (ischemia) or bleeding (hemorrhage) of the blood vessels. Waves of spreading depression also occur in the retina, though whether they are related to migraines is not clear [Hill et al. 2007].

Spreading depressions were first described by Aristides Leão in the 1940s in a series of three papers [Leao 1944a,b, 1947]. Leão was a Brazilian doing his Ph.D. at Harvard Medical School. While his original intention was to investigate the propagation of seizure discharges in the cortex, Leão chanced upon the phenomenon of spreading depression and in the process pioneered a major field of research [Somjen 2005]. SD waves have been observed to evolve into a variety of spatial patterns, such as waves rotating around a lesion in the cortex of rats [Shibata and Bureš 1972] and spiral waves in isolated chicken retina [Dahlem et al. 2010; Gorelova and Bureš 1983].

4.4 Organ: Waves of contraction in uterus near term

Uterine muscle cells or myocytes are not normally active in the life course of a woman. But close to the delivery of the fetus, these cells begin to contract spontaneously. Contraction of these myocytes is triggered via electrical action potentials. These action potentials evolve from single spikes to trains in the lead up to *parturition* [Bengtsson et al. 1984; Csapo and Kuriyama 1963; Landa et al. 1959]. However there is much that is not understood about the precise mechanism underlying the transition of the uterus from a quiescent state, almost throughout the duration of the pregnancy, to a state of rhythmic contraction of muscles observed at the onset of labor.

Understanding the origins of uterine contractions is of tremendous practical importance. Roughly 10% of all births are *preterm*, i.e., occurring prior to the completion of 37 weeks of gestation [Martin et al. 2009b]. This typically occurs because of spontaneous emergence of premature uterine contractions [Pervolaraki and Holden 2012]. Preterm births have been implicated as the cause of over a million neonatal deaths per year worldwide, and in around 50% of all cases of infant neurological damage [Pervolaraki and Holden 2012]. A clearer understanding of the mechanism of spontaneous uterine tissue contraction could therefore greatly facilitate the development of effective strategies to help curb neonatal mortality and morbidity.

Analogous to the sinus node in the heart, the spontaneous origin of contractions in the uterus has often been associated with the existence of *pacemaker* cells [Garfield and Maner 2007b; Lammers 2013]. However, despite much effort aimed at identifying the origin of uterine contractions [Duquette et al. 2005; Wray et al. 2001], including considering some candidates like interstitial Cajal-like cells (ICLCs) [McHale et al. 2006; Sergeant et al. 2000], there has, so far, not been any conclusive evidence for the existence of these pacemaker cells in the uterus. Another hypothesis that spontaneous electrical behavior is an inherent property of uterine smooth-muscle cells [Garfield and Maner 2007b] is considered to have some merit. It is known that numerous electrophysiological changes occur in the uterus during the course of pregnancy, including significant changes in the ionic currents, weight, surface area and subsequently capacitance of the myometrium [Wang et al. 1998; Yoshino et al. 1997]. However, it is not clear if and how these changes lead to the origin of spontaneous contractions [Young 2007].

Dramatic changes in the number and conductance of gap junctions close to the term [Miller et al. 1989; Miyoshi et al. 1996] have suggested an alternative paradigm for the origins of coherent uterine activity. In fact, nothing captures the importance of gap junction expression more spectacularly than the observation that chemical disruption of the gap junctions can almost immediately inhibit the oscillatory uterine contractions [Loch-Caruso et al. 2003; Tsai et al. 1998; Wang and Loch-Caruso 2002]. These findings strongly suggest that gap junctional coupling between neighboring cells has a crucial role in the development of coordinated uterine electrophysiological activity. They may also be responsible for the transition from the weak, desynchronized myometrial contractions seen in a quiescent uterus to the strong, synchronous contractions observed during labor [Miller et al. 1989; Miyoshi et al. 1996].

Analyzing the spatiotemporal details of the propagation of electrical activity, Lammers et al. have suggested that reentrant waves of excitation may occur spontaneously in isolated pregnant uterus [Lammers 1997; Lammers et al. 2008b](see Fig. 4.4 and Fig. 4.5). Whether these spirals induce transient pathological contractions like in the heart or are required to generate a stronger force to facilitate the delivery of the fetus is still an open question. This cannot be answered without developing a better understanding both at the cellular and the tissue level.

4.5   Chemical medium: Oscillations and spiral waves in Belousov–Zhabotinsky

The experimental studies on biological systems that have been discussed so far, are usually difficult to perform and often require sophisticated tools and expertise. So scientists often opt for a simpler but equally interesting class of chemical reaction experiments that have become a popular and useful model for studying complex nonlinear systems displaying excitations or oscillations [Epstein and Pojman 1998]. The prototypical behavior of such systems is the phenomenon of chemical oscillations, where the concentration of one or more species varies temporally.

Chronologically speaking, it was Bray who first reported an instance of oscillatory reactions [Bray 1921]. His results were not believed by most scientists as they were believed to violate the second law of thermodynamics [Mikhailov and Showalter 2006]. Similar scepticism ensured that Boris Belousov's observations of a periodic change in color from yellow (Ce^{4+}) to colorless (Ce^{3+}) was only published in a conference proceedings [Belousov 1959]. It was only in the sixties, subsequent to the experiments of A. M. Zhabotinsky and others that chemical oscillations were accepted to be a real phenomenon. Since Zhabotinsky rediscovered Belousov's reaction the reaction is known as the Belousov–Zhabotinsky (B-Z) reaction, in honor

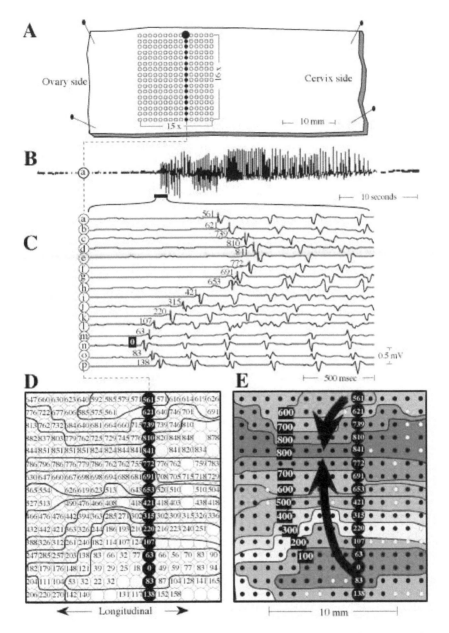

FIGURE 4.4 Reconstruction of the activity propagation map from recordings of individual spikes in rat pregnant myometrium. (A) shows the orientation of the electrode array used for recording activity in relation to the tissue preparation. (B) displays the recording from the electrode 'a' [indicated in (A)], exhibiting a characteristic bursting pattern. The initial portion of the burst time-series, recorded from all 16 electrodes in a single column, is shown magnified in (C). As the spikes occur at distinct times in the different recording sites, the times of local activation can be used for reconstructing the propagation front of electrical activity in the myometrium. The local activation times for all 240 electrodes (expressed with reference to the time of the first spike at electrode 'n' taken to be $t = 0$ ms) are shown in (D). In (E), isochrones (in steps of 100 ms) are drawn around simultaneously active regions to visualize the spread of excitation. This figure is used with permission from W. J. E. P. Lammers, Circulating excitations and re-entry in the pregnant uterus, *Pflügers Arch* (1997) 433: 287–293.

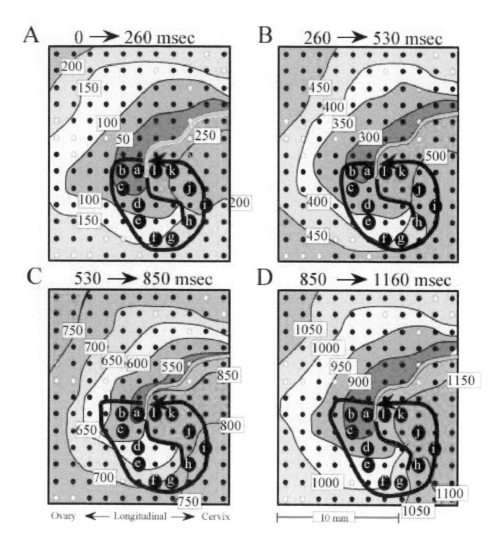

FIGURE 4.5 Experimental observation of counter-clockwise propagating spiral wave of electrical excitation in pregnant myometrium of a rat (day 19). The shades on the grayscale indicate the time corresponding to a specific isochrone, the local time of activation for each contour being indicated in the figure. This figure is used with permission from W. J. E. P. Lammers, Circulating excitations and re-entry in the pregnant uterus, *Pflügers Arch* (1997) 433: 287–293.

of the two Russian chemists [Winfree 1987].

The dramatic periodic changes in color that Belousov had observed, indicating the occurrence of chemical oscillations, were found to occur in "well-mixed" homogeneous chemical systems. While in the original experiments the periodic oscillations were transient with the reaction stopping once the reactants were fully consumed, later experiments were conducted in continuously stirred tank reactors with the reactants fed in and the products taken out continuously. Under these conditions, the oscillations would continue for as long as the supply of reactants was maintained. A detailed chemical mechanism for the Belousov–Zhabotinsky reaction was proposed by Field, Körös, and Noyes in the 1970s [Field et al. 1972] and it was finally established that chemical oscillations were a genuine phenomenon worthy of study. Over the last few decades several other oscillating chemical systems that exhibit a rich variety of dynamical behavior, have been discovered [Epstein and Pojman 1998]. A simplified version of the FKN mechanism that captures the qualitative features of the experiment was later developed and called the Oregonator, named after the place (Oregon) where it was developed. In its simplest form, the Oregonator model has two variables and has been used extensively to understand the dynamics of several phenomena observed in the Belousov–Zhabotinsky reaction.

In the case of B-Z reactions, spatiotemporal patterns are typically observed when the experimental system is not stirred. The waves correspond to spatially varying chemical concentrations, with the concentrations between the front and back of the wave (i.e, the excited region) differing from those in both the recovered and as yet unexcited regions. Typical patterns formed include target patterns and spiral waves [Epstein and Pojman 1998]. The B-Z has been a convenient model system to experimentally explore the mechanism of transition from single spiral to spatiotemporal chaos [Ouyang et al. 2000] (see Fig. 4.6 and Fig. 4.7), demonstrate techniques of control of spiral waves (see Ref. [Mikhailov and Showalter 2006] and references therein), investigate the dynamics of spirals and scrolls in heterogeneous [Vinson et al. 1997a] and anisotropic excitable media [Steinbock et al. 1995a], implement chemical logic gates [Tóth and Showalter 1995], find the paths of least distance in a complicated labyrinth [Steinbock et al. 1995b], etc.

4.6 Solid state devices: Spirals in CO oxidation on Pt(110) surface)

Another example of a chemical system that supports patterns of concentration is the catalytic surface reaction of CO oxidation on platinum. When a mixture of carbon monoxide and oxygen, both at extremely low partial pressures, is heated to 400 K in a chamber containing a single-crystal wafer of platinum [surface Pt (110)], these gases *adsorb* onto the crystal surface and react to form carbon dioxide which again goes back into the gas phase. This reaction would not happen in the gas phase, but only becomes possible when the gases get adsorbed onto the surface of a suitable catalyst. In general, during adsorption at very low pressures, the gas or liquid atoms impinges on the surface of the catalyst and forms new bonds with the atoms on the

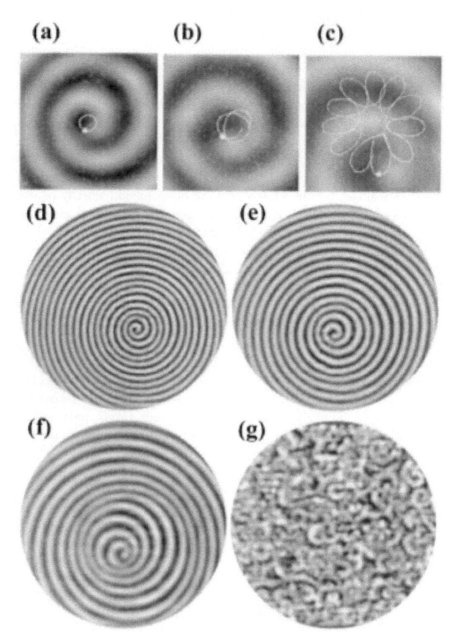

FIGURE 4.6 Transition from a stable spiral to a state of turbulence on changing the excitability by decreasing the malonic acid concentration. (a)–(c) corresponds to the zoomed in version of (d)–(f), with the white lines indicate the trajectory of the spiral tip. Malonic acid concentration of 0.40 M for (a) and (d) results in a simple spiral with a circular core. Onset of meander in (c) and (f) on decreasing malonic acid concentration to 0.30 M of the acid to 0.30 M in (c) and (f), followed by a transition to spiral turbulence in (g) for 0.10 M concentration. This figure is used with permission from Qi Ouyang, Harry L. Swinney and Ge Li, Transition from spirals to defect-mediated turbulence driven by a doppler instability, *Phys. Rev. Lett.* 84 (2000) 1047.

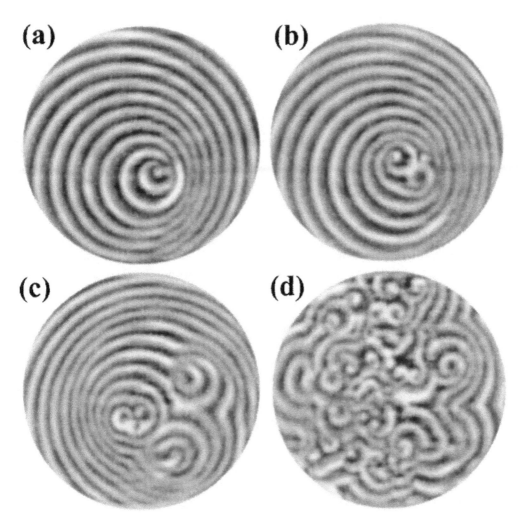

FIGURE 4.7 Illustration of the spiral breakup process leading to a state of spatiotemporal turbulence in the Belousov–Zhabotinsky reaction. A new pair of spirals is generated at time t = 0 (a), which then subsequently breaks up into multiple daughter spirals (b-c), finally forming a spatiotemporally turbulent state (d). The concentration of the malonic acid is 0.10 M and the diameter of the system is 19.5 mm. This figure is used with permission from Qi Ouyang, Harry L. Swinney and Ge Li, Transition from spirals to defect-mediated turbulence driven by a Doppler instability, *Phys. Rev. Lett.* 84 (2000) 1047.

catalyst surface. Once adsorbed to the surface of the catalyst, the atoms can change positions and react with each other almost immediately. The reaction itself happens only on a very thin layer of the surface and the products formed actually move back to the gas phase via desorption. In an open system, where the flow is continuously maintained, this reaction can be sustained without the depletion of the catalyst. This process is of great environmental significance, as it can be exploited in the making of car exhaust catalysts in order to reduce CO pollution. The concentration patterns of the adsorbed gases are imaged by photo electron emission microscopy (PEEM) [Rotermund 1997].

Since this is a surface reaction, it is an ideal system to study two-dimensional patterns, such as spiral waves (Fig. 4.8), forming far from equilibrium. They can show target patterns, traveling pulses, states with steadily rotating waves, and turbulent states corresponding with continuous creation and annihilation of spirals [Jakubith et al. 1990; Rotermund et al. 1990]. The spiral wave of oxygen concentration formed on a surface covered predominantly by CO is called an oxygen spiral(Fig. 4.9(top)). Conversely for the case of a CO spiral wave, the surface is mostly covered by oxygen(Fig. 4.9(bottom)). The length scale of these patterns is of the order of tens of micrometers and the time scales of the order of seconds. When CO oxidation takes place on perfect surfaces of single Pt crystals, its properties strongly depend on the crystallographic plane used to run the reaction. For example, CO oxidation on the Pt (110) plane produces a wider variety of dynamics including oscillations and excitations, when compared to Pt (111), which typically shows a bistable behavior. The reason for change in complexity of the dynamical behavior is due to structural changes accompanying the reaction [Imbihl and Ertl 1995].

Figure (4.10) (left) and (right) are spiral patterns of CO and O concentrations, observed in the Bär model [Hildebrand et al. 1995], which is used to study pattern formation in surface reactions. For a different parameter the spirals might become unstable and break up into a chaotic state as shown in Fig. 4.11.

4.7 Population: Waves of spreading infection in epidemics

Mankind has had to struggle with deadly infectious diseases from time immemorial. An example of a killer outbreak that comes to mind immediately is the outbreak of plague in the 14th century. "Black Death," as it is popularly known, was the scourge of Europe, killing several millions and almost halving the population of the continent. It is now understood that waves of infection spread across Europe for several years [Christakos et al. 2007]. Another infamous outbreak that spread as waves is the bubonic plague of the 19th century that spanned over decades killing several millions in India and China [Christakos et al. 2007]. Influenza is another viral disease that occurs almost every year, killing people across the world [Kaplan and Webster 1977]. Several *pandemics* (epidemics affecting large populations spread across large regions) of influenza have been reported in the past three centuries, often emerging at 10 to 50 year intervals [Potter 2001]. The pandemic of 1918-19 (frequently referred to as the "Spanish flu") traveled across the world in three separate waves and

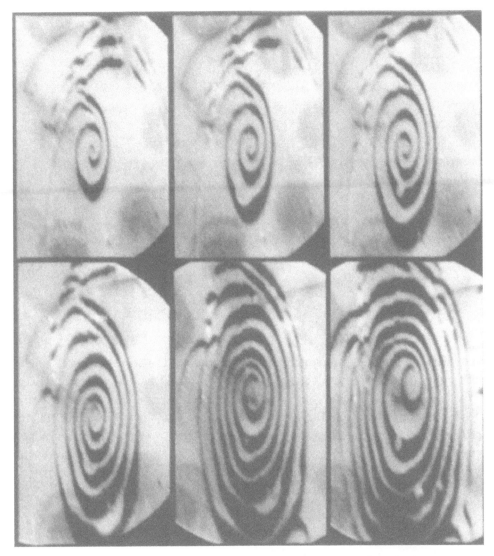

FIGURE 4.8 Series of photoemission electron microscopy (PEEM) images showing the growth of a spiral wave during oscillatory reaction of catalytic CO oxidation on Pt(110) surface (at temperature $T = 434$ K and CO and O_2 partial pressures of 2.8×10^{-5} mbar and 3.0×10^{-4} mbar, respectively). Areas covered by O appear dark in the images, while those covered by CO appear bright. Width of each image is 0.2 mm. The times corresponding to each panel (from left to right, top to bottom) are 0, 10, 21, 39, 56, and 74 s. Figure is used with permission from S. Jakubith, H. H. Rotermund, W. Engel, A. von Oertzen and G. Ertl, *Phys. Rev. Lett* 65 (1990) 3013–3016.

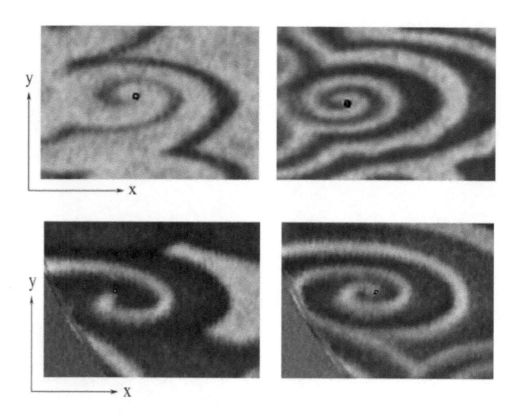

FIGURE 4.9   PEEM images of moving oxygen (top) and CO (bottom) spirals in the presence of temperature gradient along the y-axis. The initial pattern is shown on the left, while the one on the right shows the evolution of the spiral at a later time, viz., after 12.64 s for oxygen (top) and after 9.92 s for CO (bottom). The black circle indicates the position of the tip of the spiral. The top image corresponds to a temperature gradient of 1.8 K/mm while for the bottom image it is 1.3 K/mm. This figure is used with permission from P. Sadeghi and H. H. Rotermund, Gradient induced spiral drift in heterogeneous excitable media, *Chaos* 21 (2011) 013125.

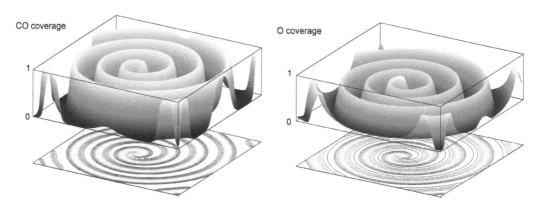

FIGURE 4.10   Pseudocolor plots of CO (left) and oxygen (right) spirals at time T=3000 iterations, obtained by simulating the model described in Ref. [Hildebrand et al. 1995] with a value of $\varepsilon = 0.07$.

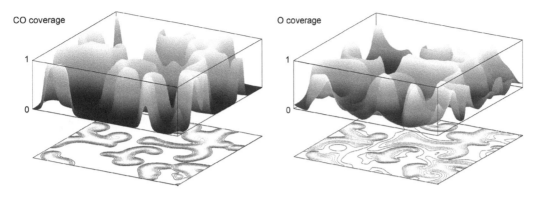

FIGURE 4.11 Pseudocolor plots of the spatiotemporally chaotic states in CO (left) and oxygen (right) concentrations at time T=10000 iterations, obtained by simulating the model described in Ref. [Hildebrand et al. 1995] with a value of $\varepsilon = 0.075$.

resulted in the death of about 50 million people [Johnson and Mueller 2002], making it one of the worst natural disasters in the history of mankind. In 2009, a novel influenza strain termed influenza A(H1N1)v, rapidly spread to different countries and became the predominant influenza virus in circulation worldwide. By April 11, 2010, it caused at least 17,798 deaths in 214 countries. Thus, understanding the spatial patterns by which diseases spread has possibly important consequences, in terms of suggesting better methods of controlling them.

Figure (4.12) shows an example of the wavelike spatial spreading of malaria disease caused by *Plasmodium falciparum* in a region of northern Bengal in India, over the period January 2005–February 2009 [Jesan 2013]. Malaria is a vector-borne infectious disease, endemic to many developing countries, that annually kills at least a million people [Murray et al. 2012]. From data consisting of records of disease incidence recorded at the different health centers located throughout the region, time-series w constructed. Subsequently, using wavelet analysis, the series corresponding to the dominant period (= 1 year) was extracted, and were compared between the locations served by the different health centers in order to determine the phase inter-relationships between the time-series. The mean relative phase angle (MRPA), i.e., the average phase angle of a given location relative to the spatial average over all regions, shown using the thicker contour lines in Fig. (4.12) indicates whether the infections in a given location are leading or lagging in phase relative to their neighboring locations. Thus, the health center with the highest MRPA represents the epicenter which is the location where the infection outbreak originates, while regions having smaller MRPAs indicate where the infection spreads subsequently. Similar methods have been used to infer traveling waves in the spreading of other epidemics, such as measles [Grenfell et al. 2001].

How is spreading of epidemics related to excitable media? Mathematical modeling of epidemics often involves considering different states of the epidemic and assigning individuals to them. Over time each individual evolves through the different states. Typical models have 2, 3, or 4 states, depending on the characteristics of the disease that is being modeled. In a three-state model, the individuals could

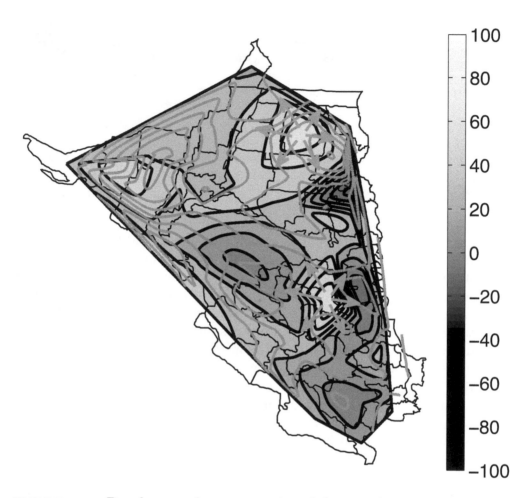

FIGURE 4.12 Pseudograyscale representation of the spreading pattern of malaria infection caused by *Plasmodium falciparum* in a region of northern Bengal, India. The two epicenters of the disease (indicated by arrows moving outward from them), from which the infections begin to spread to the rest of the region every season, are marked by the highest values of the mean relative phase angle (MRPA) [see grayscale bar shown at right] calculated from the time series of infections at different spatial locations, after reconstructing them using the wavelet spectral components corresponding to a dominant periodicity of 1 year. While the epicenter at top corresponds to a region having high population density, population distribution being indicated by the contour lines superposed on the figure, the lower epicenter does not occur close to zones of large population, indicating that factors other than population density may be responsible for it acting as a source for the infection spreading.

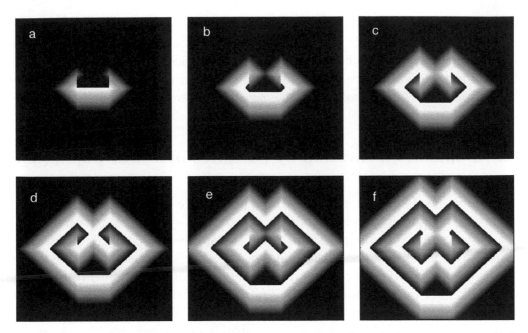

FIGURE 4.13 Pseudocolor image of the time evolution of a spiral wave using a cellular automata model, starting from a row of excited cells. The lattice size is 200×200 and each of the cells evolves through the discrete states to return to its initial value after 40 time steps.

at any instant be in any of the following states: Susceptible (S), Infected (I) or Removed (R). Thus $S(t)$ corresponds to those who can potentially be, but are not yet infected, $I(t)$ corresponds to those who are in the infected state, and $R(t)$ corresponds to those who are either removed from the dynamics due to death or immunization. Over time, the populations could evolve from one state to another. Such a model is called SIR model and was first introduced in 1927 [Kermack and McKendrick 1927]. A modification of this model is the SIRS model, where the R now corresponds to recovered, so as to become susceptible (S) again. That is, no immunity is gained by the individual on being infected once. Non-fatal instances of malaria or influenza are suitable diseases to model using the SIRS model. A patient infected with malaria can recover, but does not gain any immunity against it in the process. Hence the patient is as susceptible as before to infection by the parasite. Note that the dynamics in SIRS is conceptually akin to the cellular automaton version of the excitable medium. In the cellular automata, an excited cell returns to the initial susceptible (excitable) state passing through other discrete states corresponding to recovery, with the excitation (infection) of any cell depending on its current state and the number of excited (infected) neighbors. Not surprisingly, waves of infection observed in these models resemble the excitation wavefronts. Figure 4.13 illustrates the development of a spiral-like pattern in such a discrete state model. Each cell in this system traverses through 40 different discrete states, i.e., a cell that is excited comes back to its original state after 40 time steps.

Factors that influence the spatial spread of the disease include the mobility of the vector, the animal host and the infected human or animal. Heterogeneity in the spatial environment and the kind of interactions between the different individuals are other factors that determine the characteristics of the spatial distribution of the infection. The increase in spatially long-range connections in the modern world, primarily via air travel, can give rise to different spatial dynamics of these diseases [Marvel et al. 2013].

4.8   Population: Invasion by parasites of a host population

The interaction of a predator species with a prey to produce spatiotemporally varying patterns in populations is a hallmark of ecological systems. A typical feature of such systems is that the populations of the two species are coupled, i.e., a change in one affects the other. The most famous of the models that capture these interactions is the Lotka–Volterra coupled differential equations, mentioned in Chapter 1. Another kind of interaction dynamics occurs between the host and parasitoid populations. Parasitoids are organisms that spend a part of their life on a host species, often killing it to feed themselves. For example, insect parasitoids like wasps parasitically use the bodies of arthropods to lay their eggs. The larvae that hatch from the egg kills the host and consumes it as food. The life cycles of the host and the parasitoids often tend to be synchronized with both of them producing a single generation every year. Nicholson–Bailey is one model that describes the evolution of host and parasitoid populations using difference equations, with the time steps corresponding to the generations [Nicholson and Bailey 1935]. The female parasitoid performs a random search for a suitable host, over an environment where the distribution of hosts and parasitoids is assumed to be non-contiguous, to lay eggs in and does not differentiate between a free and an already parasitized host.

In the original form of the Nicholson–Bailey model, the host and parasitoid populations are unstable. However when details like local movement in a patchy environment are added it was found that the population dynamics became persistent [Hassell et al. 1991a]. Spatial patchiness has been observed in and is considered to be important to the persistence of natural populations in both predator-prey and host-parasitoid kinds of interactions [Hassell et al. 1991b; Pacala et al. 1990]. In their study Hassell et al. allowed for the dispersal of a fixed fraction of both host and parasitoid populations to 8 nearest patches. Provided there was a sufficient number of patches, such local dispersions led to a wide variety of dynamical behavior including crystal lattices, spiral waves, and spatially chaotic patterns [Hassell et al. 1991a]. The crystal lattice corresponds to a temporally static mosaic comprising low - and high-density patches that could be spatially irregular.

The Rosenzweig–MacArthur model is another system that displays spatial patterns of population densities. Like the Lotka–Volterra it comprises two equations, one for the the dynamics of the prey population and the other describing the predator dynamics. Additionally, it also captures the process of self-limitation of the population growth and has a non-linear term for the density/consumption relationship. Figure 4.14 describes the prey (left) and predator (right) population densities

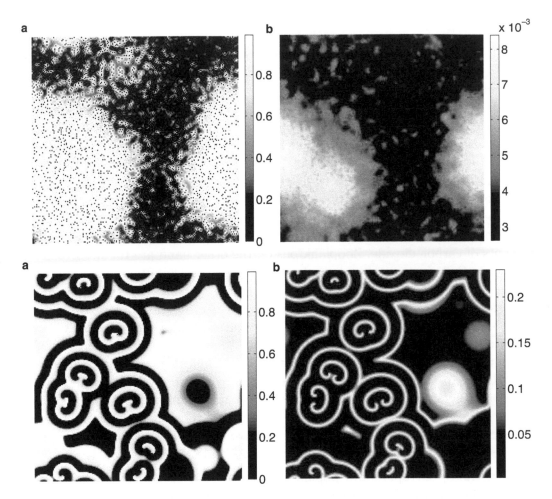

FIGURE 4.14 Population densities in the Rosenzweig–MacArthur model of prey-predator interaction on a two-dimensional lattice of size 200×200. While the prey is immobilized for both cases (a) and (c), the diffusion of the predator populations is set to (b) 0.1 and (d) 0.01.

in a square lattice of length 200. The the prey is immobilized by setting the diffusion coefficient $D = 0$. Depending on the strength of the diffusion of the predator, a spiral pattern could be observed. In this case, Fig. 4.14(a-b) corresponding to $D = 0.1$ do not form a spiral, while a $D = 0.01$ produces a spiral (Fig. 4.14(c-d)). Such a model can also describe key aspects of spatial dynamics of hunter-gatherer societies. In the hunter-gatherer stage of evolution, humans do not have access to regular food supply and depend on hunting other species for food, similar to the predator in the Rosenzweig–MacArthur model.

4.9   Ecology: Plankton growth

We will now describe another ecological phenomenon, namely phytoplankton population dynamics, that can be modeled as an excitable system. Plankton refers to an *ecological niche* of diverse aquatic organisms, ranging from bacteria to jelly fish, that typically tend to drift with the water current. In fact, the German oceanographer Victor Hensen, christened these organisms plankton after the Greek word *planktos* which means wanderer or drifter [Malchow et al. 2001]. Planktons can consist of both autotrophic phytoplanktons that produce their own food and heterotrophic zooplanktons that consume other autotrophs or heterotrophs. Phytoplankton, the plants of the plankton world, produce their own food via photosynthesis. In fact, they are imperative for life on earth to exist as they produce nearly half the amount of oxygen needed by living organisms and also act as a carbon sink by absorbing large amounts of carbon dioxide. Zooplanktons on the other hand are the herbivores and carnivores that consume the phytoplanktons. Together these planktons are a vital link in the aquatic food web as they are consumed by many other water organisms such as fish and whales. A number of factors affect the population of these planktons, including nutrient concentrations, temperature of the water, its salinity, and the intensity of sunlight. Further there can also be spatiotemporal variations in the concentration of the different constituents of the plankton community [Mackas and Boyd 1979]. In addition to the crucial ecological role of the plankton, understanding their population dynamics has important economic consequences too, especially in the context of fisheries. Commercial interests of the fishing industry provided the early motivation to understand the positive correlations existing between the availability of fish and zooplankton abundance [Steele 1977].

Sometimes, phytoplankton populations show rapid surges followed by equally swift fall in numbers. This rapid change in population can be divided into "spring blooms" and "red tides." The spring bloom typically occurs in the North Atlantic and in the subpolar waters and its spread and duration depend on several environmental factors [Truscott 1995]. During winter the mixing of water along vertical columns due to changes in temperature leads to the replenishment of nutrients at the surface that is used by the plankton for photosynthesis. On the other hand, the decreased illumination during winter and the vertical mixing of water results in the removal of plankton. This balance between growth and removal is lost during the spring when the water columns are stratified and no water mixing occurs. This causes the nutrients and the phytoplankton to be constrained to the water surface and leads to a rapid increase in plankton growth. Red tides are more localized surges of plankton population, typically occurring in coastal waters and estuaries [Pingree et al. 1975]. During red tides, the phytoplankton secrete large amounts of neurotoxins that can kill fish, sea birds, and even humans. In view of the significant effect that red tides can have on the fishing industry and on the lives of the people in coastal communities, understanding the dynamics of the rapid changes in plankton population is of practical importance.

Truscott and Brindley [Truscott and Brindley 1994a,b] developed an excitable media model to understand the origins of red tide patterns as arising from the interaction of phytoplankton and zooplankton dynamics. In this model, the fast excitation variable is the population of the phytoplankton, while the zooplankton numbers act as the recovery variable. The time for the development and termination of the red tide (typically over a week) and the duration of the outbreak (several weeks) correspond to the fast and slow timescales of the problem, respectively. Factors like temperature and pollution provide the necessary perturbation for the system to shift from the quiescent pre-tide state to the excited bloom state. A range of complex dynamics such as spirals, self-splitting and reflecting waves, has been observed in the spatially extended versions of this model, with the spread of individuals in space described by the diffusion process [Biktashev and Brindley 2004; Brindley et al. 2005]. Later variations of this model have incorporated convection terms in order to model more complicated dynamics such as evasion of pursuit [Biktashev et al. 2004].

II

Dynamics in Disordered Excitable Media

5

Localized defects and dynamics

5.1 Introduction

Spatiotemporal dynamics of the patterns that arise in homogeneous excitable media (discussed in Chapter 3) is relatively well understood. However, many examples of excitable media that are seen among living systems are heterogeneous in nature. In order to be able to explain the patterns arising in such systems one must therefore try to understand how such heterogeneities affect and alter the characteristic dynamics of homogeneous excitable systems. A particularly important example of heterogeneous excitable medium is the heart. Reentrant waves in the heart giving rise to abnormally rapid excitation can result from an impulse that rotates around an inexcitable obstacle ("anatomical reentry") [Abildskov and Lux 1995] or within a region of cardiac tissue that is excitable in its entirety ("functional reentry") [Davidenko et al. 1995; Rudy 1995]. Such deviations from the normal rhythmic functioning of the heart are not desirable, especially because they can lead to complications that may well be fatal [Cross and Hohenberg 1993; Garfinkel et al. 2000; Gray and Chattipakorn 2005; Gray et al. 1995b; Jalife et al. 1998; Winfree 1998; Witkowski et al. 1998]. Trains of local electrical stimuli are widely used to restore normal wave propagation in the heart during tachycardia. Such "antitachycardia pacing" is not always successful and may inadvertently cause tolerated tachycardias to degenerate to life-threatening spatiotemporally irregular cardiac activity such as ventricular fibrillation [Rosenthal and Josephson 1990]. The underlying mechanisms governing the success or failure of antitachycardia pacing algorithms are not yet clear. Understanding these mechanisms is essential, as a better knowledge of the processes involved in the suppression of ventricular tachycardia (VT) through such pacing might aid in the design of more effective therapies.

In the following sections we look at how the presence of conduction inhomogeneities (such as zones of slow conduction or inexcitable obstacles) affects the development of spatiotemporal patterns in excitable media in different spatial dimensions [Sinha and Christini 2002; Sinha et al. 2002]. In the next two sections we consider the specific problem of how heterogeneities affect the interaction between reentrant waves and excitations generated by external stimuli. Results of such studies have important practical consequences, e.g., in the treatment of VT by rapid pacing. It is known that several factors influence the ability of such pacing to interact with reentrant activity in the heart. The most prominent are [Josephson 1993]: (a) VT rate (for anatomical reentry this is determined by the length of the VT circuit and impulse conduction velocity around the obstacle), (b) the refractory period (i.e., the duration of time following excitation during which cardiac tissue cannot be re-excited) at the pacing site and in the VT circuit, (c) the conduction time from the pacing site to the VT circuit, and (d) the duration of the excitable gap (the region of excitable tissue in the VT circuit between the front and refractory tail of the reentrant wave [Fei et al. 1996]). A single stimulus is rarely sufficient to satisfy the large number of conditions for successfully terminating reentry. Therefore, in practice, multiple stimuli are often used — where the earlier stimuli are believed to "peel back" refractoriness to allow the subsequent stimuli to enter the circuit earlier than is possible with only a single stimulus [Josephson 1993]. In fact, simple theoretical arguments may appear to suggest that pacing should be ineffective in terminating reentrant activity in most cases. We show here that the high success rate of pacing in removing VT may be associated partly with the occurrence of local conduction heterogeneities in the heart muscle.

5.2   Termination of reentrant activity in a ring of heterogeneous excitable medium

The dynamics of pacing termination of one-dimensional reentry has been investigated in a number of studies [Glass and Josephson 1995; Nomura and Glass 1996; Qu et al. 1997; Rudy 1995; Vinet and Roberge 1994], but most of these are concerned exclusively with a homogeneous ring of cardiac cells. The termination of reentry in such a geometry (which is effectively that of the reentry circuit immediately surrounding an anatomical obstacle) occurs in the following manner (Fig. 5.1). Each stimulus splits into two branches that travel in opposite directions around the reentry circuit. The retrograde branch (proceeding opposite to the direction of the existing reentrant wave) ultimately collides with the reentrant wave, causing mutual annihilation. The anterograde branch (proceeding in the same direction as the reentrant wave) can, depending on the timing of the stimulation, lead to resetting, where the anterograde wave becomes a new reentrant wave, or termination of reentry, where the anterograde wave is blocked by the refractory tail of the original reentrant wave. From continuity arguments, it can be shown that there exists a range of stimuli phases and amplitudes that leads to successful reentry termination [Glass and Josephson 1995]. Unfortunately, the argument is essentially applicable only to a 1-D ring – the process is crucially dependent on the fact that the pacing site is on the reentry circuit itself. However, in reality, the location of the reentry circuit is typically not known when an electrical pacing device is implanted, and, it is unlikely that the pacing site will be so fortuitously located.

   In the following pages, we look at the dynamics of pacing from a site located some distance away from the reentry circuit. Such off-circuit pacing introduces the

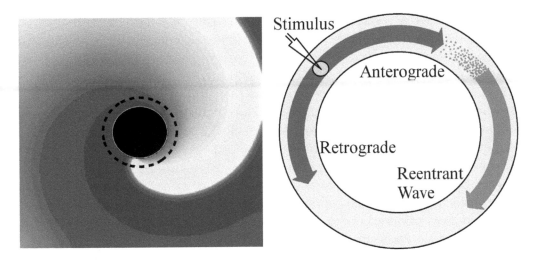

FIGURE 5.1 (left) Reentrant activity resulting from a spiral wave pinned to a non-conducting circular obstacle can be analyzed by focusing on the region immediately surrounding the perimeter of the obstacle (bounded by the broken curve in the figure). Unpinning the spiral wave by pacing from an external source to unpin the spiral wave, which is subsequently moved to the boundary by pacing, is thus equivalent to terminating the reentrant wave in the region around the obstacle that can be represented by a one-dimensional circuit. (right) The interaction of reentrant wave in a 1-D ring with waves generated as a result of stimulation from an external source. Each stimulation results in a retrograde and an anterograde wave. The former collides with the existing reentry thereby eliminating it, while the latter can, depending on the timing of the stimulation and the length of the circuit, either become the new reentrant wave (*resetting*) or can be blocked by the refractory region behind the initial reentrant wave, resulting in elimination of all activity in the ring (*termination*).

realistic propagation of the stimulus from the pacing site to the reentry circuit. Because reentrant waves propagate outwardly from, in addition to around, the circuit, the stimulus will be blocked before it reaches the circuit under most circumstances. Multiple stimuli are necessary to "peel back" refractory tissue incrementally until one successfully arrives at the reentry circuit [Josephson 1993]. Further, once a stimulus does reach the circuit, its anterograde branch must be blocked by the refractory tail of the reentrant wave, or resetting will occur and termination will fail. However, as outlined below, this is extremely unlikely to happen in a homogeneous medium.

Let us consider a reentrant circuit as a 1-D ring of length L with separate entrance and exit side branches (Fig. 5.2). This arrangement is an abstraction of the spatial geometry involved in anatomical reentry [Josephson 1993]. Further, let the pacing site be located on the entrance side branch at a distance z from the circuit. The entrance side branch is defined as the point of spatial origin ($x = 0$) to define the location of the wave on the ring. The conduction velocity and refractory period at a location a distance x away (in the clockwise direction) from the origin are denoted by $c(x)$ and $r(x)$, respectively.

For a homogeneous medium, $c(x) = c$, $r(x) = r$ (c, r are constants). Therefore, the length of the region in the ring which is refractory at a given instant is $l = cr$. For sustained reentry to occur, an excitable gap must exist (i.e., $L > c\,r$). Let us assume that restitution effects (i.e., the variation of the action potential duration as a function of the recovery time) can be neglected. Further, the circuit length L is considered to be large enough so that the reentrant activity is simply periodic. For convenience, associate $t = 0$ with the time when the reentrant wavefront is at $x = 0$ (i.e., the entrance side branch) [Fig. 5.2 (a)]. Let us assume that a stimulus is applied at $t = 0$. This stimulus will collide with the branch of the reentrant wave propagating out through the entrance side branch at $t = z/2c$ [Fig. 5.2 (b)]. The pacing site will recover at $t = r$ and if another stimulus is applied immediately it will reach the reentry circuit at $t = r+(z/c)$ [Fig. 5.2 (c)]. By this time the refractory tail of the reentrant wave will be at a distance $x = z$ away from the entrance side branch and the anterograde branch of the stimulus will not be blocked. Thus, when $z > 0$, it is impossible for the stimulus to catch up to the refractory tail in a homogeneous medium. This results in resetting of the reentrant wave rather than its termination.

Note that, if the first stimulus is given at a time $t < -z/c$, it reaches the reentry circuit before the arrival of the reentrant wave. As a result, the retrograde branch collides with the oncoming reentrant wave, while the anterograde branch proceeds to become the reset reentrant wave. Even in the very special circumstance that the reentrant wave reaches the entrance side branch exactly at the same instant that the stimulus reaches the circuit, the two colliding waves allow propagation to continue along the reentrant circuit through local depolarization at the collision site. As a result, pacing termination seems all but impossible in a homogeneous reentry circuit.

The situation changes, however, if an inhomogeneity (e.g., a zone of slow conduction) exists in the circuit [Fig. 5.2 (d)]. In this case, the above argument no longer holds because the inhomogeneity alters the electrophysiological dynamics (notably

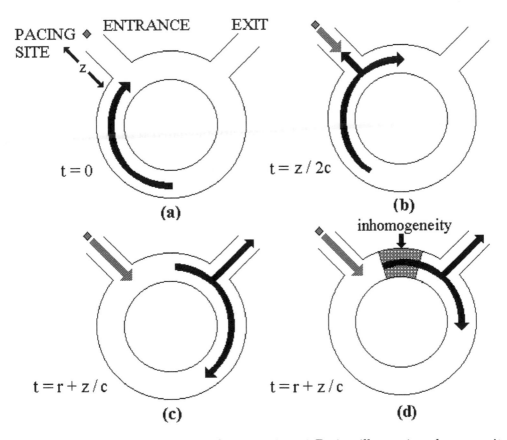

FIGURE 5.2 Schematic diagram of reentry in a 1-D ring illustrating the necessity of a region of inhomogeneity for successful termination of reentry by pacing. At $t = 0$ the reentrant wave reaches the entrance sidebranch (a). At $t = z/2c$ the reentrant wave propagating through the sidebranch and the first stimulus mutually annihilate each other (b). At $t = r + (z/c)$ the second stimulus reaches the reentry circuit by which time the refractory tail is a distance z away from the sidebranch (c). The presence of a region of inhomogeneity in the ring makes it possible that the anterograde branch of the second stimulus will encounter a refractory region behind the reentrant wave (d).

refractory period) of the excitation waves. As a result, stimuli may arrive at the circuit from the pacing site and encounter a region that is still refractory. This leads to successful block of the anterograde branch of the stimulus, while the retrograde branch annihilates the reentrant wave (as in the homogeneous case), resulting in successful termination.

The following simulation results also support the conclusion that the existence of inhomogeneity in the reentry circuit is essential for pacing termination of VT. Let us consider a cardiac impulse propagating in a continuous one-dimensional ring of tissue (ignoring the microscopic cell structure) with ring length L, which represents a closed pathway of circus movement around an anatomical obstacle (e.g., scar tissue). The initial condition is a stimulated wave at some point in the medium with transient conduction block on one side to permit wavefront propagation in a single direction only. Extra-stimuli are introduced from a pacing site.

To understand the process of inhomogeneity-mediated termination one can introduce a zone of slow conduction in the ring. Slow conduction has been observed in cardiac tissue under experimental ischemic conditions (lack of oxygen to the tissue) [Kleber et al. 1987] and the conduction velocity in affected regions is often as low as 10% of the normal propagation speed in the ventricle [de Chillou et al. 2002]. This phenomenon can be reasonably attributed to a high degree of cellular uncoupling [Rohr et al. 1998], as demonstrated by model simulations [Quan and Rudy 1990]. Here slow conduction is implemented by varying the diffusion constant D' from the value of D used for the remainder of the ring. The length and diffusion constant(D') of the zone are varied to examine their effect on the propagation of the anterograde branch of the stimulus. Varying the length of the zone of slow conduction (specifically, between 1.5 mm and 25 mm) has no qualitative effect on the results. For the simulation results shown below $D = 0.556$ cm^2/s and $D' = 0.061$ cm^2/s, corresponding to conduction velocities $c \simeq 47$ cm/s and $c' \simeq 12$ cm/s, respectively, which are consistent with the values observed in human ventricles [de Chillou et al. 2002; Surawicz and Surawicz 1995].

The reentrant wave activates the point in the ring (proximal to the zone of slow conduction) chosen to be the origin ($x = 0$), where the stimulus enters the ring, at time $t = T_0$. At time $t = T_1$, an activation wave is initiated through stimulation at $x = 0$. If this first stimulus is unable to terminate the reentry, a second stimulus is applied at $t = T_2$, again at $x = 0$. Note that, the first stimulus is always able to terminate the reentry if it is applied when the region on one side of it is still refractory — leading to unidirectional propagation. This is identical to the mechanism studied previously for terminating reentry by pacing within a 1-D ring [Glass and Josephson 1995]. However, here we are interested in the effect of pacing from a site away from the reentry circuit. In this case it is generally not possible for the first stimulus to arrive at the reentry circuit exactly at the refractory end of the reentrant wave (as discussed above). Therefore, values of T_1 have been used for which the first stimulus can give rise to both an anterograde, as well as, a retrograde branch. The only mechanism of reentry termination considered here is that which takes place through the anterograde branch of the stimulated wave being blocked in the zone of slow conduction.

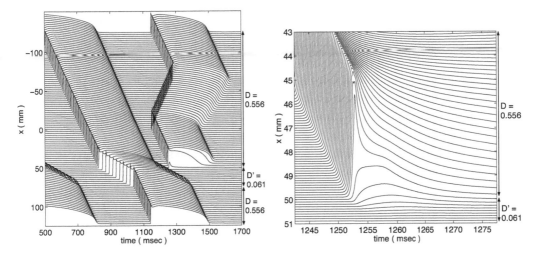

FIGURE 5.3   (left) Plot of the membrane potential, V, showing spatiotemporal propagation of a reentrant wave in a Luo–Rudy ring of length 250 mm, successfully terminated by pacing with a single stimulus. The zone of slow conduction is between $x =$
50 mm and $x = 75$ mm. In this region the diffusion constant changes from $D = 0.556$
to $D' = 0.061$ cm^2/s with an infinite gradient at the boundaries. The reentrant wave
activates the site at $x = 0$ mm at $T_0 = 718.22$ ms. The stimulus is applied at $x = 0$
mm at $T_1 = 1146.22$ ms (coupling interval $= 428.00$ ms). (right) Magnification of the
above plot for the region between $x = 43$ mm and $x = 51$ mm and the time interval
$1242.50 \leq t \leq 1277.50$ ms. Note that, inside the inhomogeneity, no action potential is
generated.

Figure 5.3 (left) shows an instance of successful termination of the reentrant wave where the anterograde branch of the stimulus applied at $T_1 = 1146.22$ ms ($T_1 - T_0 = 428.00$ ms) is blocked at the boundary of the zone of slow conduction (at time $t \simeq 1253$ ms, $x = 50$ mm). Figure 5.3 (right) shows a magnified view of the region at which conduction block occurs. The region immediately within the zone of slow conduction shows depolarization but not an action potential (i.e., the excitation is sub-threshold), while the region just preceding the inhomogeneity has successively decreasing action potential durations. The peak membrane potential attained during this sub-threshold depolarization sharply decreases with increasing depth into the inhomogeneity so that, at a distance of 0.5 mm from the boundary inside the zone of slow conduction, no appreciable change is observed in the membrane potential V. For the same simulation parameters as Fig. 5.3, a single stimulus applied at $T_1 - T_0 > 428.63$ ms is not blocked at the inhomogeneity, and a second stimulus needs to be applied at time T_2 to terminate reentry.

Different values of coupling interval ($T_1 - T_0$) and pacing interval ($T_2 - T_1$) are used to find which parameters led to block of the anterograde wave. Figure 5.4 is a parameter space diagram which shows the different parameter regimes where termination is achieved.

The conduction block of the anterograde branch of the stimulated wave is due to a dynamical effect linked to a local increase of the refractory period at the boundary of the zone of slow conduction in the reentrant circuit. To obtain an idea about the variation of refractory period around the inhomogeneity, a closely related quantity, viz., the action potential duration (APD), is measured. Figure 5.5 shows the variation of the APD as the reentrant wave propagates across the ring. The APD of an activation wave is approximated as the time-interval between successive crossing of -60 mV by the transmembrane potential V. Note that there are other methods of measuring APD which may make minor quantitative differences, but would not qualitatively change the findings. As the wave crosses the boundary into the region of slow conduction, the APD increases sharply. Within this region, however, the APD again decreases sharply. Away from the boundaries, in the interior of the region of slow conduction, the APD remains constant until the wave crosses over into the region of faster conduction again, with a corresponding sharp decrease followed by a sharp increase of the APD, around the boundary. This result is consistent with the observation by Keener [Keener 1991] that non-uniform diffusion has a large effect on the refractory period in discrete systems.

However, the APD lengthening alone cannot explain the conduction block. Figure 5.5 shows that the maximum APD, at the boundary of the inhomogeneity, is ~ 345 ms, yet conduction failed for a coupling interval of 428.00 ms. The blocking of such a wave (i.e., one that is initiated at a coupling interval that is significantly longer than the previous APD) implies that APD actually underestimates the effective refractory period. While it is believed that such "post-repolarization refractoriness" does not occur in homogeneous tissue, it has been observed experimentally for discontinuous propagation of excitation [Jalife and Delmar 1991; Jalife et al. 1983]. Post-repolarization refractoriness is caused by residual repolarizing ionic current that persists long after the transmembrane potential has returned almost

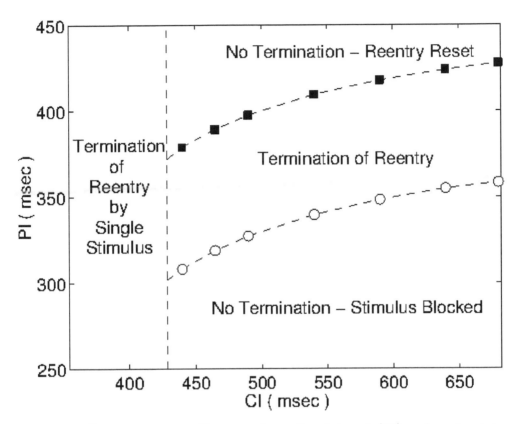

FIGURE 5.4 Parameter space diagram of coupling interval (CI) and pacing interval (PI) at which termination occurs in the 1-D Luo–Rudy ring of length 250 mm with a zone of slow conduction between $x = 50$ mm and 75 mm ($D = 0.556$ cm^2/s, $D' = 0.061$ cm^2/s). The VT period around the ring is 692.46 ms. The line connecting the circles represents the critical pacing interval value below which the second stimulus gets blocked by the refractory tail of the first stimulus. The region between the circles and squares represents the regime in which the second stimulus is blocked in the antero-grade direction in the zone of slow conduction (leading to successful termination) for a boundary with infinite gradient. For longer PI (the region above the squares), the second stimulus propagates through the inhomogeneity and only resets the reentrant wave. Note that the APD of the reentrant wave at $x = 0$ is approximately 340.6 ms. For CI < 355.29 ms, the first stimulus is blocked at $x = 0$, while for CI > 428.63 ms, the anterograde branch of the first stimulus is not blocked at the zone of slow conduction (leading to resetting of the reentrant wave). For intermediate values of CI (i.e., between $355.29 - 428.63$ ms), the first stimulus is able to successfully terminate reentry.

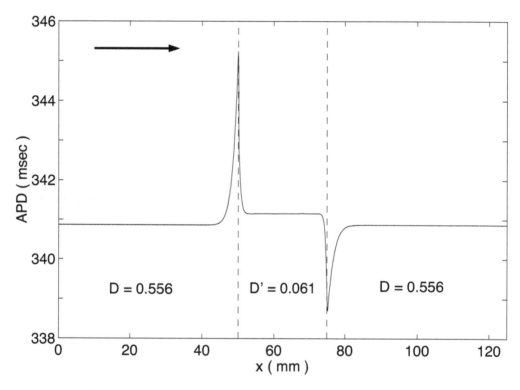

FIGURE 5.5   Variation of the action potential duration (APD) for the reentrant wave as it proceeds from $x = 0$ to $x = 125$ mm in the ring. The arrow indicates the direction of wave propagation. The dashed lines between $x = 50$ mm and 75 mm enclose the region where the diffusion constant changes from $D = 0.556$ to $D' = 0.061$ cm^2/s with an infinite gradient at the boundaries.

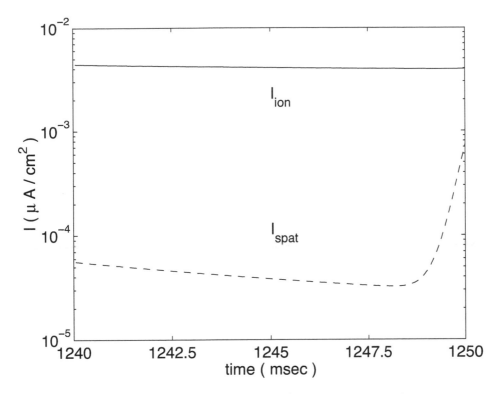

FIGURE 5.6 The ionic current I_{ion} (solid curve) and the spatial electrotonic current $I_{spat} = C_m D \nabla^2 V$ (broken curve) at $x = 50$ mm (the boundary of the inhomogeneity). Note that, initially, the latter is two orders of magnitude smaller than the former. At $t \simeq 1248.5$ ms, I_{spat} begins to increase due to current inflow from the upstream excited neighboring region. The high value of I_{ion} throughout this time indicates that the boundary of the inhomogeneity has not yet recovered from the passage of the previous wave by the time the next stimulated wave arrives.

to its quiescent value. As seen in Fig. 5.6, this indeed occurs in the system — at the time when the electrotonic current of the stimulated wave begins entering the inhomogeneity, the ionic current has not fully recovered from the previous wave, and is actually working against the electrotonic current ($I_{ion} > 0$). If the stimulated wave is early enough, this ionic current mediated post-repolarization refractoriness is enough to prevent the electrotonic current from depolarizing the zone of slow conduction.

This prolongation of the refractory period at the border of the inhomogeneity is the underlying cause of the conduction block leading to successful termination of reentry, as can be seen in Fig. 5.7. Here, the behavior at the inhomogeneity boundary when the wave is blocked (CI = 428.00 ms) is compared with the case when the wave does propagate through the inhomogeneity (CI = 428.64 ms). As already mentioned, in the case of block, the depolarization of the region immediately inside the zone of slow conduction is not sufficient to generate an action potential. Instead, the electrotonic current increases the transmembrane potential V to ~ -55 mV and

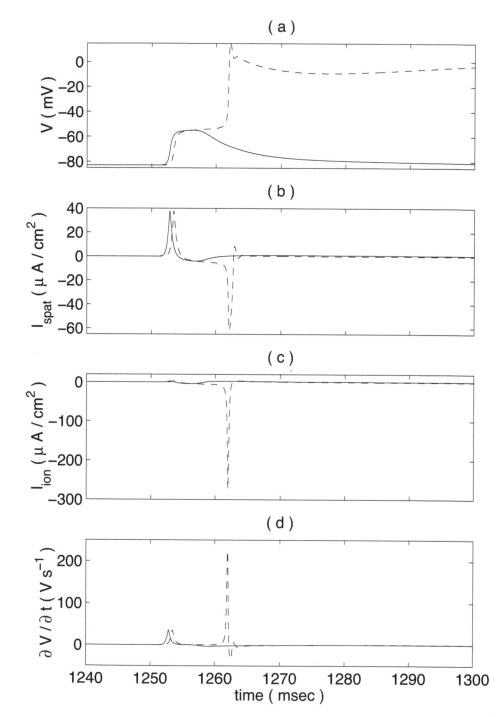

FIGURE 5.7 The (a) transmembrane potential V, (b) the spatial (electrotonic) current $I_{spat} = C_m D\nabla^2 V$, (c) the ionic current I_{ion}, and (d) the rate of change of transmembrane potential $\partial V/\partial t = D\nabla^2 V - (I_{ion}/C_m)$, at $x = 50$ mm for coupling intervals CI = 428.00 ms (solid curves) and 428.64 ms (broken curves) [$D = 0.556$ cm^2/s, $D' = 0.061$ cm^2/s]. In the former case, the initial depolarization is insufficient to generate an action potential and the excitation wavefront is blocked. In the latter case, the region has recovered sufficiently so that the stimulation is able to generate an action potential and the wave propagates through the inhomogeneity.

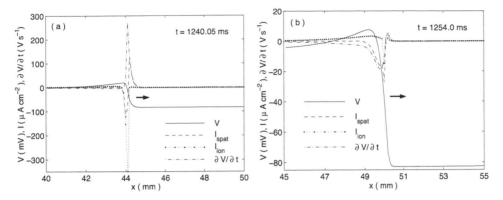

FIGURE 5.8 The spatial variation of transmembrane potential V, the spatial current $I_{spat} = C_m D \nabla^2 V$, the ionic current I_{ion}, and the rate of change of transmembrane potential $\partial V / \partial t = D \nabla^2 V - (I_{ion}/C_m)$, in (a) a homogeneous region of tissue (at $t = 1240.05$ ms) and (b) at the border of slow conduction (at $t = 1254.00$ ms). The zone of slow conduction is between $x = 50$ mm and $x = 75$ mm ($D = 0.556$ cm^2/s, $D' = 0.061$ cm^2/s) and the coupling interval is 428.00 ms. In (a) the excitation wavefront propagates normally, while in (b) the wavefront is blocked at the boundary of the inhomogeneity. The arrows indicate the direction of propagation of the wavefront.

then slowly decays back to the resting state. On the other hand, if the stimulation is applied only 0.64 ms later, V (as in the earlier case) rises to ~ -55 mV and then after a short delay (~ 8 ms) exhibits an action potential. This illustrates that in the earlier case the region had not yet recovered fully when the stimulation had arrived.

Figure 5.8 shows in detail how the difference in diffusion constant in the inhomogeneity leads to conduction block. (For similar analysis of the qualitatively different dynamics of conduction block that occurs as a wave moves from a region of impaired coupling to normal tissue, see [Wang and Rudy 2000]). In the region with normal diffusion constant [Fig. 5.8 (a)], the initial stimulation for a region to exhibit an action potential is provided by current arriving from a neighboring excited region. This spatial "electrotonic" current $I_{spat} = C_m D \nabla^2 V$ is communicated through gap junctions which connect neighboring cardiac cells. As soon as the "upstream" neighboring region is excited with a corresponding increase in potential V, the difference in membrane potentials causes an electrotonic current to flow into the region under consideration (positive deflection in I_{spat}). This causes the local potential to rise and subsequently, an outward current to the "downstream" neighboring non-excited region is initiated (negative deflection in I_{spat}). Thus, the initial net current flow into the region is quickly balanced and is followed, for a short time, by a net current flow out of the region, until the net current I_{spat} becomes zero as the membrane potential V reaches its peak value. The change in V due to I_{spat} initiates changes in the local ionic current I_{ion}. This initially shows a small positive hump, followed by a large negative dip (the inward excitatory rush of ionic current), which then again sharply rises to a small positive value (repolarizing current) and gradually goes to zero. The net effect of the changes in I_{spat} and I_{ion} on the local membrane potential V is reflected in the curve for $\partial V/\partial t = D\nabla^2 V - (I_{ion}/C_m)$. This shows a rapid increase to a large positive value and then a decrease to a very small negative value as the membrane potential V reaches its peak value followed by a slight decrease to the plateau phase value of the action potential.

At the boundary of the inhomogeneity, however, the current I_{spat} is reduced drastically because of the low value of the diffusion constant D' at the inhomogeneity [Figs. 5.7 (b) and 5.8 (b)]. Physically, this means that the net current flow into the cell immediately inside the region of slow conduction is much lower than in the normal tissue. This results in a slower than normal depolarization of the region within the inhomogeneity. Because the region is not yet fully recovered, this reduced electrotonic current is not sufficient to depolarize the membrane beyond the excitation threshold. Therefore, the depolarization is not accompanied by the changes in ionic current dynamics needed to generate an action potential in the tissue [Fig. 5.7 (c)]. The resulting $\partial V/\partial t$ curve therefore shows no significant positive peak. Hence, the initial rise in V is then followed by a decline back to the resting potential value as is reflected in the curve for $\partial V/\partial t$ in Fig. 5.7 (d).

When the stimulation is given after a longer coupling interval $(T_1 - T_0)$, i.e., after the cells in the inhomogeneous region have recovered more fully, a different sequence of events occurs. As in the case shown in Fig. 5.8 (e), the depolarization of the region immediately inside the zone of slow conduction is extremely slow because of the reduced electrotonic current I_{spat}. However, because the tissue has had more time to recover, the electrotonic current depolarizes the membrane potential beyond the threshold and the ionic current mechanism responsible for generating the action potential is initiated (as illustrated by the broken curves in Fig. 5.7). As a result, the excitation is not blocked but propagates through the inhomogeneity, although with a slower conduction velocity than normal because of the longer time required for the cells to be depolarized beyond the excitation threshold.

To ensure that these results are not model dependent, especially on the details of ionic currents, a modified FitzHugh–Nagumo-type excitable media model of ventricular activation proposed by Panfilov [Panfilov 1998; Panfilov and Hogeweg 1993] has also been investigated. The details about the simulation of the Panfilov model are identical to those given in Refs. [Sinha et al. 2001, 2002]. As this two-variable model lacks any description of ionic currents and does not exhibit either the restitution or dispersion property of cardiac tissue, quantitative changes in termination requirements are indeed observed. For example, the parameter regions at which termination occurs for the Panfilov model are different from those shown in Fig. 5.4 for the Luo–Rudy model. Nevertheless, despite quantitative differences, these fundamentally different models shared the requirement of an inhomogeneity for termination, thereby supporting the model independence of the above results.

There are some limitations of this study, the most significant being the use of a 1-D model. However, studies using 2-D excitable media models of anatomical reentry [Sinha et al. 2002] have show similar results (discussed later in the chapter). Also, a monodomain formulation of the excitable media has been used instead of considering a bidomain model which would have involved separate sets of equations describing the dynamics in the intracellular and extracellular spaces, respectively.

However, for very low amplitude stimulation (as used above) the two formulations give identical results. Also, in order to simulate ischemic tissue where slow conduction occurs a higher degree of cellular uncoupling has been used. There are, of course, other ways of simulating ischemia [Shaw and Rudy 1997], e.g., by increasing the external K^+ ion concentration [Xie et al. 2001a]. But this is transient and does not provide a chronic substrate for inducing irregular cardiac activity. Further, ischemic zone is only one of a number of possible types of inhomogeneity that can occur in biological tissue. For example, the existence of a region having longer refractory period will lead to the development of patches of refractory zones in the wake of the reentrant wave. If the anterograde branch of the stimulus arrives at such a zone before it has fully recovered, it will be blocked [Abildskov and Lux 1995].

Despite these limitations, the results described above offer insight into pacing termination of reentrant activity as they involve general (i.e., model independent) mathematical arguments which have been verified by numerical simulations. Given the clear dynamical similarity between the bidirectional block required for reentry termination and the unidirectional block required for reentry initiation [Rudy 1995], it is noteworthy that the inhomogeneity-mediated mechanism of termination suggested here is consistent with the widely accepted theory that initiation of reentry is facilitated by circuit inhomogeneities [Rudy 1995; Zhang et al. 1998] (but does not necessarily require a zone of slow conduction, in particular [Girouard and Rosenbaum 2001]). The key observation is that inhomogeneities are required in the reentrant circuit for successful termination of structural heterogeneity induced reentry (as in certain types of tachycardia) when external stimulation arrives from a distant source. It is possible that more effective pacing algorithms for treating arrhythmia can be developed which explicitly consider any pre-existing inhomogeneities on the heart surface obtained by cardiac electrical mapping.

5.3 Termination of reentry in two-dimensional heterogeneous excitable medium

In this section we look at how inhomogeneities play a critical role in successful termination of reentry in a two-dimensional excitable medium. Extending the study of reentry termination to higher dimensions has immense practical significance in the context of treating cardiac arrhythmias that are related to structural disorder in the heart. The underlying substrate for anatomical reentry (i.e., reentrant waves circulating around an existing conduction or ionic heterogeneity) in the ventricles could be scar tissue that serves as a conduction barrier. This is often the result of myocardial-infarction induced ischemia, i.e., oxygen deprivation of cardiac tissue due to obstruction of the coronary artery supplying oxygenated blood to that area. Experimental studies on patterns of activation during the acute phase of myocardial ischemia have shown reentrant activity around and across the ischemic border zone (e.g., in tissue which has healed after undergoing an episode of myocardial infarction [El-Sherif et al. 1981]). This type of activity often takes the form of "figure-of-eight reentry," which consists of two adjacent reentry circuits. Propagation in these cir-

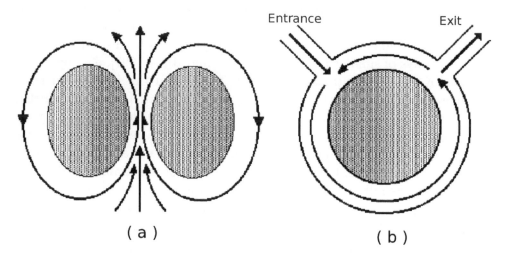

FIGURE 5.9 (a) Schematic diagram of "figure-of-eight" reentry around two anatomical obstacles (shaded regions represent the ischemic central zone). Normal tissue and ischemic border zone are not differentiated in this figure. The arrows indicate the direction of the two counter-propagating rotating waves. (b) One half of the spatial arrangement involved in figure-of-eight reentry is used to set up a quasi-1-D analog model. This abstract model of anatomical reentry has an entrance and an exit sidebranch to signify the region of cardiac myocardium surrounding the circuit.

cuits occurs within normal tissue and the ischemic border zone, while the anatomical barriers are the scar tissue comprising the ischemic central zone [Fig. 5.9]. This situation may be fatal if the VT rate is too rapid or if, as occurs in a large number of cases, the VT degenerates into ventricular fibrillation [Rosenthal and Josephson 1990].

To study pacing termination of reentry, one can look specifically at geometries of anatomical obstacles which give rise to a figure-of-eight pattern. For instance, in two dimensions, this can be modeled by using a pair of rectangular inexcitable patches ("scar tissue") at the center of the simulation domain with a narrow channel for wave propagation between them. In order to investigate the effect of pacing from a location at some distance from the reentry circuit, the pacing site is placed away from the immediate vicinity of the patches. The one-dimensional model that we looked at earlier can be seen to be derived from this geometry where the reentry circuit around only one of the inexcitable regions is being considered. As the pacing site is located away from the circuit, an entrance and an exit sidebranch have to be considered explicitly in the one-dimensional model [Fig. 5.9 (b)]. Real scar tissue is unlikely to have regular shapes like squares or circles — rather they will have irregular contours. However, if one considers only the portion of the reentry circuit immediately surrounding the inexcitable region, then this can be modeled by a smooth circuit such as a circle. The one-dimensional model represents this kind of circuit, which doesn't have any sharp edges. To ensure that the introduction of angular extremities does not change the qualitative picture, one can look at scars

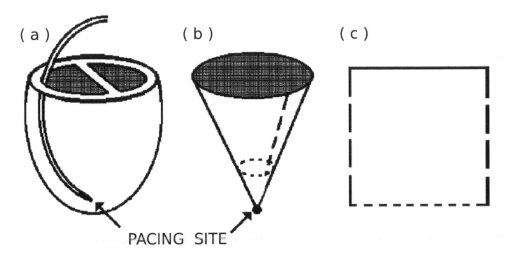

(a)   (b)   (c)

PACING SITE

FIGURE 5.10 (a) Schematic diagram of a typical cardiac pacing arrangement with an electrode from the ICD attached to the endocardial surface of the right ventricle. The electrode (shaded black) is at the apex of the ventricle (pacing site). (b) The ventricles can be represented by a cone with the electrode from the ICD attached at the apex — the pacing site. The broken lines indicate the way the cone is "opened up" to form the 2-D simulation domain shown in (c). The long dashes in (b) represent the line along which the cone is cut to "open" it — the corresponding vertical edges in the 2-D representation therefore have periodic boundary conditions. The line formed by the short dashes truncate the cone a fixed distance above the apex. The region beyond this line (including the pacing site) is not explicitly included in the 2-D plane shown in (c) — propagation through this region could be simply accounted for via a fixed delay time.

in the shape of squares in the two-dimensional model simulations. If the presence of sharp edges in the square patches of inexcitable tissue do not qualitatively affect the simulation results, one may conclude that the conclusions of such a study may be valid even when considering inexcitable regions with realistic irregular contours.

A few studies have looked at the role of inhomogeneities in two-dimensional excitable media [Abildskov and Lux 1995; Panfilov and Keener 1993]. Ref. [Abildskov and Lux 1995] examined the role of a region of increased refractoriness placed in the reentry circuit, while Ref. [Panfilov and Keener 1993] investigated the effects of high-frequency stimulation of excitable media with an inexcitable obstacle. Both studies noted conditions in which reentrant patterns are eliminated. However, by building upon the 1-D model results presented earlier and connecting it to 2-D models, in this section the critical role of inhomogeneities in pacing termination of reentry will be established more firmly.

The model study described here uses the excitation kinetics proposed by Panfilov [Panfilov and Hogeweg 1993]. The spatial grid consists of a square lattice with $L \times L$ points with values of L ranging from 128 to 256. For the top and bottom of the 2-D simulation domain no-flux (Neumann) boundary conditions are used because the ventricles are electrically insulated from the atria (top) and the waves converge and annihilate at the region around the apex of the heart (see Fig. 5.10). For the sides, periodic boundary conditions are used to represent the "wrap-around" nature of the ventricle. Two rectangular patches of inexcitable region (with conductivity constant zero) are placed about the center of the simulation domain with a narrow passage for wave propagation between them. No-flux boundary conditions are used at the interface between the active medium and the inexcitable obstacle. This arrangement is used to represent anatomical obstacles (e.g., scar tissue) around which "figure-of-eight" reentry can occur, as depicted schematically in Fig. 5.9. The initial condition is a wavefront initiated at some point in the narrow conducting channel between the two inexcitable regions. A conduction block immediately below the point of initiation (implemented by making the corresponding region refractory) permits the propagation of the wave along a single direction (upward) only. The wave proceeds along the narrow channel and then around the two anatomical obstacles, setting up a figure-of-eight reentry pattern.

We can now investigate the effectiveness of different pacing algorithms in terminating figure-of-eight reentry. Here pacing is simulated by a planar wave initiated at the base of the simulation domain. This represents a wave that has propagated a fixed distance up the ventricular wall from a local electrode at the apex — similar to a wave initiated at the tip of an inverted cone before propagating up its walls [Fig. 5.10 (b)]. Two different pacing patterns are applied. The first uses pacing bursts at a cycle length which is a fixed percentage of the tachycardia cycle length (determined by the size of the obstacles). This corresponds to "burst pacing" used in ICDs [Horwood et al. 1995]. The pacing cycle length as well as the number of extra-stimuli in each pacing burst (4 to 8) are varied to look at their effects on the reentrant wave. The second pacing pattern used is decremental ramp pacing in which the interval between successive extra-stimuli is gradually reduced (corresponding to "ramp pacing" used in ICDs [Horwood et al. 1995]).

For a homogeneous cardiac medium, both ramp and burst pacing are unsuccessful at terminating reentry. An example of the time evolution of such a procedure with a 4-stimulus pacing burst is shown in Fig. (5.11). Pacing trains containing up to 8 successive pulses and a comprehensive range of combinations of coupling and pacing intervals were considered but none resulted in successful termination. The reason for the failure of pacing to terminate reentry is similar to that in the 1-D model. As the pacing wave approaches the reentry circuit, it is intercepted by the outwardly propagating wavefront originating from the reentry circuit, resulting in mutual annihilation (not shown). During the ensuing refractory period a pacing wave cannot approach the circuit at all. If a stimulus is applied immediately after the end of the refractory period, the wave propagates into the simulation domain (see the panels corresponding to $t = 495\text{-}550$ ms in Fig. 5.11), mutually annihilating with the reentrant wave at some point in the circuit (as occurs in the regions between the anatomical barriers and the vertical domain boundaries in the time between the $t = 550$ ms and 605 ms snapshots of Fig. 5.11). However, the branch of the stimulated wave proceeding along the narrow passage between the inexcitable obstacles continues on its path (see the panel corresponding to $t = 605$ ms in Fig. 5.11).

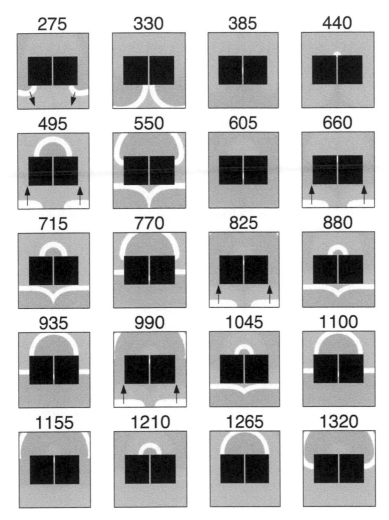

FIGURE 5.11 Snapshots of waves (in white) at different times (labelled in ms above each plot) for the excitability e, upon application of a 4-stimulus pacing burst (applied at 462 ms, 627 ms, 792 ms, and 957 ms). Upon first appearance of each wave, its direction is indicated by a black arrow. The initial condition is a wave rotating around the rectangular obstacles (in black) as seen in the first four panels. The simulation domain (128 mm × 128 mm) is completely homogeneous with a conductivity constant $D = 1$. The two inexcitable obstacles are each 48.5 mm long and 43 mm wide, while the channel between the two obstacles has a width of 2.5 mm. Pacing failed to terminate the reentrant wave.

When stimulation is halted, this intrachannel impulse gives rise again to a reentrant wave (e.g., the panels corresponding to $t = 1155$-1320 ms in Fig. 5.11). This merely resets the reentry rather than terminating it. Only by blocking the branch of the stimulus propagating along the narrow channel is it possible to achieve successful termination. As shown in Fig. (5.12), termination via such intrachannel blocking is possible in the presence of inhomogeneity. This is accomplished by placing a small zone (7.5 mm in length) of slow conduction in the narrow channel between the two non-conducting patches. The conductivity in this region is 0.05 times smaller than the rest of the tissue. A 4-stimulus pacing burst (with a pacing interval of 165 ms) is found to successfully block the anterograde branch of the extra-stimulus traveling through the narrow channel and hence terminate the reentry (see the panels corresponding to $t = 1155$-1320 ms in Fig. 5.12). The timing of the stimulus relative to the reentrant wave is important, as applying the pacing at a slightly later time (but with the same pacing interval) leads to failure of termination (Fig. 5.13).

Cardiac tissue is an example of anisotropic excitable medium, i.e., the action potential propagates faster along the direction of the muscle fibers (longitudinal) than perpendicular to them (transverse). Although this is also a type of inhomogeneity, the spatial conductivity gradient is much more gradual than that of the inhomogeneous region used in the simulation shown in Fig. 5.12. As we have seen for one-dimensional models, the parameter space in which successful termination can occur decreases with the increasing smoothness of the transition at the boundary of the inhomogeneity. This implies that there is a critical spatial gradient of conductivity below which the inhomogeneity will not block the anterograde branch of the applied stimulus. Because anisotropy is a very gradual inhomogeneity, anterograde wave propagation is not disrupted enough by the anisotropy to lead to block and termination of propagation. This conclusion is supported by 2-D simulations in which the effect of anisotropy is investigated by making the conductivities along the vertical and the horizontal axes differ with a ratio of 1:0.3, a value that is consistent with cardiac tissue [Keener and Panfilov 1997]. As with the homogeneous case, in the anisotropic case none of the pacing methods tried were successful in terminating reentry (one example is shown in Fig. 5.14).

For the 2-D studies, in addition to the Panfilov model, two other 2-variable modified FitzHugh–Nagumo type models: the Aliev–Panfilov model (parameters are those given in Ref. [Aliev and Panfilov 1996]), which simulates the restitution property of cardiac tissue, as well as the model proposed by Kogan et al. (parameters are those given in Ref. [Kogan et al. 1991]), which simulates the restitution and dispersion properties of cardiac tissue. In both cases the results are qualitatively similar to that obtained in the case of the Panfilov model and support the model independence of the 2-D results. A possible limitation of the above study is that the heart has a three-dimensional structure which is not being considered here. The rotation of the axis of anisotropy along the thickness of the myocardium governs the propagation of impulses along the ventricle, this may have important consequences for reentrant activity and interaction with external stimulation.

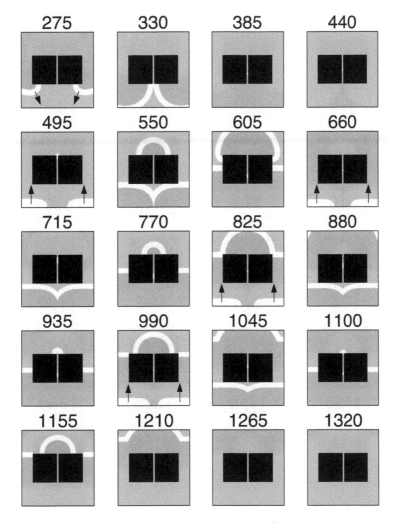

FIGURE 5.12 Snapshots of waves at different times (labeled in ms above each plot) for the excitability e, upon application of a 4-stimulus pacing burst (applied at 462 ms, 627 ms, 792 ms, and 957 ms). The simulation domain has a region of slow conduction in the narrow channel between the two inexcitable obstacles, of length 7.5 mm and a conductivity constant $D' = 0.05$. The fourth stimulus is blocked at the boundary of the inhomogeneity, leading to successful termination by $t \sim 1250$ ms.

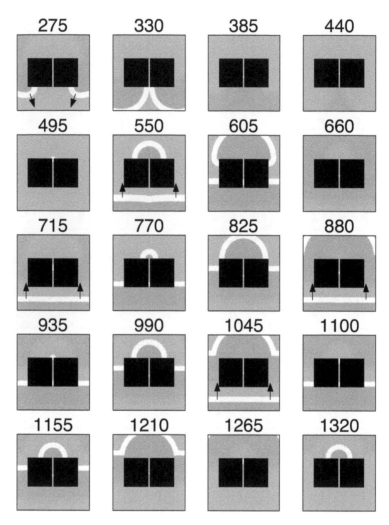

FIGURE 5.13 Snapshots of waves at different times (labeled in ms above each plot) for the excitability e, upon application of a 4-stimulus pacing burst (applied at 495 ms, 660 ms, 825 ms, and 990ms). The simulation domain has a region of slow conduction in the narrow channel between the two inexcitable obstacles, of length 7.5 mm and a conductivity constant $D' = 0.05$. The distinction from Fig. 5.12 is in the relative timing of the stimulus with respect to the reentrant wave — which results in failure of termination. The pacing interval (=165 ms) is the same in both figures. Pacing failed to terminate the reentrant wave.

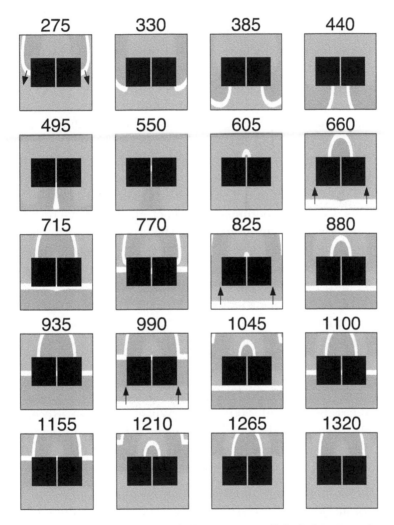

FIGURE 5.14 Snapshots of waves at different times (labeled in ms above each plot) for the excitability e, upon application of a 4-stimulus pacing burst (applied at 462 ms, 627 ms, 792 ms, and 957 ms). The simulation domain has an anisotropy ratio of 1:0.3 for the conductivity constants, with the fast axis oriented along the up-down direction. Pacing failed to terminate the reentrant wave.

5.4   Effect of obstacles on long-term stability of spiral waves: Fractal basin of boundaries

As is evident from the earlier discussions, the role of heterogeneities in excitable media appear to be crucial for the initiation of a rotating spiral wave (RS), as well as the initiation and maintenance of spiral turbulence (ST) resulting from multiple breakup of spiral waves. Unfortunately, the precise mechanisms by which reentrant activity is affected by conduction inhomogeneities in excitable media such as ventricular tissue is still not well understood [Weiss et al. 2000]. While it is known that spiral waves can appear as a result of reentrant activity being triggered under certain conditions around a pre-existing non-conducting obstacle (i.e., anatomical reentry) [Agladze et al. 1994; Wiener and Rosenblueth 1946], they can also arise in the absence of any structural disorder [Allessie et al. 1977]. The interaction of such spiral waves arising through functional reentry with an anatomical obstacle can be quite complex especially in the spatiotemporally chaotic state associated with spiral turbulence. Indeed, experiments with obstacles in cardiac tissue have yielded a variety of results. Certain experiments [Ikeda et al. 1997] have reported that while small obstacles do not affect spiral waves, as the size of the obstacle is increased, a wave can get pinned to the obstacle. Other experiments have discussed the role of an anatomical obstacle as an anchoring site for spiral waves, which can lead to the conversion of ST into RS [Kim et al. 1999; Pertsov et al. 1993; Valderrabano et al. 2000]. Davidenko et al. [Davidenko et al. 1992] have found that, when they induced spiral waves in cardiac tissue preparations "... in most episodes, the spiral was anchored to small arteries or bands of connective tissue, and gave rise to stationary rotations. In some cases the core drifted away from its site of origin and dissipated at the tissue border." Other studies have shown that an obstacle, in the path of a moving spiral wave, can break it and lead to many competing spiral waves [Ohara et al. 2001; Valderrabano et al. 2001, 2003; Wu et al. 1998]. Recent experiments by Hwang et al. [Hwang et al. 2005] have suggested that multistability of spirals with different periods in the same cardiac-tissue preparation can arise because of the interaction of spiral tips with small-scale inhomogeneities.

Conduction inhomogeneities in the ventricle include scar tissues, resulting from an infarction, or major blood vessels. Some theoretical studies of the effects of tissue inhomogeneities have been carried out by using model equations for cardiac tissue; however, they have not addressed the issues that are going to be discussed below. The interaction of an excitation wave with piecewise linear obstacles has been studied by Starobin et al. [Starobin and Starmer 1996] to understand the role of obstacle curvature in the pinning of such waves. Xie et al. [Xie et al. 1998] have considered spiral waves around a circular obstacle and given a plausible connection of the ST-RS transition to the size of the obstacle. Panfilov et al. [Panfilov 2002; ten Tusscher and Panfilov 2003b, 2005] have shown that a high concentration of randomly distributed non-excitable cells can suppress spiral breakup. Conduction inhomogeneities can also play a very important role in pacing termination of cardiac arrhythmias [Sinha et al. 2002]; in particular, it is easier to remove a spiral wave once it is pinned to an obstacle, as described in Refs. [Biktashev and Holden 1998; Takagi et al. 2004], than to control a state with spiral turbulence.

Here we describe a study which systematically investigates the effects of conduction inhomogeneities in excitable media models [Shajahan et al. 2007]. The results show that the position of obstacles of different sizes determines whether ST can be suppressed or not. This *sensitive dependence* on the sizes and positions of obstacles is a manifestation of a complex, fractal-like boundary [Lai and Winslow 1994; Sommerer and Ott 1993] between the domains of attraction of ST and RS. Inhomogeneities in other model parameters, such as those which govern the ratio of time scales of the excitation kinetics, also show similar features.

The results are described below have been obtained using the Panfilov [Panfilov 1998; Panfilov and Hogeweg 1993] and Luo–Rudy I [Luo and Rudy 1991; Shajahan et al. 2005] models of excitable systems. The former is well suited for extensive numerical studies because of its relative simplicity, while the latter, being realistic, allows one to check that the results obtained are qualitatively correct and not artifacts of a simple model.

The Panfilov model [Panfilov 1998; Panfilov and Hogeweg 1993], discussed earlier in Chapter 2, is described by:

$$\begin{aligned} \partial V/\partial t &= \nabla^2 V - f(V) - g; \\ \partial g/\partial t &= \epsilon(V,g)(kV - g), \end{aligned} \qquad (5.1)$$

where V is the transmembrane potential (activation variable) and g is the recovery variable. The initiation of action potential is encoded in $f(V)$, which is piecewise linear: $f(V) = C_1 V$, for $V < e_1$, $f(V) = -C_2 V + a$, for $e_1 \leq V \leq e_2$, and $f(V) = C_3(V-1)$, for $V > e_2$. The physically appropriate parameters given in Refs. [Panfilov 1998; Panfilov and Hogeweg 1993] are $e_1 = 0.0026$, $e_2 = 0.837$, $C_1 = 20$, $C_2 = 3$, $C_3 = 15$, $a = 0.06$, and $k = 3$. The function $\epsilon(V,g)$ determines the dynamics of the recovery variable: $\epsilon(V,g) = \epsilon_1$ for $V < e_2$, $\epsilon(V,g) = \epsilon_2$ for $V > e_2$, and $\epsilon(V,g) = \epsilon_3$ for $V < e_1$ and $g < g_1$ with $g_1 = 1.8$, $\epsilon_1 = 0.01$, $\epsilon_2 = 1.0$, and $\epsilon_3 = 0.3$. As in Refs. [Panfilov 1998; Panfilov and Hogeweg 1993], dimensioned time T is defined to be 5 ms times dimensionless time and 1 spatial unit to be 1 mm. The dimensioned value of the conductivity constant D is 2 cm^2/s. To make sure that the features observed are not artifacts of the Panfilov model, they are also shown to occur for the realistic Luo–Rudy I model [Luo and Rudy 1991].

The Panfilov model equations are solved in d spatial dimensions by using the forward-Euler method in time t, with a time step $\delta t = 0.022$, and a finite-difference method in space, with step size $\delta x = 0.5$ and five-point and seven-point stencils, respectively, for the Laplacian in $d=2$ and $d=3$. The spatial grids consist of square or simple-cubic lattices with side L mm, i.e., $(2L)^d$ grid points, with $L=200$. The LRI model has been solved using a forward-Euler scheme with $\delta t = 0.01$ ms, a finite-difference method in space, with $\delta x = 0.0225$ cm, and a square simulation domain with 400×400 grid points, i.e., $L=90$ mm has been used.

For both models no-flux (Neumann) boundary conditions are used on the edges of simulation domain and on the boundaries of obstacles. Conduction inhomogeneities are introduced in the medium by setting the diffusion constant D equal

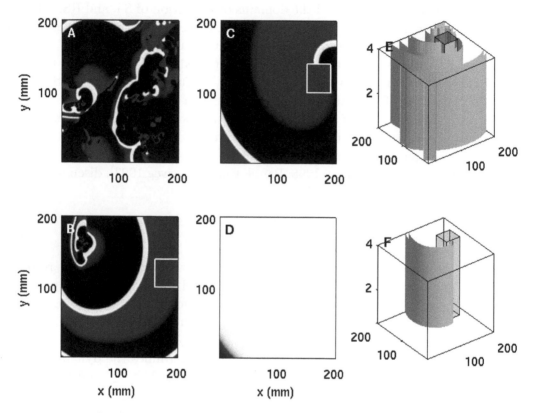

FIGURE 5.15 Spiral turbulence (ST) in a two-dimensional medium with Panfilov excitation dynamics. Transmembrane potentials for two dimension (pseudo-grayscale plots A-F) and three dimension (isosurface plots G and H). Two-dimensions 200 mm $\times$ 200 mm domain and a 40 mm $\times$ 40 mm square obstacle with left-bottom corner at (x, y). (A) no obstacle -ST; (B) $(x = 160, y = 100)$ ST persists; (C) $(x = 150, y = 100)$ ST replaced by RS (one rotating anchored spiral); (D) $(x = 140, y = 100)$ spiral moves away (medium quiescent). Three-dimensional analogs of (B) and (C): $(200 \times 200 \times 4)$ domain; an obstacle of height 4 mm and a square base of side 40 mm at (E) $(x = 140, y = 120, z = 0)$ and (F) $(x = 140, y = 110, z = 0)$.

to zero in regions with obstacles; in all other parts of the simulation domain D is a nonzero constant. The dimensioned value of D is 2 cm^2/s for the Panfilov model and between 0.5 cm^2/s for the LRI model. Most results are for an inhomogeneity that is a square region with lateral dimensions l, 10 mm $\leq l \leq$ 40 mm; however, illustrative simulations with circular or irregularly shaped inhomogeneities have also been carried out. Three-dimensional simulations have been done with an obstacle of height 4 mm and a square base of side 40 mm, i.e., 8 and 80 grid points, respectively.[†] Inhomogeneities in which the kinetic parameter ϵ_1 governing one of the time-scales of the Panfilov model varies over the simulation domain have also been considered.

The initial conditions used are such that, in the absence of inhomogeneities, they lead to a state that displays spatiotemporal chaos and spiral turbulence. For the Panfilov model one starts with a broken-wavefront initial conditions. From this broken wavefront a spiral wave develops with a core in the center of the simulation domain and, in the absence of inhomogeneities, evolves to a state with broken spiral waves and turbulence (Fig. (5.15) A). The spirals continue to break up even after 35000 ms for the parameter values used. For the LRI model the initial conditions are shown in Fig. (5.16) A, which develop, in the absence of an obstacle, into the spiral-turbulent state shown in Fig. (5.16) B.

In the presence of an obstacle the spiral turbulence (ST) state of Fig. (5.15) A can either remain in the ST state or evolve into a quiescent state (Q) with no spirals or the RS state with one rotating spiral anchored at the obstacle. Before exploring these possibilities, let us consider the criteria used to decide whether a given state of the system is of type ST, RS, or Q. In the Panfilov model, if the spiral wave continue to form and break up even up to 3500 ms, the state is identified as ST (Fig. (5.15) B); if, by contrast, a single spiral wave anchors to the obstacle and rotates around it at least for ten rotations ($\simeq$ 3500 ms for the Panfilov model with a 40 $\times$ 40 mm^2 obstacle) this is called an RS state (Fig. (5.15) C) (once it anchors, this rotation of the spiral wave continues even after 100 rotation periods); lastly, if the spiral wave moves away from the simulation domain and is absorbed at the boundaries within 3500 ms, the state is considered to be Q (Fig. 5.15 D).

For the LRI model, if the spiral formation and breakup continues up to 2200 ms, the state is identified as ST (Fig. 5.16 C); if the spiral wave gets anchored to the obstacle and completes 4 rotation periods ($\simeq$ 2200 ms for the obstacle being considered) the state is RS (Fig. 5.16 D); and the state Q (Figs. 5.15 E and F) is achieved if the spiral wave moves away from the simulation domain within 2200 ms.

Cardiac tissue can have conduction inhomogeneities at various length scales. Even minute changes in cell or gap-junctional densities might act as conduction inhomogeneities [Hwang et al. 2005]; these are of the order of microns. Scar tissues or blood vessels can lead to much bigger obstacles; these are in the mm to cm range so they can be studied effectively by using the PDEs mentioned above. Here we

---

[†]Note that, for a detailed understanding of the three-dimensional case one must also consider the effects of rotational anisotropy of muscle fibers in cardiac tissue [Fenton and Karma 1998].

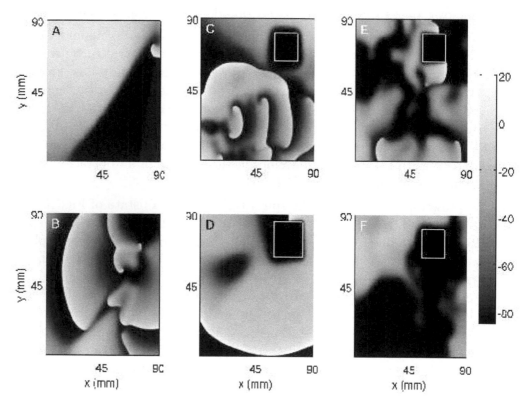

FIGURE 5.16   Spiral turbulence in a two-dimensional medium with Luo–Rudy excitation dynamics. Pseudo-grayscale plots in a 90×90 mm^2 illustrate how the initial condition (A) evolves, in the absence of obstacles, to (B) via the generation of spiral waves and their subsequent breakup. In the presence of a square obstacle of side l placed with its bottom-left corner at (x, y) the following is observed: (C) l=18 mm and $(x = 58.5$ mm, $y = 63$ mm) ST persists; (D) l=22.5 mm and $(x = 58.5$ mm, $y = 63$ mm) RS (one spiral anchored at the obstacle); for l=18 and $(x = 54$ mm,$y = 63$ mm) spirals disappear leaving the medium quiescent (E) at 800 ms and (F) at 1000 ms.

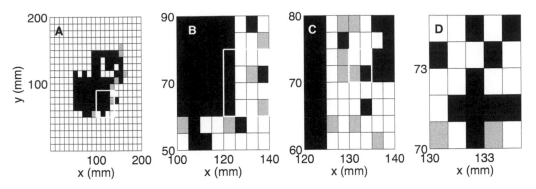

FIGURE 5.17 Sensitive dependence of dynamics on obstacle location in the Panfilov model. The effect of a 40 × 40 mm^2 obstacle in a 200 × 200 mm^2 domain shown by small squares (side l_p) the colors of which indicate the final state of the system when the position of the bottom-left corner of the obstacle coincides with that of the small square (white, black, and gray denote ST, RS, and Q, respectively). (A) for l_p=10 mm. Fractal-like interfaces between ST, RS, and Q are observed by magnifying small sub domains (white boundaries in A, B, and C with (B) l_p=5 mm, (C) l_p=2.5 mm, and (D) l_p= 1 mm).

focus on such large obstacles. As in the experiments of Ikeda et al. [Ikeda et al. 1997], the position of the obstacle is fixed and the spiral-wave dynamics is studied as a function of the obstacle size. For this a square obstacle of side l is introduced in the two-dimensional ($d = 2$) Panfilov model. With the bottom-left corner of the obstacle at the point (50 mm, 100 mm), one finds that spiral turbulence (ST) persists if $l \leq (40 - \Delta)$ mm, a quiescent state (Q) with no spirals is obtained if $l=$ 40mm, and a state with a single rotating spiral (RS) anchored at the obstacle is obtained if $l \geq (40 + \Delta)$ mm. To obtain these results l has been varied from 2 to 80 mm in steps of $\Delta= 1$ mm. Hence there is a clear transition from spiral turbulence to stable spirals, with these two states separated by a state with no spirals.

The final state of the system depends not just on the size of the obstacle but also on how it is placed with respect to the tip of the initial wavefront. For example, even a small obstacle placed close to the tip [l=10 mm obstacle placed at (100 mm, 100 mm)] can prevent the spiral from breaking up, whereas a bigger obstacle placed far away from the tip [$l= 75$ mm, placed at (125 mm, 50 mm)] does not affect the spiral.

To understand in detail how the position of the obstacle changes the final state, the spatiotemporal dynamics of the Panfilov model is investigated on a two-dimensional square domain with length 200 mm (i.e., 400 × 400 grid points) within which a square obstacle with l=40 mm is placed. Figure 5.17 A shows the simulation domain divided into small squares of side l_p mm (l_p=10 mm in Fig.5.17 A). The color of each small square indicates the final state of the system when the position of the lower-left corner of the obstacle coincides with that of the small square: white, black, and gray indicate, respectively, ST, RS, and Q. In Figs. 5.17 B and C the rich, fractal-like structure of the interfaces between the ST, RS, and Q regions is

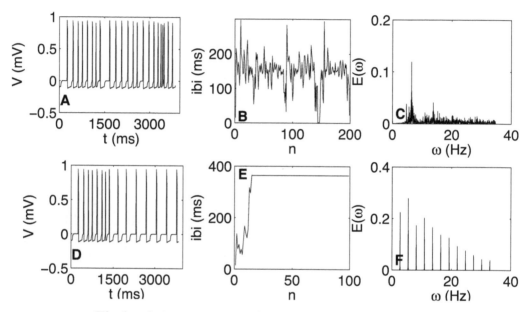

FIGURE 5.18   The local time series, interbeat interval IBI, and power spectrum of the transmembrane potential $V(x, y, t)$ at a representative point (x, y) in the tissue. When the obstacle is at (160 mm,100 mm) a spiral turbulent state ST is obtained with the time series (A), and interbeat interval (B) showing non-periodic chaotic behavior and a broad-band power spectrum (C). However, with the bottom-left corner of the obstacle at (150 mm,100 mm), the spiral wave gets attached to the obstacle after 9 rotations ($\simeq$ 1800 ms); this is reflected in the time series (D) and the plot of the interbeat interval (E); after transients the latter settles on to a constant value of 363 ms; the power spectrum (F) shows discrete peaks with a fundamental frequency $\omega_f = 2.74$ Hz and its harmonics. Initial transients over the first 50,000 δt are removed before data for calculating the power spectrum is obtained.

shown by successively magnifying small subdomains encompassing sections of these interfaces (white boundaries in Figs. 5.17 A and B) and reducing the sizes of the small squares into which the subdomain is divided. Clearly very small changes in the position of the obstacle can change the state of the system from ST to Q or RS, i.e., the spatiotemporal evolution of the transmembrane potential depends very sensitively on the position of the obstacle.

The time series of the transmembrane potential $V(x, y, t)$ taken from a representative point (x, y) in the simulation domain illustrates the changes that occur when one moves from the ST to the RS regime in Fig. 5.17. Such time series are shown in Fig. 5.18. For example, when the obstacle is placed with its bottom-left corner at (160 mm, 100 mm), the system is in the spiral turbulent state ST. The time series of V from the point (51 mm, 50 mm) clearly shows non-periodic, chaotic behavior. The times between successive spikes in such time series, or interbeat intervals (IBI), are plotted versus the integers n, which label the spikes, in Fig. 5.18 B; this also shows the chaotic nature of the state ST. Figure 5.18 C shows the power spectrum $E(\omega)$ of the time series in Fig. 5.18 A; the broad-band nature of this power spectrum provides additional evidence for the chaotic character of ST. By combining Figs. 5.18 (A-C) with the pseudo-grayscale plots of Figs. 5.15 A and B, one can conclude that ST is not merely chaotic but exhibits *spatiotemporal chaos*. Indeed, it has been shown that the Panfilov model, in the spiral turbulence regime, has several positive Lyapunov exponents whose number increases with the size of the simulation domain; consequently the Kaplan–Yorke dimension also increases with the system size (e.g., see Fig. 12.1 in Chapter 12); this is a clear indication of spatiotemporal chaos.

If the position of the obstacle s changed slightly, viz., if it is moved such that the lower-left corner is in the position (150 mm, 100 mm) of the simulation domain, the spiral eventually gets attached to the obstacle. For this case the analogs of Figs. 5.18 A-C are shown, respectively, in Figs. 5.18 D-F. The time series in Fig. 5.18 D shows that the transmembrane potential displays some transients up to about 2000 ms but then it settles into periodic behavior. This is also mirrored in the plot of IBI versus n in Fig. 5.18 E in which the transients asymptote to a constant value for the IBI (363 ms) which is characteristic of periodic spikes. Not surprisingly, the corresponding power spectrum in Fig. 5.18 F consists of discrete spikes at frequencies $\omega_m = m\omega_f$, where m is a positive integer and ω_f is the fundamental frequency ($\omega_f = 2.74$ Hz). A simple rotating spiral anchored at the obstacle (Fig. 5.15 C) will clearly result in such a periodic time series in the state RS.

The analogs of Figs. 5.18 A-C for the quiescent state Q are not shown as the transmembrane potential V just goes to zero after an initial period of transients. The durations for which the transients last, say in Fig. 5.18 D, vary greatly depending on the position of the obstacle relative to the spiral tip. Transient times ranging from 300 ms to 2000 ms have been seen in simulations.

Similar results are obtained for the three-dimensional Panfilov and the two-dimensional Luo–Rudy I models: Illustrative pictures from simulations of spiral turbulence (ST) and a single rotating spiral (RS) anchored at the obstacle are shown in Figs. 5.15 and 5.16, respectively. From these and similar figures, one can

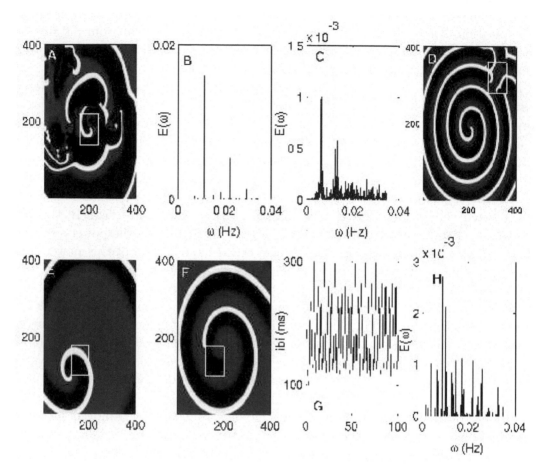

FIGURE 5.19 Inhomogeneities in the parameter ϵ_1 of the Panfilov model result in the coexistence of different types of spatiotemporal behavior in the same system. With ϵ_1^{out}=0.01 and ϵ_1^{in}=0.02 (see text), spatiotemporal chaos is seen outside the inhomogeneity but quasiperiodic behavior is observed inside it (A); the latter is illustrated by the power spectrum of $V(x,y,t)$ with discrete peaks (B) and the former by a broad-band power spectrum (C). With ϵ_1^{out}=0.03 and ϵ_1^{in}=0.01 and the left-bottom corner of the inhomogeneity placed at (x=140 mm, y=140 mm), single and broken spiral waves coexist in same medium (D), whereas, with the inhomogeneity at (x=60 mm, y=50 mm), a single rotating spiral gets anchored to the inhomogeneity (E, F) with quasiperiodic behavior illustrated by the interbeat interval (G) and the power spectrum (H). The power spectrum (H) shows six frequencies ($\omega_1 = 4.06, \omega_2 = 5.56, \omega_3 = 6.57, \omega_4 = 7.05, \omega_5 = 8.58$, and $\omega_6 = 9.07$ Hz) not rationally related to each other; all other frequencies can be expressed as $\sum_{i=1}^{6} n_i\omega_i$, where the n_i are integers. Initial transients over the first 50,000 δt are removed before data for calculating power spectra is obtained.

conclude that the final state, ST, RS, or Q, depends not only on the size of the obstacle but also on its position. Obstacles of different shapes, e.g., circles, irregular shapes, and two squares separated from each other, lead to similar results.

The study also explores the effects of inhomogeneities in parameters such as ϵ_1, which controls the recovery time at large values of the inactivation or recovery variable and intermediate values of the activation variable in the Panfilov model [Panfilov and Hogeweg 1993]. As ϵ_1 is increased, the absolute refractory period of the action potential decreases and this in turn decreases the pitch of the spiral wave (Fig. 5.20). In a homogeneous simulation domain (of size say 200×200 mm^2) values of $\epsilon_1 > 0.03$ produce a single periodically rotating spiral. As ϵ_1 is lowered, e.g., if $\epsilon_1 < 0.02$, quasiperiodic behavior is seen; this is associated with the meandering of the tip of a simple rotating spiral. Even lower values of ϵ_1, say $\epsilon_1 = 0.01$, lead to spatiotemporal chaos.

Let us now consider an inhomogeneous simulation domain in which all parameters in the model except ϵ_1 remain constant over the whole simulation domain. A square inhomogeneity is now introduced, inside which ϵ_1 assumes the value ϵ_1^{in} and outside which it has the value ϵ_1^{out}. Different choices of ϵ_1^{in} and ϵ_1^{out} lead to the behavior summarized below.

With a square patch of size 40×40 mm^2, $\epsilon_1^{in} = 0.02$, and $\epsilon_1^{out} = 0.01$, a spatiotemporally chaotic state is obtained for most positions of this inhomogeneity. But there are certain critical positions of this inhomogeneity for which all spirals are completely eliminated (e.g., when the left-bottom corner of the inhomogeneity is at $x=70$ mm, $y=120$ mm the spiral moves out of the simulation domain). For yet other positions of the inhomogeneity, spatiotemporal chaos is obtained outside the inhomogeneity but inside it quasiperiodic behavior is seen (Figs. 5.19 A-C). However, with $\epsilon_1^{in} = 0.01$ and $\epsilon_1^{out} = 0.03$, spiral breakup occurs inside the inhomogeneity and coexists with unbroken periodic spiral waves outside it (Fig. 5.19 D), as previously noted by Xie et al. [Xie et al. 2001b]. Even in this case, for certain positions of the inhomogeneity, a single spiral wave gets anchored to it (Figs. 5.19 E, F) as in the case of a conduction inhomogeneity (Fig. 5.15 C). However, the temporal evolution of V at a representative point in Fig. 5.19 E is richer than it is in Fig. 5.15 C: $V(x, y, t)$, with $x=51$ and $y=50$, displays the interbeat interval of Fig. 5.19 G; the associated power spectrum shows six fundamental frequencies, not rationally related to each other, and their combinations; this indicates strong quasiperiodicity of $V(x, y, t)$. So, even an inhomogeneity in the excitability of the medium can cause the ST-RS or ST-Q transitions discussed above for the case of conduction inhomogeneities. Furthermore, an inhomogeneity in excitability can also lead to rich temporal behaviors as shown in Figs. 5.19 E-H.

The results described above suggest that spiral turbulence in excitable media models depends sensitively on the size and position of inhomogeneities in the medium. In particular, with the inhomogeneity at a particular position, the state of the spiral wave changes from ST to RS as the size of the obstacle increases. Also, for an obstacle with fixed size, this transition also depends upon the position of the obstacle. Two important questions arise from this: (1) What causes the sensitive dependence of such spiral turbulence on the positions and sizes of conduction in-

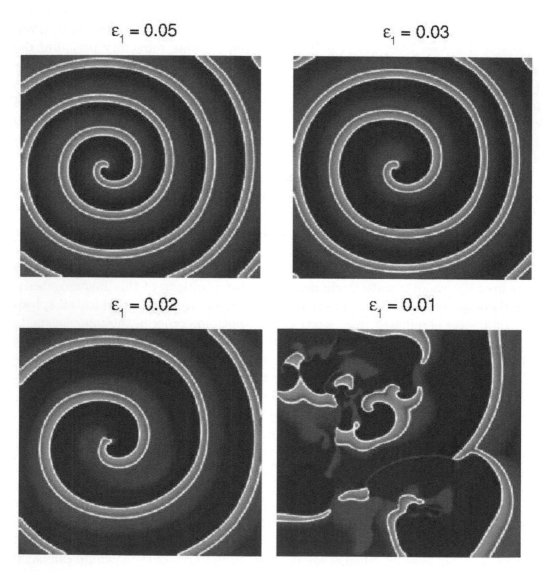

FIGURE 5.20 Pseudocolor plots of the activation field for the Panfilov model in a two-dimensional excitable medium as the parameter ϵ_1 controlling one of the recovery dynamics timescales is decreased. The pitch of the spiral wave is seen to increase as the refractory period increases on decreasing ϵ_1 from 0.05 to 0.03 and then to 0.02, until spiral breakup is observed when ϵ_1 is decreased to 0.01.

homogeneities? (2) What are the implications of this theoretical study for cardiac arrhythmias and their control? As spiral turbulence (ST) and a single rotating spiral (RS) have been linked to fibrillation and tachycardia, it is possible that the latter phenomena may also depend sensitively on the positions and sizes of conduction inhomogeneities. This critical sensitivity appears to be a natural consequence of the spatiotemporal chaos associated with spiral turbulence. Indeed, even for much simpler, low-dimensional dynamical systems it is often the case that a *fractal* basin boundary [Lai and Winslow 1994; Sommerer and Ott 1993] separates the basin of attraction of a strange attractor from the basin of attraction of a fixed point or limit cycle; thus a small change in the initial condition can lead either to chaos, associated with the strange attractor, or to the simple dynamical behaviors associated with fixed points or limit cycles.

The PDEs considered here are infinite-dimensional dynamical systems; the complete basin boundaries for these are not easy to determine; however, it is reasonable to assume that a complex, fractal-like boundary separates the basins of attraction of spatiotemporally chaotic states (e.g., ST) and those with simpler behaviors (e.g., RS or Q). In the study described above the initial conditions are not changed; instead the dynamical system is altered slightly by moving the position, size, or shape of a conduction inhomogeneity. This too affects the long-time evolution of the system as sensitively as does a change in the initial conditions. It is suggestive that by changing the position of a conduction inhomogeneity, one may convert spiral breakup to single rotating spiral or vice versa as depicted graphically in Figs. 5.17 and 5.18. Even more intriguingly, conditions leading to the quiescent state Q can be found at the boundary between these two types of behavior (Fig. 5.17). Thus the study obtains all the analogs of possible qualitative behavior found in experiments, namely, (1) ST might persist even in the presence of an obstacle, (2) it might be suppressed partially and become RS, or (3) it might be eliminated completely.

These results also have implications for antitachycardia-pacing and defibrillation algorithms used for the suppression of cardiac arrhythmias. It suggests that optimal pacing algorithms might well have to be tailor made for the specific inhomogeneities existing in different hearts. Indeed, clinicians often adapt their hospital procedures for the treatment of arrhythmias on a case-by-case basis to account for cardiac structural variations between patients [Christini and Glass 2002]. The modeling study points out the need for further systematic experimentation on the effects of obstacles on cardiac arrhythmias.

5.5 Effect of obstacles on dynamics of scroll waves: A novel breakup mechanism

As already mentioned, while patterns in homogeneous active media have been the subject of intense theoretical and experimental investigations for several decades [Meron 1992; Winfree 1972; Zykov 1986], natural systems possess significant inhomogeneities. Therefore, the dynamics of waves in disordered excitable media have come under increasing scrutiny in recent times [Panfilov and Keener 1993;

Sinha and Christini 2002; Sinha et al. 2002; Xie et al. 1998]. The heterogeneities considered range from partially or wholly inexcitable obstacles [Pumir et al. 2010; Sinha and Christini 2002; Sinha et al. 2002], gradients of excitation or conduction properties [Sridhar et al. 2010], and anisotropy in the speed of propagation [Krinsky and Pumir 1998]. Most such studies have focused on two-dimensional systems and the results show unexpected complexity, such as fractal basins of attraction for different dynamical states corresponding to pinned waves, spatiotemporal chaos and complete termination of activity, as discussed in the previous section [Shajahan et al. 2003, 2007, 2009]. Nevertheless, these "planar" models do not completely describe real systems which are necessarily three-dimensional.

Adding an extra dimension is equivalent to considering the thickness of the system, so that one can in principle distinguish between phenomena on the surface and those in the bulk. More importantly, three-dimensional disordered media can exhibit novel dynamical phenomena that do not appear in lower dimensions. A frequently occurring pattern of activity in such systems is the scroll wave, which is a higher dimensional generalization of the spiral wave.[Fig. 5.21 (a)]. It can be visualized as a set of contiguous rotating spirals whose phase singularities describe a continuous line (filament) along the rotation axis perpendicular to the plane of the spirals [Winfree 1987] [Fig. 5.21 (c)]. Scroll waves have been experimentally observed in many natural systems including chemical waves in the Belousov–Zhabotinsky reaction [Pertsov et al. 1990; Welsh et al. 1983; Winfree 1973], patterns of aggregation during *Dictyostelium* morphogenesis [Steinbock et al. 1993; Weijer 1999], and electrical waves in heart muscles [Efimov et al. 1999]. Indeed scroll waves have been implicated in several types of arrhythmia, i.e., disturbances in the natural rhythmic activity of the heart that can be potentially fatal. Under certain conditions these three-dimensional waves can develop instabilities and break up into multiple scroll fragments. The ensuing spatiotemporally chaotic state of excitation is associated with the complete loss of coherent cardiac activity in life-threatening arrhythmia such as ventricular fibrillation [Gray et al. 1998; Witkowski et al. 1998]. Therefore a deeper understanding of the mechanisms that lead to chaotic activity through breakup of scroll waves is of great practical significance.

At present almost all proposed scenarios for scroll wave breaking involve complex filament dynamics [Fenton et al. 2002]. The question of whether there can be other mechanisms, especially one involving breakup far from the singularities, is of special significance when considering disordered media [Panfilov 2002]. This is because heterogeneities such as inexcitable obstacles can anchor rotating waves and stabilize filament dynamics preventing the usual breakup scenario. While interaction of scroll waves with inexcitable obstacles has recently been investigated [Jimenez et al. 2009; Majumder et al. 2011; Pertsov and Vinson 1994; ten Tusscher and Panfilov 2003b; Vinson et al. 1994], the existence of several types of disorder in the systems considered in these studies does not clearly reveal the exact mechanisms for wavebreaks that are involved. By focusing on a simplified situation of a defect with regular geometry one can ask whether there can be a purely obstacle-induced breakup mechanism for scroll waves [Sridhar et al. 2013a].

The specific model system being considered here comprises an isotropic three-

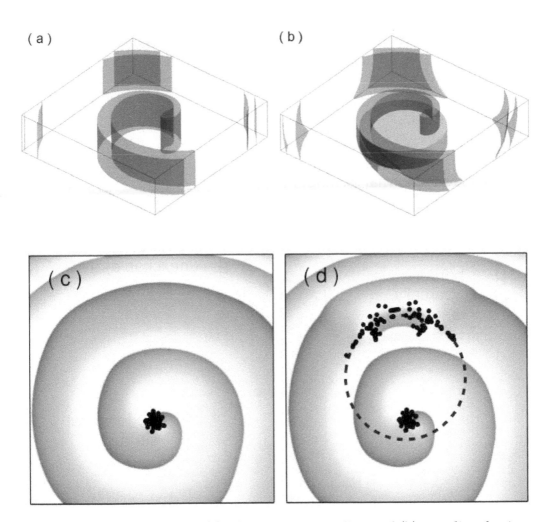

FIGURE 5.21 Scroll waves in (a) a homogeneous medium and (b) a medium having an inexcitable cylinder-shaped obstacle that extends only partly through the bulk of the medium. (c-d) Pseudocolor images of spiral waves observed in a cross-sectional plane perpendicular to the axis of rotation, for the systems shown in (a) and (b), respectively. The black dots correspond to the intersection of the plane with the filament at different times, indicating the trajectory of the phase singularity in the plane. In (d), the plane chosen is just above the upper bounding surface of the obstacle whose circumference is indicated using broken lines. The appearance of additional black dots along the broken circle in (d) indicates a wavebreak induced by the obstacle at its boundary, far from the existing scroll filament.

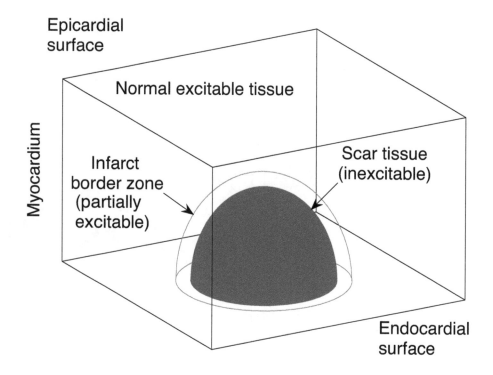

FIGURE 5.22   Schematic diagram of an inexcitable obstacle surrounded by a border zone comprising partially excitable cells embedded in cardiac tissue composed of fully excitable cells. The obstacle extends only partly through the thickness of the medium, and cannot be seen by observing the top surface alone. In the context of the heart, such situations can arise after myocardial infarction, when a region of scar tissue is formed inside the bulk of the myocardium and cannot be detected by imaging the epicardial surface.

dimensional medium with an inexcitable obstacle of uniform cross-section embedded within it, which does not span the entire thickness of the medium [Fig. 5.21 (b)]. This is motivated by the physiological observation that an inexcitable obstacle may be located deep in the bulk of heart tissue (Fig. 5.22), where it cannot be detected through electrophysiological imaging of the surface [Peters et al. 1997, 1998]. A novel dynamical transition is seen, which does not appear in the absence of a defect, nor in the effectively two-dimensional situation when the obstacle spans the thickness of the medium. We observe that under certain circumstances, the wave can break near the bounding edge of the obstacle and far from the scroll wave filament [Fig. 5.21 (d)]. This happens when the wavefront velocity decreases sharply at the edge as a result of sudden change in its curvature (which is related to the 'source-sink' relation expressing the balance of electrical currents during wave propagation [Fast and Kleber 1997]). The novelty of the mechanism presented here lies in the fact that the breakup does not involve the filament, unlike the previously proposed pathways for the onset of chaos in three-dimensional excitable media. It is especially relevant for disordered media as the transition to chaos is essentially defect induced because the wave is *stable* (i.e., does not break or result in the generation of additional filaments) in the *absence* of the inexcitable obstacle.

To simulate spatiotemporal activity in three-dimensional excitable tissue, models having the following generic form are used:

$$\frac{\partial V}{\partial t} = -\frac{I_{ion}(V, g_i)}{C_m} + \nabla.D\nabla V, \tag{5.2}$$

where V is the activation variable, typically, the potential difference across a cellular membrane (measured in mV), C_m ($= 1\ \mu\text{F cm}^{-2}$) is the transmembrane capacitance, D represents the inhomogeneous intracellular coupling, I_{ion} ($\mu\text{A cm}^{-2}$) is the total current density through ion channels on the cellular membrane, and g_i describe the dynamics of gating variables of different ion channels. The specific functional form for I_{ion} varies for different biological systems. For the results reported here the Luo-Rudy I (LR I) model [Luo and Rudy 1991] that describes the ionic currents in a ventricular cell is used with the following modifications. The slow inward Ca^{2+} channel conductance G_{si} is taken to be ≤ 0.04 mS cm^{-2} and the maximum K^+ channel conductance G_K is increased to 0.705 mS cm^{-2} [Xie et al. 2001b]. This reduces the slope of the restitution curve and enables us to study scroll wave dynamics in a stable regime where the waves do not break up spontaneously in the absence of an obstacle [ten Tusscher and Panfilov 2003a]. It has been explicitly verified that the results are not sensitively dependent on model-specific details (e.g., description of ion channels) as similar effects have been observed in the simpler phenomenological model proposed by Panfilov [Panfilov 1998].

The equations are solved using a forward-Euler scheme with time-step $\delta t = 0.01$ ms in a three-dimensional simulation domain having $L \times L \times L$ points. For most results reported here $L = 400$, although size independence has been verified by performing the simulations with different L (viz., up to 600). The space step used is $\delta x = 0.0225$ cm, with a standard 7-point stencil for the Laplacian describing the spatial coupling [Press et al. 1995; Sinha et al. 2001]. No-flux boundary conditions are applied on the boundary planes of the simulation domain. To simulate the electrically non-conducting nature of the inexcitable obstacle, no-flux boundary conditions are imposed along its walls. The coupling D is set to zero inside the obstacle, while outside it is 0.001 cm^{2}s^{-1}. We have considered obstacles of different shapes, including cylinders and cuboids, and have not observed qualitative differences between them. Obstacles are characterized by their height L_z (beginning from the base of the simulation domain and ranging between 0 to L) and cross-sectional area (viz., πR^2 for cylinders where R is the radius and $L' \times L'$ for a cuboid). The initial scroll wave, with its filament aligned along the L_z-axis, is obtained by breaking a three-dimensional planar wavefront. Qualitatively similar results are achieved with waves having curved filaments. Note that the implementation of the non-conducting property of the inexcitable obstacle is important to establish the exclusively three-dimensional feature of the breakup mechanism shown here. This is because wavefronts interacting with heterogeneities that act as electrical sinks (e.g., Ref. [Jie et al. 2008]) may result in conduction block independent of the specific geometry of the setup [Sinha and Christini 2002; Sinha et al. 2002].

The most important result obtained from the simulations is that a sufficiently large pinning defect can promote wavebreaks in an otherwise stable rotating wave (Figs. 5.23 and 5.24). At the initial stage, when a broken wave is anchored by the obstacle, the partial extension of the latter through the bulk of the medium produces differential rotation speeds of the scroll wave along the L_z-axis. Far above the obstacle the period is closer to T_{free}, that of a free scroll wave unattached to any obstacle, while at the base of the obstacle the period $T_{pinned} \sim R/c$ is substantially slower depending upon the radius R of the obstacle (c being the average propagation speed normal to the wavefront). This results in a helical winding of the scroll wave around the obstacle until a steady state is achieved when the rotation period becomes the same across the domain [Fig. 5.23 (a)]. The pitch of the wound scroll wave depends on the obstacle size, with larger radius resulting in a tighter winding, i.e., smaller pitch. Note that the filament of the resulting scroll wave, which stretches from the top of the obstacle to the upper boundary of the simulation domain, remains the same as that of a free scroll wave [see Figs. 5.21 and 5.24 (a)]. When the wavefront is close to the filament it has velocity components only along the plane perpendicular to the L_z-axis. However, when the wave crosses the obstacle boundary, it develops a velocity component parallel to the L_z-axis as the front travels down the vertical sides of the obstacle. Figure 5.25 shows that both velocity components exhibit a large decrease as the wave crosses the obstacle boundary, with the component parallel to L_z-axis having the greater reduction. The magnitude of this change in velocity depends on the pitch of the helical winding, and hence on the circumference of the obstacle, with larger obstacles resulting in greater velocity differences. As seen in Fig. 5.23 (b), this sudden change in velocity components as the front crosses the edge of the obstacle upper surface may result in detachment of a succeeding wave from the obstacle surface. This is manifested as a "tear" on the wavefront surface as it breaks generating a new filament [Fig. 5.24 (b)]. Further evolution of the system produces more complex wave fragments as additional filaments are generated and which interact with each other [Figs. 5.23 (c-d) and 5.24 (c-d)],

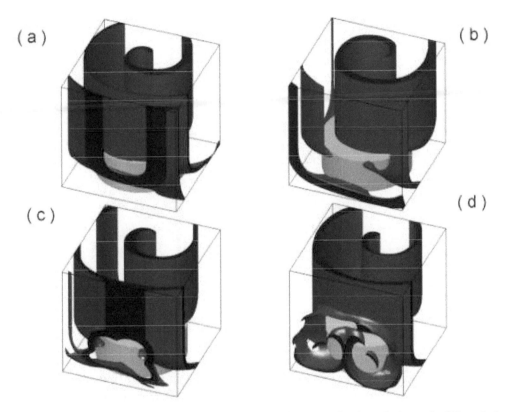

FIGURE 5.23 Breakup of a scroll wave induced by a cylindrical obstacle ($R = 3.4$ cm, $L_z = 2.7$ cm). (a) By $T = 275$ ms after the initial wavebreak, the resulting scroll wave has wound itself around the obstacle. (b) A wavefront detaches from the surface of the obstacle, at the boundary of its upper surface, by $T = 300$ ms. This broken wavefront eventually evolves into new scroll wave filaments (c-d). The period of a free (unpinned) scroll wave in this parameter regime is $T_{free} = 65.13$ ms.

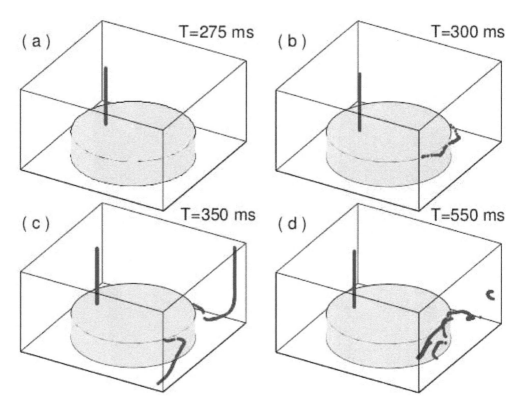

FIGURE 5.24 Filament dynamics during obstacle induced breakup of a scroll wave. (a) At $T = 275$ ms after the initial wavebreak results in generation of a scroll wave rotating around the obstacle, there is a single filament (corresponding to the rotating scroll wave) in the system, oriented parallel to the axis of the cylindrical obstacle. (b) By $T = 300$ ms, another filament has developed away from the original filament (which continues to exist unperturbed by the wavebreak phenomenon occurring at the obstacle boundary). The new filament appears as a result of a "tearing" of the wavefront as it travels past the bounding edge of the inexcitable obstacle. (c) The newly formed filament subsequently collides with the domain boundary, splitting into two separate filaments moving independently of each other. (d) Further interactions between the filaments, the obstacle, and the domain boundaries result in the generation of even more filaments over time, eventually leading to spatiotemporal chaos.

eventually leading to spatiotemporal chaos (Fig. 5.26).

To better understand the mechanism by which the pinned scroll wave breaks, consider the cross-sectional view of the system parallel to the L_z-axis [Fig. 5.27 (a-c)]. Figure 5.27 (a) indicates the direction of propagation as the waves move first along the upper surface and then down the vertical sides of the obstacle (shaded dark in the figures). The wave W_1 slows down as it moves past the boundary of the upper surface of the obstacle This results from the sudden change in the velocity components of the front mentioned earlier and is associated with an increased curvature of the wave [Mikhailov et al. 1994]. Thus, the wave W_2 closely following W_1 encounters an incompletely recovered region at the edge of the obstacle [Fig. 5.27 (b)]. The resulting propagation block of W_2 in the direction parallel to the L_z-axis (immediately adjacent to the obstacle boundary) dislodges the wave from the surface of the obstacle and generates a singularity at the point where the scroll wave breaks [Fig. 5.27 (c)]. This phenomenon can be further enhanced by filament meandering that results in Doppler-effect induced changes in the propagation speed of successive waves. As previously mentioned, larger obstacles result in sharper differences in the velocity (and curvature) of the front as it crosses the obstacle, which suggests increased probability of wavebreak (measured as the fraction of different initial conditions that eventually lead to breakup) with increasing radius R. This is indeed observed in Fig. 5.27 (d) supporting the mechanism outlined above. Note that this increased probability is a function of increasing obstacle size (and not an artifact of decreased gap between the obstacle and the system boundary) by making the obstacle bigger while keeping the ratio of its size to that of the system a constant.

The significance of the results reported here lies in the fact that almost all previously reported mechanisms by which scroll waves break up leading to spatiotemporal chaos necessarily involves the dynamics of the filament, e.g., negative filament tension in low excitability regime or filament twist instabilities in the presence of cardiac fiber rotation [Fenton et al. 2002]. By contrast, the novel transition route to chaos described here is a result of changes in the wavefront velocity components at the edge of a pinning obstacle, far from the existing filament. While there exist a few other mechanisms which can lead to scroll waves breaking, e.g., decreased cell coupling, these are not exclusively three-dimensional phenomena and are known to be involved in two-dimensional spiral wave breakup [Fenton et al. 2002].

We stress that the mechanism described here has no two-dimensional analog. Wavebreaks created through interaction between high-frequency excitation fronts and an inexcitable obstacle, observed in two-dimensional media [Agladze et al. 1994; Panfilov and Keener 1993], are fundamentally different from the phenomenon described here, as the three-dimensional analog of the former situation would correspond to the obstacle interacting with scroll waves originating from a filament that is *far* from the obstacle (and not pinned by it). Note that, in the geometric setup discussed in detail here, this is not the case. As the filament is attached to the obstacle, the two-dimensional situation corresponding to it will be a source of high-frequency waves contiguous with the obstacle. We have explicitly verified through simulations in two-dimensional media that rapid waves generated by either point or

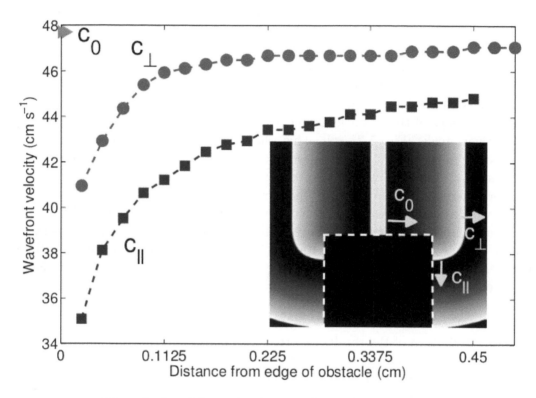

FIGURE 5.25   The velocity of the excitation wavefront after it moves past the bounding edge of the obstacle, shown as a function of increasing distance from the edge. The initial propagation speed of the wave before it crosses the boundary c_0 is indicated. The two velocity components of the front, $c_\perp$ and $c_\parallel$, emerge just after the wavefront traverses the boundary. The large relative change with respect to c_0 very close to the edge indicates a sudden slowing down of the wave at the obstacle boundary. As the wave propagates away from the obstacle, its velocity increases asymptotically toward c_0. The inset shows a pseudocolor plot of the transmembrane potential V for a cross-section of the three-dimensional system along a plane parallel to the L_z axis passing through the center of the obstacle. The different velocity components of a traveling excitation front generated by a line electrode parallel to L_z axis are indicated.

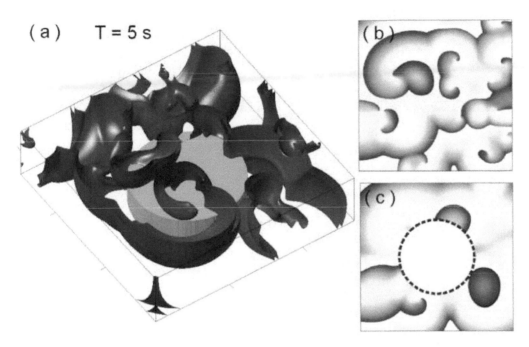

FIGURE 5.26 (a) Fully developed spatiotemporal chaos following scroll wave breakup induced by a cylindrical obstacle ($R = 3.4$ cm, $L_z = 1.8$ cm) after a duration of $T = 5$ s following the initiation of the broken wavefront in a rectangular simulation domain having lateral dimensions of 13.5 cm and a height of 3.6 cm. (b-c) Pseudocolor images of the cross-sectional planes perpendicular to the L_z-axis corresponding to the (b) top ($z = L$)and (c) bottom ($z = 0$) surfaces of the simulation domain. In (c), the perimeter of the obstacle is indicated using broken lines.

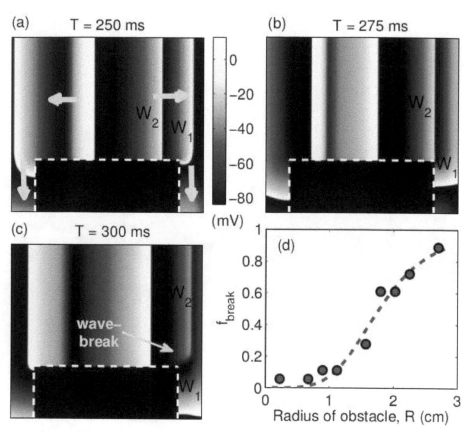

FIGURE 5.27   (a-c)Pseudocolor plots of the transmembrane potential V for a cross-section of the system shown in Fig. 5.23, along a plane parallel to the L_z-axis. Boundary of the obstacle (shaded) is marked by broken lines. The arrows in (a) indicate the direction of propagation for wavefronts. The wave W_1 slows down at the boundary of the upper surface of the obstacle. As a result, the wave W_2 closely following W_1 encounters a region that has not fully recovered from its prior excitation by W_1, leading to (c) the detachment of W_2 from the surface of the obstacle and formation of a singularity. (d) The fraction of initial conditions resulting in scroll wave breakup by $T = 3$ s, $f_{breakup}$, shown as a function of the obstacle size. The fraction for each obstacle size is estimated by using 18 distinct initial scroll wave configurations generated from a three-dimensional planar wavefront and determining the number of cases where new filaments are subsequently formed at the obstacle boundary. The broken line is a sigmoid fit to the data.

line-electrode positioned next to the obstacle do not generate wavebreaks.

The role played by a three-dimensional obstacle in inducing the breakup of an otherwise stable rotating wave is extremely pertinent for understanding the genesis of certain cardiac arrhythmias as an ageing heart gradually accumulates defects through increased instances of local tissue necrosis [Zipes and Jalife 2004]. While obstacles in three-dimensional media that do not extend through the entire thickness of the system have been considered earlier, these studies focused on the depinning transition of scroll waves in the presence of drift inducing parameter gradients [Pertsov and Vinson 1994; Vinson et al. 1994]. In contrast, here it is seen that such defects can give rise to complex dynamics including transition to chaos. The mechanism presented above also provides a qualitative dynamical framework for explaining recent observations in chemical systems of scroll waves wrapping around an inexcitable obstacle [Jimenez and Steinbock 2012] and the prediction of obstacle-induced breakup of scroll waves can be directly tested in a similar experimental setup [Jimenez and Steinbock 2012].

Thus, the study described in this section shows that the presence of an inexcitable obstacle in three-dimensional excitable media can result in scroll wave breakup far from the filament through a novel physical mechanism. The helical winding of wave around a pinning defect causes sudden changes in the velocity components of the wavefront as it crosses the boundary of the obstacle, with an associated increase in the curvature of the wave in the plane parallel to the axis of its rotation. The resulting enhanced interaction between successive waves at the bounding edge of the obstacle increases the probability of a wavefront detaching from the surface of the obstacle giving rise to new filaments. These wavebreaks can eventually lead to spatiotemporal chaos, manifested as fibrillation in the heart. These results may have consequences for understanding the critical role of defects (such as, inexcitable regions of necrotic tissue) embedded deep inside the bulk of cardiac muscle in the genesis of life-threatening arrhythmia.

6

Spatial gradients and dynamics

In Chapter 5, we considered the effect of a localized heterogeneity, specifically a region or a defect having different conduction properties from that of its surrounding medium, on the dynamics of two-dimensional and three-dimensional spatial patterns of excitation. Another kind of heterogeneity often encountered in naturally occurring excitable systems is the spatial gradient. A gradient typically involves the variation of one or many system properties over some spatial extent. Several types of gradients have been observed in naturally occurring excitable systems. Dynamics of spiral and scroll waves are significantly influenced by these gradients. For example, variation of a system property along the gradient causes a spiral wave to drift. Drift is a kind of non-meandering rotation characterized by a significant linear translation motion. While spiral drift has been observed in several contexts such as, in the presence of an electric field [Krinsky et al. 1996; Steinbock et al. 1992] and high frequency stimulation of the medium [Gottwald et al. 2001b], gradient induced drift is especially relevant in the context of biological excitable media. In a heterogeneous excitable medium like the heart, drift can be a significant component of the spiral wave dynamics, so much so that it is believed to be a possible underlying mechanism for arrhythmias like polymorphic ventricular tachycardia [Garfinkel and Qu 2004; Gray et al. 1995a]. Characterized by an aperiodic electrocardiogram, polymorphic ventricular tachycardia may also be a precursor to fully disordered activity that characterizes potentially fatal ventricular fibrillation [Cobbe and Rankin 2005]. Thus understanding the phenomenon of spiral drift has potential clinical significance [Fenton et al. 2002].

We first describe the different kinds of gradients that have been experimentally observed in natural systems. We then define the concept of drift and discuss how it is characterized in experiments and simulations. Next we describe the various kinds of gradient induced drift in model excitable medium. We then explore the connection between spiral drift observed in cardiac models and the theory of mother-rotor fibrillation in the heart. Finally, we conclude with a small section on the other kinds of non-gradient perturbations that lead to spiral drift.

6.1 Spatially varying heterogeneities

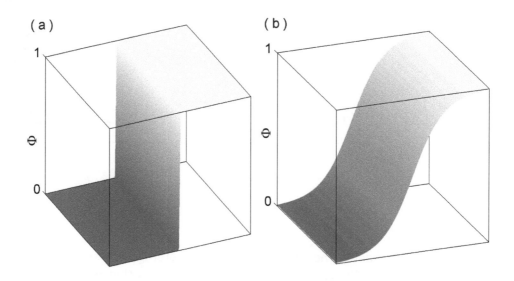

FIGURE 6.1   Spatial variation in a parameter Φ describing either local dynamics or coupling in an excitable medium, showing either an abrupt step-like change (a) or a smoother gradient (b) along a specific direction.

Naturally occurring excitable systems are often neither homogeneous nor isotropic. Typically one or more of the properties of these systems would show either continuous or abrupt variation in space. Biological excitable media in particular show a spatial dependence on properties such as excitability (manifested as the expression of ion channels), density of fibroblasts and Interstitial Cajal-like cell (in uterus), etc. Studies that have investigated the effect of heterogeneities in excitable media have typically considered either a step-like inhomogeneity or a smoothly varying spatial gradient. In a step-like heterogeneity, one or more properties of the system jump discontinuously to another value at a specific location as shown in Fig. 6.1 (a). Here Φ could be any property of the system such as excitability, conduction, or temperature that has a spatial dependence. Smooth gradients, on the other hand, occur when properties of the medium such as excitability vary gradually across the spatial extent of the medium(Fig. 6.1 (b)). Gradients can also occur when electrotonic interactions flatten abrupt discontinuities in the properties of the medium.

Such heterogeneities can significantly influence the dynamics of spatial patterns in the medium. Abrupt changes in a medium parameter have been shown to induce spiral drift along the border in both simulations [Krinsky 1968] and experiments [Fast and Pertsov 1990; Markus et al. 1992]. More recently it has been shown that step-like heterogeneities can induce spiral breakup [Xie et al. 2001a] even in systems where the restitution curve is shallow enough to not cause spontaneous breakup of spirals. Spatial gradients too have been shown to induce spiral drift in both simple and realistic models of cardiac tissue [Panfilov and Vasiev 1991; Rudenko and Panfilov 1983; ten Tusscher and Panfilov 2003a].

6.2 Gradients in the lab

Before we discuss the phenomenon of gradient-induced spiral drift, let us take a quick look at the different kinds of gradients that have been observed in the experimental and clinical contexts. In the cardiac tissue, electro-physiological gradients are known to arise due to the spatial variation in the distribution of ion-channel expression [Akar and Rosenbaum 2003]. Gradients in intercellular coupling can also occur as a result of a spatial variation in gap junction conductances that electrically couple adjacent cells [Bub et al. 2002]. Gradients in APD within the ventricles, differences in APD across right and left ventricles or across the two atria, are typical examples of heterogeneity occurring in the normally functioning heart [Schram et al. 2002]. But under conditions of ischemia or disorders like long QT syndrome, there can occur an increase in heterogeneity that can potentially disturb the electrical activity of the heart.

Cardiac ventricular cells form fibers whose orientation varies gradually through the heart wall. This rotational anisotropy can result in the accumulation of twist in the scroll filaments. The resultant twisting gradient can lead to the scroll waves drifting [Berenfeld and Pertsov 1999]. Another biological example is the occurrence of gradients in the resting membrane potential in smooth muscle fibers of the animal digestive tract [Sarna 2008]. These gradients are thought to be vital for *peristalsis*, which is the term for the contraction of smooth muscles that enables the directed movement of food through the digestive tract toward the anus. Externally imposed gradients in intensity of light have been used to precisely control spiral wave dynamics in chemical excitable systems like the Belousov–Zhabotinsky reaction [Markus et al. 1992; Zhang et al. 2005b]. Thermal gradients are another example of an externally imposed heterogeneity in chemical excitable systems such as Belousov–Zhabotinsky [Vinson et al. 1997b] and CO oxidation on the surface of platinum (110) crystals [Sadeghi and Rotermund 2011].

6.3 Drifting spirals

In this section we shall look at the effect of gradients on the dynamics of spatial patterns of excitation, specifically the phenomenon of spiral drift. Spiral waves can display a range of dynamics such as steady stationary rotation, non-stationary meander, and drift [Garfinkel and Qu 2004]. We characterize the dynamics of a spiral wave in terms of the motion of its core. As explained in Chapter 3, spiral core is the trajectory of the wave tip which is a phase singularity. During rigid rotation, the core of the spiral traces a circle with a fixed period. Meander is the result of the presence of multiple frequencies leading to a quasiperiodic motion of the spiral core. A whole zoo of meandering patterns have been observed and characterized in excitable media, such as inward petals, outward petals, etc. [Alonso and Panfilov 2008; Barkley 1994; Li et al. 1996; Winfree 1991] Drift is characterized by the directed translational motion of the spiral core and is quite different from meander where the translational component is not along a single direction.

6.3.1   Why do spirals drift?

The governing parameter of spiral drift is the refractory period, which varies along the excitability gradient. Typically drift velocity of the spiral can have both longitudinal (parallel or anti-parallel to the gradient) and transverse (perpendicular to the gradient) components. So even if the gradient is along one direction, its effect on the spiral dynamics, manifested as its drift, has components along both the direction of the gradient and perpendicular to it. This is because the morphology of the spiral wave is such that it breaks the symmetry and hence its motion is not restricted to a direction parallel to the gradient. It has been observed that changing the chirality (handedness) of the spiral changes the sign of the transverse component (V_T), while the longitudinal component (V_L) remains unchanged. However Voignier et al. showed experimentally that there might be other mechanisms, not yet understood, that can alter the direction of the drift of the transverse component [Voignier et al. 1999].

In an excitability gradient the spiral rotation period and velocity vary along the axis. Thus along the gradient different regions can sustain spirals of different periods. For example, let us assume that the spiral tip is initially at '1' (see Fig. 6.2(A)). The directed spatial variation in the recovery period of the medium (recovery period determines the period and velocity of the spiral) ensures that the spiral tip after one rotation does not return to 1. The tip rotates quickly through a region of the shorter refractory period, but traces a wider arc as it passes through the section where the recovery time is longer. This causes the tip to precess in the direction of the longer recovery period. Thus after some time the spiral tip has moved to location '2'. The mechanism described here explains the kind of drift observed typically in experiments. In fact, almost all experimental or simulated observations of the interactions of a spiral with a gradient describe drift to regions of longer spiral rotation period or lower excitability [Panfilov and Vasiev 1991; Rudenko and Panfilov 1983; Sadeghi and Rotermund 2011; ten Tusscher and Panfilov 2003a]. This means that for a spiral of a given chirality, the longitudinal component of drift velocity V_L has the same sign irrespective of the type of gradient or its strength. However kinematic studies that approximate the motion of spirals as dynamics of a curve, have suggested the possibility of drift to regions with higher excitability [Hakim and Karma 1999; Mikhailov et al. 1994]. These studies have considered gradients in local excitability of the medium and have shown that the direction of drift changes above a critical value of the strength of the gradient. Are there gradients that can show drift to regions of higher excitability? If yes, what is the implication of such a drift for the dynamics and stability of the spiral wave? What is the biological consequence of such a drift? These are some of the questions that we will explore in the following sections.

6.4   Normal and anomalous drift

In this section we will describe a study performed to ascertain whether drift to regions of higher excitability is indeed possible [Sridhar et al. 2010]. The drift of spiral waves is studied in the presence of gradients in a simple model of heterogeneous excitable medium. In this study linear gradients in the (a) distribution of ion-channel expression and (b) intercellular coupling have been considered. Electrophysiological heterogeneities in cardiac tissue may arise, in general, through spatial variation in the distribution of ion-channel expression in excitable tissue [Akar and Rosenbaum 2003]. There can also be gradients in the intercellular coupling as a result of the inhomogeneous distribution of the conductances of gap junctions that connect neighboring cells [Bub et al. 2002]. Spirals which generally have stationary non-meandering rotation, tend to drift in the presence of such linear gradients. As mentioned earlier, this drift motion has both longitudinal (along or against the direction of the gradient) and transverse (perpendicular to the gradient) components. Spiral drift can thus be classified as either *normal* or *anomalous* depending on longitudinal direction of the drift. As noted earlier, the sign of the transverse velocity V_T, does not change.

- **Normal drift** of the spiral corresponds to motion anti-parallel to the gradient. This is the kind of drift that has been typically observed in experiments and simulations and is manifested as drift to regions of slower rotation or smaller intercellular coupling. Since the motion is against the gradient the longitudinal velocity V_L for normal drift has a negative sign.
- **Anomalous drift** on the other hand moves the spiral to a region of where it has a shorter rotation period or stronger intercellular coupling. The sign of V_L in this case is positive.

Reduction in either the ion-channel density or the intercellular coupling can impede wave propagation and can be qualitatively regarded as decreasing the excitability of the medium. On the other hand, a drift to regions of larger ion-channel expression or intercellular coupling can be considered to be drift toward region of higher excitability.

6.5 Implementation of gradient

As described in Chapter 2, a generic model of excitable media that describes the dynamics of transmembrane potential V in cardiac tissue has the form

$$\partial V/\partial t = \nabla \gamma D \nabla V + \alpha I_{ion}(V, g_i), \qquad (6.1)$$

$$\partial g_i/\partial t = F(V, g_i). \qquad (6.2)$$

Here, I_{ion} is the total ionic current traveling through the channels on the cellular membrane, D accounts for the intercellular coupling and g_i describes the dynamics of gating variables for the various ion-channels. In order to study the effects of a heterogeneous distribution of ion-channel expression and intercellular coupling, the parameters α and γ, representing the spatial variation in ion-channel expression and conduction properties (respectively) for an inhomogeneous medium are introduced. Parameter α directly scales the value of ion-channel expression in Eq. (6.1), while

γ scales the diffusion coefficient as $D = D_0 + \gamma(x)$. The activation-inactivation dynamics is given by the Barkley model [Barkley et al. 1990], which is described in Chapter 2. The spatial heterogeneity of ion-channel expression and cellular coupling are assumed to have linear functional form, viz., $\alpha(x) = \alpha_0 + \Delta\alpha\, x$ and $\gamma(x) = \gamma_0 + \Delta\gamma\, x$. The variable $x\ (= -d/2, \ldots, d/2)$ represents the spatial position along the principal direction of the inhomogeneity gradient, where d is the length of the domain measured along this direction and the origin (i.e., $x = 0$) is at the midpoint of the simulation domain. At this point, $\alpha = \alpha_0$, $\gamma = \gamma_0$, and $\Delta\alpha, \Delta\gamma$ measure their rate of change along the gradient. For all the figures in this chapter, $\alpha_0 = 1.15$, $\gamma_0 = 1.3$, and $\epsilon = 0.02$.

6.6   Observation of anomalous drift

To investigate the role of heterogeneity in spiral drift, spatial gradients in α or γ are individually considered (keeping the other parameter constant). Anomalous drift of the spiral, i.e., a drift toward regions with shorter period or higher intercellular coupling, is indeed observed [Fig. 6.2 (A,C)] after extensive numerical simulations that scan over the (a, b) parameter space of the Barkley model. In both the cases shown in Fig. 6.2 (A,C) the drift is toward the region of higher excitability (in the qualitative sense) because increase in either α or γ enhances wave propagation. For comparison, in Fig. 6.2 (B, D) the normal drift of the spiral, i.e., toward regions of lower excitability, is shown. This is seen for a set of (a, b) values which is farther from the boundary with the sub-excitable region of the Barkley model [Pumir et al. 2010] than the (a, b) parameter set for which anomalous drift is observed in Fig. 6.2 (A,C).

To analyze the genesis of anomalous drift, the effect of the parameters γ and α on the spiral wave in a *homogeneous* medium (i.e., no gradient) is considered. As γ is only a scaling factor for the diffusion coefficient (see Eq. 6.1), the period of the spiral wave does not depend on it. Thus, neither normal nor anomalous drift is associated with a significant change in the period. However, due to the discrete nature of wave propagation in real systems such as cardiac tissue, there is a small decrease in the period when the spiral wave moves toward regions having higher cellular coupling during anomalous drift.

Figure 6.3 (A) shows the variation of the spiral period as a function of the parameter α, which decreases as α increases.[†] Thus, for normal drift in the presence of α gradient, the period of the spiral increases as the core moves towards lower α regions. On the other hand, a decrease in the period is observed for the case of anomalous drift towards regions having higher values of α. In contrast, the wavelength does not appear to be a determinant of the drift. For anomalous drift in presence of a gradient in γ, the wavelength increases (as $\sqrt{\gamma}$), while it decreases for

---

[†]For a, b parameters where anomalous drift is observed, the period and wavelength of spiral wave exhibits a more rapid divergence with decreasing α compared to the parameter regime showing normal drift.

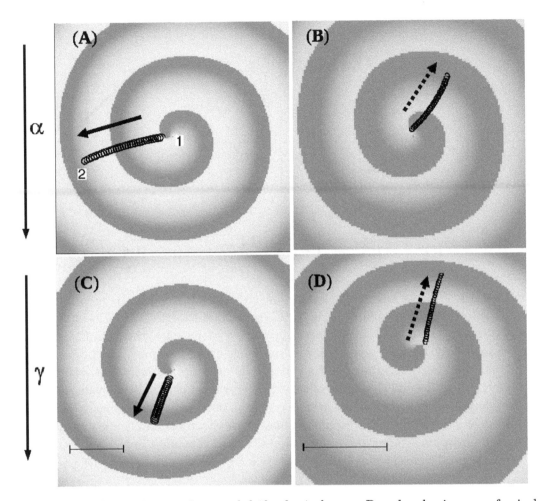

FIGURE 6.2 Anomalous and normal drift of spiral wave. Pseudocolor images of spiral wave at the instant when the gradient in ion-channel expression α (top row) and cellular coupling (bottom row) is applied. (A, C) Anomalous drift toward increasing values of α (A) or γ (C), the direction being shown by solid arrows. (B, D) Normal drift in α (B) or γ (D) gradient. Parameter values are $a = 0.82$, $b = 0.13$ (for A, C) and $a = 1.02$, $b = 0.15$ (for B, D). The gradients applied are (A, B) $\Delta\alpha = 0.005$, $\Delta\gamma = 0$, and (C, D) $\Delta\alpha = 0$, $\Delta\gamma = 0.040$. In all cases, the gradient is along the vertical direction, with α or γ increasing from top to bottom. In (B, D) the region around the core is magnified to make the wavelength of the spiral comparable to that in (A,C). The trajectories shown correspond to 100 time units and are obtained using the algorithm given in Ref. [Fenton and Karma 1998]. The bar indicates a scale of 25 space units.

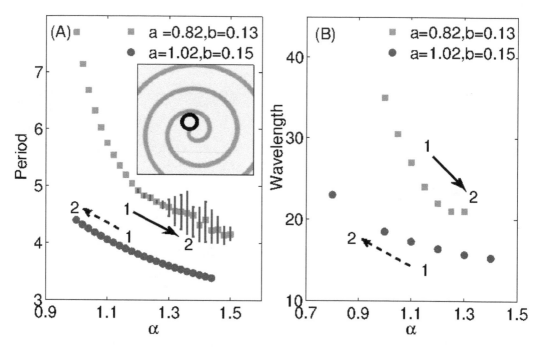

FIGURE 6.3 The variation of spiral period (A) and wavelength (B) as a function of the parameter α. The symbols "1" and "2" correspond to the values of α in the region around the initial and final positions (respectively) of the spiral waves in Fig. 6.2, with the same sets of Barkley model parameters being used. The solid and broken arrows represent the directions of anomalous and normal drift, respectively, in the presence of a gradient in α. Results shown are obtained by averaging over multiple values recorded from symmetrically placed points in the simulation domain to smooth variations arising from spiral wave meandering at high values of α. Error bars are indicated when the standard deviation is larger than the symbol size used. The inset in (A) is a pseudocolor image of a spiral wave showing its trajectory in a *homogeneous* medium with parameters $a = 0.82$, $b = 0.13$, and $\alpha = 1.275$.

an α gradient, as it moves toward regions of higher α [Fig. 6.3 (B)].

Next, the effect of the magnitude of spatial gradient in α or γ on the velocity of spiral drift is investigated. Figure 6.4 shows the longitudinal component of the drift velocity, v_L, i.e., along the gradient, as a function of the spatial variation in α or γ. Note that positive v_L corresponds to anomalous, while, negative v_L corresponds to normal drift of the spiral wave. Figure 6.4 shows that, for normal drift, increasing either of the gradients results in a monotonic increase of v_L (broken lines). However, in the case of anomalous drift as a result of α gradient, a *non-monotonic* behavior in v_L, which first increases but then decreases and becomes negative [Fig. 6.4 (A)] is observed. Thus, the anomalous drift of the spiral toward shorter period in α gradient is seen only for small $\Delta\alpha$. For higher $\Delta\alpha$, there is a reversal of direction and the spiral exhibits normal drift. On the other hand, Fig. 6.4 (B) shows that for a gradient in γ, the anomalous drift is observed for the entire range of $\Delta\gamma$ that is investigated.

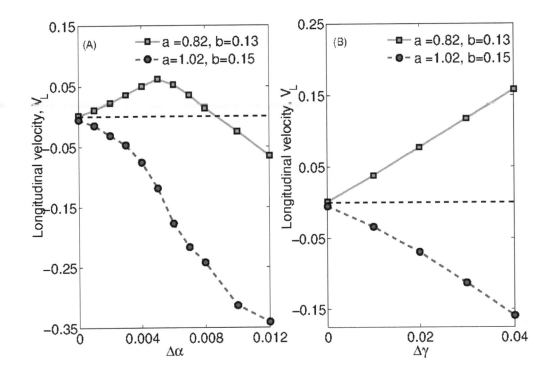

FIGURE 6.4 Drift velocity depends on the gradient in parameters α and γ. (A) Non-monotonic variation (solid curve) of the longitudinal component of spiral wave drift velocity v_L as a function of the gradient in ion-channel expression, $\Delta\alpha$, for a model system with parameters $a = 0.82, b = 0.13$. Positive values of v_L indicate anomalous drift. For a different set of parameters ($a = 1.02, b = 0.15$), normal drift is seen for the entire range of gradients used (broken curve). (B) Variation of v_L with the gradient in cellular coupling, $\Delta\gamma$. Solid and broken curves represent the anomalous and normal drift seen for the two parameter sets mentioned earlier (respectively), and are observed throughout the range of gradients used.

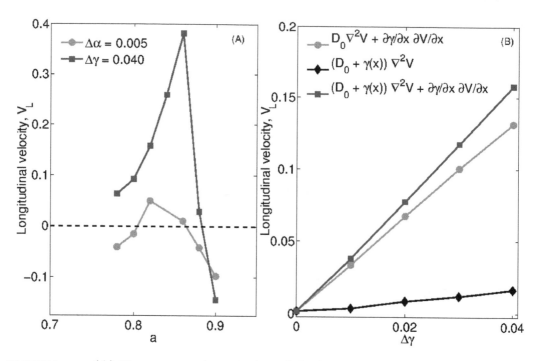

FIGURE 6.5 (A) Non-monotonic variation of the longitudinal component of drift velocity, v_L, as a function of the model parameter a ($b = 0.13$). The two curves correspond to media having a constant gradient in excitability (circles: $\Delta\alpha = 0.005$, $\Delta\gamma = 0$) and cellular coupling (squares: $\Delta\alpha = 0$, $\Delta\gamma = 0.040$). (B) The contribution to v_L from the different components in the diffusion term as a function of the gradient in cellular coupling, $\Delta\gamma$. The circles and diamonds correspond to the linear and second-order contributions, and, squares correspond to the complete Laplacian term, respectively. All data points shown are for $a = 0.82$, $b = 0.13$.

The effect of the local kinetics on anomalous drift is investigated by varying the Barkley model parameter a [Fig. 6.5 (A)]. Increasing a (keeping b fixed) decreases the activation threshold of the medium, and thus makes the system more excitable. It can be observed that for both α and γ gradients, the variation of v_L as a function of a is non-monotonic. For the cellular coupling (γ) gradient, the presence of anomalous regime clearly correlates with excitability. The drift is anomalous at lower excitability, but becomes normal at higher excitability. However, for the gradient in α, the anomalous drift occurs only over an intermediate range of a. For lower and higher excitability, the drift becomes normal. Note that arguments put forth in Ref. [Hakim and Karma 1999] suggest that in the *large core limit* (corresponding to very low excitability), the longitudinal component of drift velocity resulting from a parameter gradient in α should disappear.

6.7 Mechanism of drift

A complete understanding of the phenomenon of anomalous drift in the α gradient is still lacking. However, it is possible to understand the anomalous drift for the cellular coupling (γ) gradient, by relating it to other drift phenomena in excitable media. Note that the Laplacian term in Eq. (6.1) can be expanded as,

$$\nabla\gamma(x)D\nabla V = (D_0 + \gamma(x))\nabla^2 V + \partial\gamma/\partial x \, \partial V/\partial x. \qquad (6.3)$$

Therefore, the heterogeneous cellular coupling $\gamma(x)$ contributes to both the gradient ($\partial\gamma/\partial x \, \partial V/\partial x$), as well as second order ($\gamma(x)\nabla^2 V$) terms. The relative contributions of these terms to the longitudinal component of drift velocity is shown in Fig. 6.5 (B). The principal effect on v_L is due to the $\partial\gamma/\partial x \, \partial V/\partial x$ term, while $\gamma(x)\nabla^2 V$ accounts only for about 10% of the observed drift. This observation allows for the following explanation for the phenomenon of anomalous drift in the presence of a gradient in γ. If the contribution of the $\gamma(x)\nabla^2 V$ term in the Laplacian is ignored, the spatial operators in Eq. (6.3) are seen to be identical to those in equations describing drift of a spiral wave in the presence of an electric field [Krinsky et al. 1996]. The latter, in turn, is similar to the Laplacian describing the drift of radially symmetric filaments of a scroll ring in three-dimensional excitable media [Alonso and Panfilov 2008; Panfilov and Rudenko 1987]. It has been shown that the drifts observed in these two kinds of systems are induced by the long wavelength instabilities [Henry 2004; Henry and Hakim 2002]. As explained in Chapter 3, long wavelength instabilities occur far away from the spiral tip with the distance between successive wavefronts varying spatially. Figure 6.5 (B) shows that the gradient $\partial\gamma/\partial x \, \partial V/\partial x$ term, which determines the drift in an electric field and that of scroll wave filaments, also determines the drift as a result of γ gradient. Therefore, it can be inferred that the anomalous drift direction (toward stronger cellular coupling) is also a result of the same long wavelength instabilities determining the drift of scroll wave filaments. This suggests that the occurrence of scroll expansion in three-dimension implies the existence of anomalous drift in γ gradient in two-dimensions. Conversely, observation of anomalous drift might suggest parameter regions where scroll wave expansion is possible.

6.8 Anomalous drift and mother rotor fibrillation

The phenomenon of anomalous drift in the presence of a gradient in excitability has important implications for understanding the mechanisms underlying the formation of mother rotors. The hypothesis of a dominant rotor was first proposed by Sir Thomas Lewis in 1920 [Ideker and Rogers 2006; Lewis 1920] as a possible mechanism for atrial fibrillation. More recently, experiments have suggested that one or few mother rotors can act as rapidly rotating sources that drive ventricular fibrillation [Jalife 2000; Nash et al. 2006; Samie et al. 2001]. The initiation of fibrillation by mother rotors depends on the heterogeneity in the recovery periods of the medium. The rotor emits high frequency waves that can propagate into some regions of the heart tissue, but are blocked by others that have longer recovery periods. The conduction block can result in the breaking of the wavefront into multiple "daughter" fragments that result in an irregular activation pattern associated with VF.

The paradox in this phenomenon arises because experimentally, only drifts to regions of larger refractory and hence longer rotation period (normal drift) have been observed. On the other hand, the formation of the mother rotor requires that the spiral move to a region of shorter refractory period. What then is the mechanism behind the drift of the rotor to a region where it rotates faster?

The phenomenon of anomalous drift described in the previous section is one plausible mechanism for the formation of the high frequency wave sources. Although a simple two-variable model was used in the analysis, this study can be easily extended to biologically realistic models, such as Luo–Rudy I or TNNP [Luo and Rudy 1991; ten Tusscher et al. 2004] as these models have the same form as Eq. (6.1). Further, inhomogeneities inferred from direct experimental measurements [Schram et al. 2002] can be used to model the heterogeneity, even though gradients in ion-channel expression may not affect all ion channels and do not always affect excitability. It might be possible to infer the parameter range in realistic models where anomalous drift may occur by using the relation between the cellular coupling gradient-induced drift and scroll ring expansion. Note that the latter phenomenon has recently been seen in the Luo–Rudy I model [Alonso and Panfilov 2008].

6.9   Other conditions for drift

While in the previous sections we described spiral drift in the presence of spatial gradients, the phenomenon as such can also be caused by other kinds of perturbations. Here we briefly summarize a few other situations where spirals have been observed to drift.

Electric fields have been shown to induce drift in spirals observed in Belousov–Zhabotinsky chemical experiments [Steinbock et al. 1992]. It has been observed that the direction of drift depends on whether the spiral is sparse or dense, with the former drifting toward the negative electrode and the latter toward the positive one [Krinsky et al. 1996]. If the electric field E is applied along the direction parallel to X-axis, it basically contributes an advective term $E\partial V/\partial X$ to the reaction-diffusion equation; for example, Eq.(1) in Ref. [Krinsky et al. 1996]. The resulting equation for the time variation of transmembrane potential is then:

$$\partial V/\partial t = \nabla \gamma D \nabla V + I_{ion}(V, g_i) + cE\partial V/\partial X, \tag{6.4}$$

Note that if γ does not have a spatial dependence Eq. (6.4) is identical to Eq. (6.3). It has been suggested that two different mechanisms contribute to the opposite effects on sparse and dense spirals. While periodic changes in core radius causes sparse spirals to drift toward negative electrode, drift of dense spirals to the positive electrode is because of the periodic changes in velocity [Krinsky et al. 1996].

Another mechanism that leads to spiral drift is the periodic stimulation of the medium. It has been shown theoretically [Gottwald et al. 2001b] that such regular perturbation of the medium can induce spirals to drift to the boundary. The periodic perturbation actually changes the excitability of the medium significantly and converts the meandering motion of the spiral into drift. They further showed that for moderately sparse spirals, perturbations with a period $T_{pacing} > T_{free}$ applied very close to the core is sufficient to produce a nonzero drift in the spiral. Experimentally, drift of spirals via periodic perturbation has also been observed in monolayers of neonatal rat myocytes [Agladze et al. 2007].

6.10   Conclusion

In this chapter, we have discussed the ubiquity of gradients in natural systems. We have also detailed the effect of these gradients on the dynamics of spatial patterns. The dominant effect on the spiral wave is the induction of drift because of the variation in recovery properties of the medium along the gradient. While drift has been experimentally verified in many systems it has always been found to be directed toward a region which only supports spirals with longer periods. While there has been some theoretical understanding of the uni-directionality in drift, the possibility of movement of spiral core to a region of shorter period is interesting especially in the context of occurrence of mother rotors that sustain fibrillation.

Wishing to see whether such a drift is at all possible, we discussed a study where spiral drift is investigated in simple models of smoothly varying heterogeneous excitable medium. We discussed regimes that support spiral wave drift toward regions with shorter rotation period (corresponding to larger expression of ion-channels) and/or stronger cellular coupling do exist. Both can be broadly considered to be drift toward a more excitable region. Note, however, that this is a qualitative statement, as quantitative definition of excitability is not straightforward. In general excitability depends on the threshold of excitation, on ability of a wavefront to generate local depolarization currents, on curvature effects for two and three-dimensional propagation, etc.

Spiral waves are not only relevant for cardiac tissue, but are also observed in many other excitable media. It may be possible to relate the observation of anomalous drift with results of models of cyclic catalysis in replicating entities [Boerlijst and Hogeweg 1995], which predict drift toward regions of shorter periods. From a clinical perspective, anomalous drift is important as it may result in fibrillation by promoting wavebreaks away from the spiral core. Spiral drift in presence of a cellular-coupling gradient may be a key factor giving rise to abnormal wave activity in regions of the heart where conductivity changes, e.g., at Purkinje-muscle cell junctions or in an infarct border zone [Pumir et al. 2005]. It can also be studied experimentally and numerically in many model systems, such as heterogeneous monolayers of neonatal rat cardiomyocytes [Biktashev et al. 2008].

To conclude, in this chapter, we have explored the phenomenon of spiral drift, specifically considering the case of gradients that induce spiral drift toward regions having shorter spiral rotation period or stronger cellular coupling. This drift appears to be related to regimes where expansion of three-dimensional scroll wave filaments is observed. Further, such anomalous drift of spiral waves may increase the likelihood of complex spatiotemporal patterns in excitable medium, e.g., turbulent electrical activity in the heart.

7

Distributed heterogeneities and dynamics

7.1 Introduction

The presence of disorder in the structural composition in many natural systems can affect the transitions between the dynamical regimes they exhibit [Binder and Young 1986; Mézard et al. 1987]. Such disorder is frequently seen in biological systems, including the uterine muscle which has a heterogeneous composition comprising electrically excitable myocytes (smooth muscle cells) that form the bulk of the tissue and a small fraction of electrically passive cells, like fibroblasts and telocytes. The passive cells are much fewer compared to the myocytes and experimental evidence suggests that the (frozen) pattern of their connections to excitable cells is highly irregular [Popescu et al. 2007], and can be treated as a quenched disorder in the system. In this chapter, we will discuss the role of disorder on the transitions between different dynamical regimes, specifically, the emergence of oscillations from a quiescent state, in extended heterogeneous systems of excitable and passive cells, induced by varying the strength of the coupling between cells. The random distribution of passive cells provides a quenched disorder in important biological contexts, such as the appearance of contractions in the pregnant uterus.

7.2 Genesis of coherent activity by increased coupling

We shall focus on heterogeneous systems whose individual components are continuous dynamical systems, in particular, excitable and passive elements. Local coupling between these different types of elements can result in oscillations and the emergence of collective rhythmic activity in such systems have been investigated. This is motivated by a puzzling observation in uterine physiology, viz., synchronized oscillations that give rise to labor during childbirth occur in the uterus even though it has been shown that none of the cells in the uterine tissue can oscillate spontaneously in isolation. Thus, the periodic activity of the uterus is distinct from several other types of biological oscillations, such as the rhythmic pumping action of the heart, which are coordinated by specialized elements known as "pacemakers" (e.g., cells in the sino-atrial node of the heart). Although in the uterus there occurs a transition from disordered activity during gestation to synchronized electrical activity giving rise to coherent contraction that ultimately leads to birth, there is no experimental evidence for the presence of such specialized coordinating elements in this organ. A novel explanation for the emergence of coherent activity in this system is through the increased coupling among heterogeneous dynamical elements. For this purpose, a lattice model of a disordered excitable system is used (disorder being in the form of a variable number of passive cells connected to each excitable cell). On increasing the strength of coupling between the elements comprising the system, a transition from a quiescent state to coherent activity via several non-trivial collective dynamical states is observed. These results help in causally connecting two apparently unrelated experimental observations: (i) coupling between uterine cells increases remarkably through the course of pregnancy and (ii) oscillatory activity is rare and extremely weak during the early stages of pregnancy but increases in frequency and strength as one approaches labor.

Rhythmic behavior is central to the normal functioning of many biological processes [Glass 2001] and the periods of such oscillators span a wide range of timescales controlling almost every aspect of life [Gillette and Sejnowski 2005; Golubitsky et al. 1999; Hakim and Brunel 1999; Winfree 2000]. Synchronization of spatially distributed oscillators is of crucial importance for many biological systems [Pikovsky et al. 2003]. For example, disruption of coherent collective activity in the heart can result in life-threatening arrhythmia [Keener and Sneyd 1998]. In several cases, the rhythmic behavior of the entire system is centrally organized by a specialized group of oscillators (often referred to as *pacemakers*) [Chigwada et al. 2006] as in the heart, where this function is performed in the sino-atrial node [Tsien et al. 1979]. However, no such special coordinating agency has been identified for many biological processes. A promising mechanism for the self-organized emergence of coherence is through coupling among neighboring elements. Indeed, local interactions can lead to order without an organizing center in a broad class of complex systems [Grégoire and Chaté 2004; Vicsek et al. 1995].

7.3 Modeling distributed heterogeneity

The pregnant uterus (discussed in Chapter 4) is distinguished from other excitable biological systems in that its principal function is critically dependent on coherent rhythmic contractions that, unlike in the heart, do not appear to be centrally coordinated from a localized group of pacemaker cells [Blackburn 2007]. In fact, the uterus remains quiescent almost throughout pregnancy until at the very late stage when large sustained periodic activity is observed immediately preceding the expulsion of the fetus [Garfield and Maner 2007a]. In the USA, in more than 10 % of all pregnancies, rhythmic contractions are initiated significantly earlier, causing preterm births [Martin et al. 2009a], which are responsible for more than a third of all infant deaths [MacDorman et al. 2007]. The causes of premature rhythmic activity are not well understood and at present there is no effective treatment for preterm labor [Garfield and Maner 2007a].

In this section, the emergence of coherence will be discussed using the modeling approach that stresses the role of coupling in a system of heterogeneous entities [Singh et al. 2012]. Importantly, recent studies have not revealed the presence of pacemaker cells in the uterus [Shmygol et al. 2007]. The uterine tissue has a heterogeneous composition, comprising electrically excitable smooth muscle cells (uterine myocytes), as well as electrically passive cells (fibroblasts and interstitial Cajal-like cells [ICLCs]) [Duquette et al. 2005; Popescu et al. 2007]. Cells are coupled in tissue by gap junctions that serve as electrical conductors. In the uterine tissue, the gap junctional couplings have been seen to markedly increase during late pregnancy and labor, both in terms of the number of such junctions and their conductances (by an order of magnitude [Miller et al. 1989; Miyoshi et al. 1996]), which is the most striking of all electrophysiological changes the cells undergo during this period. The observation that isolated uterine cells do not spontaneously oscillate [Shmygol et al. 2007], whereas the organ rhythmically contracts when the number of gap junctions increases, strongly suggests a prominent role of the coupling. The above observations have motivated the model (discussed here) for the onset of spontaneous oscillatory activity and its synchronization through increased coupling in a mixed population of excitable and passive elements. While it has been shown earlier that an excitable cell connected to passive cells can oscillate [Chen et al. 2009; Jacquemet 2006; Kryukov et al. 2008], it is observed that coupling such oscillators with different frequencies (because of varying numbers of passive cells) can result in the system having a frequency *higher* than its constituent elements. A systematic characterization has also been done for the dynamical transitions occurring in the heterogeneous medium comprising active and passive cells as the coupling is increased, revealing a rich variety of synchronized activity in the absence of any pacemaker. Finally, it is observed that the system has multiple coexisting attractors characterized by distinct mean oscillation periods, with the nature of variation of the frequency with coupling depending on the choice of initial state as the coupling strength is varied. These results provide a physical understanding of the transition from transient excitations to sustained rhythmic activity through physiological changes such as increased gap junction expression [Garfield et al. 1998].

The dynamics of excitable myocytes can be described by a model having the form

$$C_m \dot{V}_e = -I_{ion}(V_e, g_i)$$

where V_e(mV) is the potential difference across a cellular membrane C_m ($= 1 \ \mu$F cm^{-2}) is the membrane capacitance, I_{ion} (μA cm^{-2}) is the total current density through ion channels on the cellular membrane and g_i are the gating variables, describing the different ion channels. The specific functional form for I_{ion} varies in different models. To investigate the actual biological system first a detailed, realistic description of the uterine myocyte given by Tong et al. [Tong et al. 2011] is considered. However, during the systematic dynamical characterization of the spatially extended system, for ease of computation the phenomenological FitzHugh–Nagumo

(FHN) system [Keener and Sneyd 1998] is used, which exhibits behavior qualitatively similar to the uterine myocyte model in the excitable regime. In the FHN model, the ionic current is given by

$$I_{ion} = F_e(V_e, g) = AV_e(V_e - \alpha)(1 - V_e) - g,$$

where g is an effective membrane conductance evolving with time as

$$\dot{g} = \epsilon(V_e - g),$$

$\alpha(= 0.2)$ is the excitation threshold, $A(= 3)$ specifies the fast activation kinetics and $\epsilon(= 0.08)$ characterizes the recovery rate of the medium (the parameter values are chosen such that the system is in the excitable regime and small variations do not affect the results qualitatively). The state of the electrically passive cell is described by the time-evolution of the single variable V_p [Kohl et al. 1994]:

$$\dot{V}_p = F_p(V_p) = K(V_p^R - V_p),$$

where the resting state for the cell, V_p^R is set to 1.5 and $K(= 0.25)$ characterizes the timescale over which perturbations away from V_p^R decay back to it. The interaction between a myocyte and one or more passive cells is modeled by:

$$\dot{V}_e = F_e(V_e, g) + n_p\, C_r(V_p - V_e), \tag{7.1a}$$
$$\dot{V}_p = F_p(V_p) - C_r(V_p - V_e), \tag{7.1b}$$

where $n_p(= 1, 2, \ldots)$ passive elements are coupled to an excitable element via the activation variable $V_{e,p}$ with strength C_r. Here, for simplicity, it is assumed that all passive cells are identical having the same parameters V_p^R and K, as well as starting from the same initial state. It is observed that the coupled system comprising a realistic model of uterine myocyte and one or more passive cells exhibits oscillations (Fig. 7.1 (a)) qualitatively similar to the generic FHN model (Fig. 7.1 (b)), although the individual elements are incapable of spontaneous periodic activity in both cases. In Fig. 7.1 (a-b), the range of n_p and excitable-passive cell couplings for which limit cycle oscillations of the coupled system are observed is indicated with a pseudocolor representation of the period (τ). One can also look at how a system obtained by diffusively coupling two such "oscillators" with distinct frequencies (by virtue of having different n_p) behaves upon increasing the coupling constant D between V_e (Fig. 7.1 (c)). A surprising result here is that the combined system may oscillate *faster* than the individual oscillators comprising it.

To investigate the onset of spatial organization of periodic activity in the system a two-dimensional medium of locally coupled excitable cells is considered, where each excitable cell is connected to n_p passive cells [Fig. 7.1 (d)], n_p having a Poisson distribution with mean f. Thus, f is a measure of the density of passive cells relative to the myocytes. The results described here are for $f = 0.7$; it has been verified that qualitatively similar behavior is seen for various values of $f \geq 0.5$. The dynamics of the resulting medium is described by:

$$\frac{\partial V_e}{\partial t} = F_e(V_e, g) + n_p\, C_r(V_p - V_e) + D\nabla^2 V_e,$$

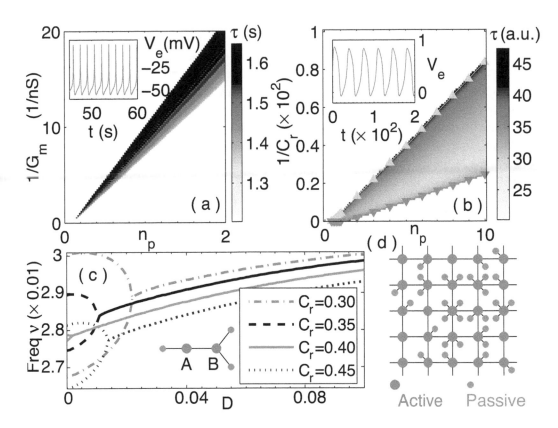

FIGURE 7.1 Oscillations through interaction between excitable and passive elements. A single excitable element described by (a) a detailed ionic model of a uterine myocyte and (b) a generic FHN model, coupled to n_p passive elements exhibits oscillatory activity (inset) with period τ for a specific range of gap junctional conductances G_m in (a) and coupling strengths C_r in (b). The triangles (upright and inverted) enclosing the region of periodic activity in (b) are obtained analytically by linear stability analysis of the fixed point solution of Eq. (7.1a). (c) Frequency of oscillation for a system of two "oscillators" A and B (each comprising an excitable cell and n_p passive cells with $n_p^A = 1$ and $n_p^B = 2$) coupled with strength D. Curves corresponding to different values of C_r show that the system synchronizes on increasing D, having a frequency that can be *higher* than either of the component oscillators. (d) Uterine tissue model as a two-dimensional square lattice, every site occupied by an excitable cell coupled to a variable number of passive cells.

where D represents the strength of coupling between excitable elements (passive cells are not coupled to each other). Note that, in the limit of large D the behavior of the spatially extended medium can be reduced by a mean-field approximation to a single excitable element coupled to f passive cells. As f can be non-integer, n_p in the mean-field limit can take fractional values [as in Fig. 7.1 (a-b)].

The system is discretized on a square spatial grid of size $L \times L$ with the lattice spacing set to 1. For most results $L = 64$, although values of L up to 1024 have been used to verify that the qualitative nature of the transition to global synchronization with increasing coupling is independent of system size. The dynamical equations are solved using a fourth-order Runge–Kutta scheme with time-step $dt \leq 0.1$ and a standard 5-point stencil for the spatial coupling between the excitable elements. While periodic boundary conditions are used for the results, it is verified that no-flux boundary conditions do not produce qualitatively different phenomena. Frequencies of individual elements are calculated using FFT of time-series for a duration 2^{15} time units. The behavior of the model for a specific set of values of f, C_r, and D is analyzed over many (~ 100) realizations of the n_p distribution with random initial conditions.

7.4   Dynamical transitions by increasing coupling

To quantitatively analyze the dynamical transitions as the intercellular coupling is increased, the differences in the oscillatory behavior of individual elements in the simulation domain are observed. In Fig. 7.2 (a) it is seen that at low D elements can have different periods, indicating the co-existence of multiple oscillation frequencies in the medium. This is explicit from the power spectral density of local activity at different sites [Fig. 7.2 (b)], which shows that there are multiple clusters in the domain, each being characterized by a principal frequency, ν [Fig. 7.2 (c)]. As all elements belonging to one cluster have the same period, this behavior is referred to as *cluster synchronization* (CS). Note that quiescent regions of non-oscillating elements, indicated in white in Fig. 7.2 (c), coexist with the clusters. As the coupling is increased the clusters merge [Fig. 7.2 (d)], thereby reducing the spread in the distribution of oscillation frequencies present in the medium, $P(\nu)$, eventually resulting in a single frequency for all oscillating elements (as seen for $D = 0.3$). As there are still a few local regions of inactivity, this behavior is termed *local synchronization* (LS). Further increasing D results in *global synchronization* (GS) characterized by *all* elements in the medium oscillating at the same frequency.

Dynamical transitions observed upon increasing the coupling between neighboring excitable elements can be interpreted as being coordinated by waves traveling over increasingly longer range in the system. Figure 7.3 (first row) shows spatial activity in the system at different values of D after long durations ($\sim 2^{15}$ time units) starting from random initial conditions. As the coupling D between the excitable elements is increased, a transition from highly localized, asynchronous excitations to spatially organized coherent activity that manifests as propagating waves is observed. Similar traveling waves of excitation have indeed been experimentally

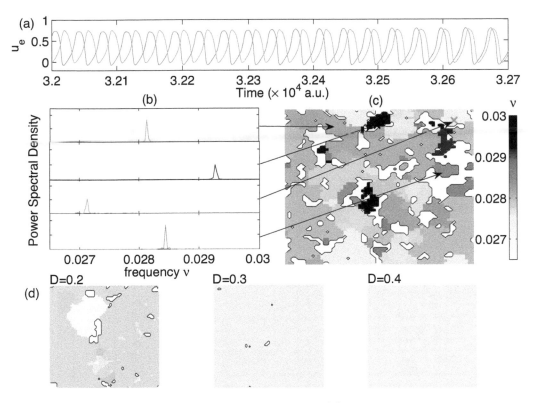

FIGURE 7.2 Synchronization via cluster merging. (a) Time-series of fast activation variable u_e for two excitable cells in the domain exhibiting distinct oscillation frequencies. (b) Power spectral density of u_e from four different sites [location shown in (c)] in a two-dimensional simulation domain with $L = 64$ ($f = 0.7, C_r = 1, D = 0.1$). (c) Pseudocolor plot indicating multiple clusters, each consisting of oscillators synchronized at a distinct frequency, i.e., cluster synchronization (white corresponding to absence of oscillation). (d) Increasing D from 0.1 in (c) to 0.2 in the left panel results in decreasing the number of clusters with distinct oscillation frequencies. Increasing D further to 0.3 results in local synchronization where all oscillators have the same frequency with a few patches showing absence of oscillation. When $D = 0.4$, all elements in the domain oscillate with the same frequency (i.e., global synchronization).

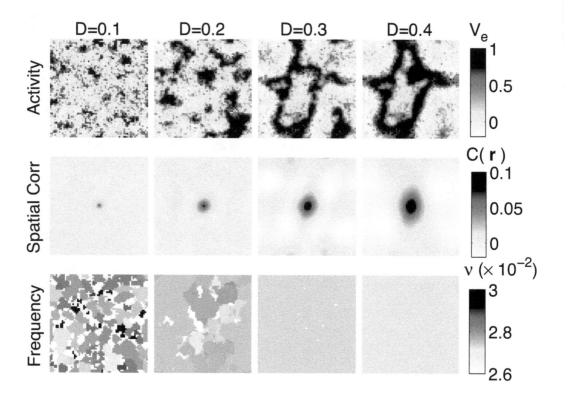

FIGURE 7.3 Emergence of synchronization via propagation of activity waves with increased coupling. Snapshots (first row) of the activity V_e in a two-dimensional simulation domain ($f = 0.7, C_r = 1, L = 64$) for increasing values of coupling D (with a given distribution of n_p). The corresponding time-averaged spatial correlation functions $C(\mathbf{r})$ are shown in the middle row. The size of the region around $\mathbf{r} = 0$ (at center) where $C(\mathbf{r})$ is high provides a measure of the correlation length scale which is seen to increase with D. The last row shows pseudocolor plots indicating the frequencies of individual oscillators in the medium (white corresponding to absence of oscillation). Increasing D results in decreasing the number of clusters with distinct oscillation frequencies, eventually leading to global synchronization characterized by spatially coherent, wavelike excitation patterns where all elements in the domain oscillate with same frequency.

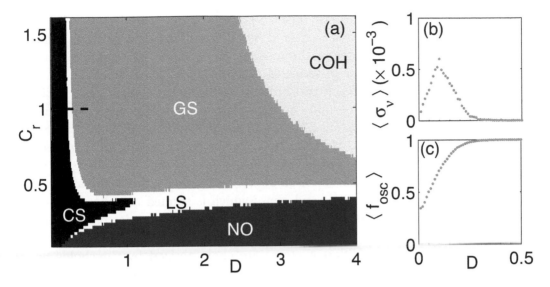

FIGURE 7.4 (a) Different dynamical regimes of the uterine tissue model (for $f = 0.7$) in $D - C_r$ parameter plane indicating the regions having (i) complete absence of oscillation (NO), (ii) cluster synchronization (CS), (iii) local synchronization (LS), (iv) global synchronization (GS), and (v) coherence (COH). (b-c) Variation of (b) width of frequency distribution $\langle \sigma_\nu \rangle$ and (c) fraction of oscillating cells $\langle f_{osc} \rangle$ with coupling strength D for $C_r = 1$ [i.e., along the broken line shown in (a)]. The regimes in (a) are distinguished by thresholds applied on order parameters $\langle \sigma_\nu \rangle$, $\langle f_{osc} \rangle$ and $\langle F \rangle$, viz., NO: $\langle f_{osc} \rangle < 10^{-3}$, CS: $\langle \sigma_\nu \rangle > 10^{-4}$, LS: $\langle \sigma_\nu \rangle < 10^{-4}$ and $\langle f_{osc} \rangle < 0.99$; GS: $\langle f_{osc} \rangle > 0.99$ and COH: $\langle F \rangle > 0.995$. Results shown are averaged over many realizations.

observed *in vitro* in myometrial tissue from the pregnant uterus [Lammers et al. 2008a]. The different dynamical regimes observed during the transition are accompanied by an increase in spatial correlation length scale (Fig. 7.3, middle row) and can be characterized by the spatial variation of frequencies of the constituent elements (Fig. 7.3, last row). For low coupling ($D = 0.1$), multiple clusters each with a distinct oscillation frequency ν coexist in the medium (CS). Note that there are also quiescent regions of non-oscillating elements indicated in white. With increased coupling the clusters merge, reducing the variance of the distribution of oscillation frequencies eventually resulting in a single frequency for all oscillating elements (LS, seen for $D = 0.3$). On increasing the coupling to even higher values ($D = 0.4$), a single wave traverses the entire system resulting in GS where *all* elements in the medium are oscillating at the same frequency. These results thus help in causally connecting two well-known observations about electrical activity in the pregnant uterus: (a) there is a remarkable increase in cellular coupling through gap junctions close to onset of labor [Miller et al. 1989] and (b) excitations are initially infrequent and irregular, but gradually become sustained and coherent toward the end of labor [Blackburn 2007].

The above observations motivate the following order parameters that allow to quantitatively segregate the different synchronization regimes in the space of model parameters D and C_r [Fig. 7.4 (a)]. The CS state is characterized by a finite width of the frequency distribution as measured by the standard deviation, σ_ν, and the fraction of oscillating elements in the medium, $0 < f_{osc} < 1$. Both LS and GS states have $\sigma_\nu \to 0$, but differ in terms of f_{osc} (< 1 in LS, $\simeq 1$ in GS). Figure 7.4 (b-c) shows the variation of the two order parameters $\langle \sigma_\nu \rangle$ and $\langle f_{osc} \rangle$ with the coupling D, $\langle \ \rangle$ indicating ensemble average over many realizations. Varying the excitable cell-passive cell coupling C_r together with D allows the exploration of the rich variety of spatiotemporal behavior that the system is capable of [Fig. 7.4 (a)]. In addition to the different synchronized states (CS, LS, and GS), a region where there is no oscillation (NO) characterized by $f_{osc} \to 0$, and a state where all elements oscillate with the same frequency and phase which is termed as coherence (COH) is also observed. COH is identified by the condition that the order parameter $F \equiv \max_t[f_{act}(t)] \to 1$ where $f_{act}(t)$ is the fraction of elements that are active ($V_e > \alpha$) at time t. In practice, the different states are characterized by thresholds whose specific values do not affect the qualitative nature of the results.

To further characterize the state of the system, the mean frequency $\bar{\nu}$ is determined by averaging overall oscillating cells for any given realization of the system. Figure 7.5 (a) reveals that several values of the mean frequency are possible at a given coupling strength. When the initial conditions are chosen randomly for each value of the coupling (broken curve in Fig. 7.5 (a)), the mean frequency decreases with increasing D. On the other hand, $\bar{\nu}$ is observed to *increase* with D when the system is allowed to evolve starting from a random initial state at low D, and then adiabatically increasing the value of D. The abrupt jumps correspond to drastic changes in the size of the basin of an attractor at certain values of the coupling strength, which can be investigated in detail in future studies. This suggests a multistable attractor landscape of the system dynamics, with the basins of the multiple attractors shown in Fig. 7.5 (d) [each corresponding to a characteristic spatiotemporal pattern of activity shown in Fig. 7.5 (e)] having differing sizes. They represent one or more plane waves propagating in the medium and are quite distinct from the disordered patterns of spreading activity (Fig. 7.5 (b-c)) seen when random initial conditions are used at each value of D. The period of recurrent activity in the uterus decreases with time as it comes closer to term [Garfield et al. 1998] in conjunction with the increase in number of gap junctions, an observation consistent with the result shown in Fig. 7.5(a) when considering a gradual increase of the coupling D.

As previously mentioned, the above results are for a fixed value of f, the mean number of passive cells per excitable cell. To investigate how varying the density of passive cells affects the spatial coherence in activity, a special case of the passive cell distribution is considered to define another spatially extended, two-dimensional lattice model for the uterine tissue. Here, an excitable cell, located at each lattice site, can be connected to either one or no passive cells (Fig. 7.6, left). This has the simplifying feature that the individual lattice sites either don't oscillate or oscillate at the same frequency in isolation. In this scenario, the passive cell density, f, which is the same as the fraction of oscillators in the lattice, varies between 0 and 1. When $f = 1$, the system corresponds to a *homogeneous* oscillatory medium. As the coupling between the excitable elements is increased, it is observed that for high

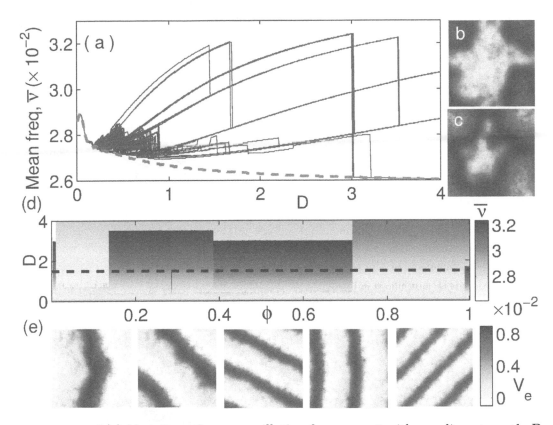

FIGURE 7.5 (a) Variation of mean oscillation frequency $\bar{\nu}$ with coupling strength D in the uterine tissue model ($f = 0.7$) for 400 different initial conditions at $C_r = 1$. Continuous curves correspond to gradually increasing D starting from a random initial state at low D, while broken curves (overlapping) correspond to random initial conditions chosen at each value of D. (b-c) Snapshots of activity in the medium at $D = 1.5$ for a random initial condition seen at intervals of $\delta T = 5$ time units. (d) Variation of the cumulative fractional volumes ϕ of the basins for different attractors corresponding to activation patterns shown in (b-c) and (e), as a function of the coupling strength D. (e) Snapshots of topologically distinct patterns of activity corresponding to the five attractors at $D = 1.5$ [shown by a broken line in (d)] when D is increased.

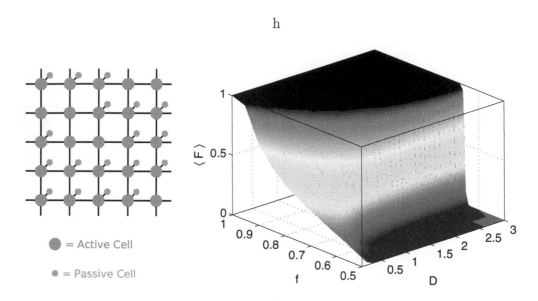

FIGURE 7.6 (left) A two-dimensional square lattice model for uterine tissue where individual excitable cells (located at each site) can couple to either one or no passive cell. (right) Variation of the mean value of the order parameter, $\langle F \rangle$, characterizing coherence (COH) shown as a function of the passive cell density (f) and the coupling strength D between the excitable cells, for $C_r = 0.6$.

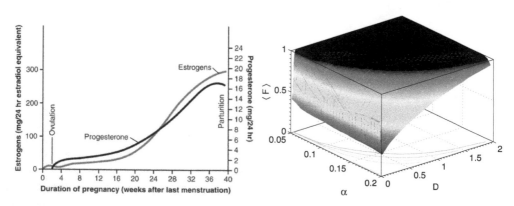

FIGURE 7.7 (left) Rate of production of human pregnancy-related hormones estrogen and progesterone over the course of pregnancy (adapted from Ref. [Guyton and Hall 2011]). (right) Variation of the mean value of the order parameter, $\langle F \rangle$, characterizing coherence (COH) shown as a function of the excitation threshold (α) and the coupling strength D between the excitable cells, for $C_r = 0.6$.

value of f the system becomes coherent (COH) as in the lattice model used earlier. At low passive cell density, increasing D results in cessation of oscillation (NO) [Fig. 7.6 (right)]. The transition to coherence can also be observed as a function of increasing passive cell density. For lower coupling, this exhibits a gradual rise, while at higher coupling there is an abrupt change in the order parameter characterizing coherence. This can be explained as a result of the system dynamics approaching that expected in the mean-field limit as the coupling D is increased.

An important biological factor that is believed to regulate the onset of uterine activity is the secretion of different hormones, such as estrogen and progesterone [Guyton and Hall 2011]. Estrogen increases the excitability of the myometrium, while progesterone reduces it [Mesiano 2004], so that altering the balance between the two can result in the uterus being quiescent or undergoing contractions. As seen from Fig. 7.7 (left), the rate of secretion of both these hormones increases during the course of pregnancy. However, close to term, the progesterone rate falls slightly while that of estrogen keeps increasing. This presumably results in a large increase in the myometrium excitability resulting in stimulation of uterine contractions. In the model discussed here, the role of such hormones can be incorporated by simply altering the value of the threshold of the excitable cells, α. Increasing ratio of estrogen to progesterone production rates can be modeled as reducing α which has the effect of making the medium more excitable. Figure 7.7 (right) shows that coherence is achieved by either increasing excitability (i.e., reducing α) or increasing the coupling strength D between excitable cells or both. Thus, the role of hormones essentially amplifies quantitatively the coherence that is achieved through increased coupling.

7.5   Implications for uterine contractions

The results described above explain several important features known about the emergence of contractions in uterine tissue. Previous experimental results have demonstrated that the coupling between cells in the myometrium increases with progress of pregnancy [Miller et al. 1989]. This suggests that the changes in the system with time amounts to simultaneous increase of D and C_r, eventually leading to synchronization as shown in Fig. 7.4 (a). Such a scenario is supported by experimental evidence that disruption of gap-junctional communication is associated with acute inhibition of spontaneous uterine contractions [Tsai et al. 1998]. The mechanism of synchronization discussed here is based on a very generic model, suggesting that the model results apply to a broad class of systems comprising coupled excitable and passive cells [Bub et al. 2002; Pumir et al. 2005]. A possible extension will be to investigate the effect of long-range connections [Falcke and Engel 1994; Sinha et al. 2007].

To conclude, the study discussed here shows that coherent periodic activity can emerge in a system of heterogeneous cells in a self-organized manner and does not require the presence of a centralized coordinating group of pacemaker cells. A rich variety of collective behavior is observed in the system under different conditions;

in particular, for intermediate cellular coupling, groups of cells spontaneously form clusters that oscillate at different frequencies. With increased coupling, clusters merge and eventually give rise to a globally synchronized state marked by the genesis of propagating waves of excitation in the medium. The model predicts that a similar set of changes occurs in the uterus during late stages of pregnancy.

7.6 Diffuse fibrosis

Before we conclude this chapter, we will discuss, in the context of wave propagation in the heart, another scenario that is of interest, namely the case of the spatial distribution of small inexcitable cells or obstacles in the medium. Unlike in the case of the passive cell distribution that we described in the previous sections, here a certain fraction of the cells in the medium are themselves inexcitable or dead. Under normal circumstance these cells account for only a small percentage of the tissue and play an important role of anchoring cardiac muscle cells via connective tissues like collagen and fibril material that help provide the heart with a structure. But with age or after an instance of heart attack, the percentage of these connective tissues can increase drastically, in some cases constituting up to 40% of the total number of cells in the heart. While this increase in fibrotic tissue has been associated with higher susceptibility to cardiac arrhythmias and in worst cases even sudden cardiac death [Burstein and Nattel 2008; Everett et al. 2007; Morita et al. 2014], the exact mechanism of the origin of such arrhythmia is still elusive.

Panfilov et al. in a series of papers explored the effect of a random distribution of inexcitable obstacles on the propagation of waves in both two- and three-dimensional tissue using both simplistic and realistic models of the heart [Panfilov 2002; ten Tusscher and Panfilov 2003b, 2005, 2007]. In these studies, a varying fraction of the total number of cells in the excitable medium was replaced by inexcitable obstacles modeled using no flux boundary conditions. They observed that the effect of these heterogeneities on wave propagation (both plane waves and spirals) were substantially different in two and three dimensions. The presence of inexcitable obstacles can lead to spiral formation and breakup. In two dimensions, a far lesser number of inexcitable cells are required, compared to three dimensions, to create a conduction block. Further the vulnerable window that gives rise to wavebreaks and spirals is much larger in the case of two dimensions. Unlike in two dimensions, the wavebreaks due to diffuse fibrosis do not develop into spirals easily in the case of three dimensional propagation. While the period of rotation for spirals increased with the fraction of inexcitable obstacles, the increase in the case of two dimensions was larger than that observed in the case of propagation in three dimensions.

Connection topology, networks, and dynamics

8.1 Introduction

The patterns characterizing the spatiotemporal dynamics of excitable media, such as spiral waves as well as spatiotemporal chaos arising from successive breakup and collisions between such waves, are observed in a wide variety of natural systems. They range from chemical systems such as the Belousov–Zhabotinsky reaction [Zaikin and Zhabotinsky 1970] and the heart in the throes of life-threatening arrhythmias [Gray et al. 1998] to neocortical slices from the mammalian brain [Huang et al. 2004]. In the latter system, it has been speculated that the observed spiral waves may provide a spatial framework for persistent cortical oscillations [Huang et al. 2004]. Theoretical work on persistent activity that can occur in excitable media has tended to focus on either the existence of specialized clusters of cells capable of oscillatory self-activation, e.g., the cells in the sinus node of the heart, or the presence of noise in the environment of the system that acts as a source of stochastic stimulation [García-Ojalvo and Schimansky-Geier 1999]. However, one can argue that, in such cases the activity in the medium is not truly self-sustained as the signal from the pacemaker region or the noise is akin to external intervention that is necessary for the initiation and persistence of spiral waves [Hou and Xin 2002; Jung and Mayer-Kress 1995a]. If such patterns are to be seen as spontaneously emerging from arbitrary initial conditions, then the pattern formation should be an outcome of the internal structure of the system alone. Furthermore, small variations in this structure may result in transitions between different dynamical regimes characterized by distinct spatiotemporal patterns.

To infer the possible nature of such structures let us recall that the brain differs from many other spatially extended excitable systems fundamentally in the topology of connections between its constituent elements. For example, in the heart, myocardial cells are electrically connected through gap junction proteins with neighboring myocytes forming a regular three-dimensional lattice. To give a visual analogy of this structure, it is rather like a wall having several layers, each made up of row upon row of closely fitting bricks. Exciting any region in this arrangement will propagate outward from the source of activation as a wave, which upon reaching the (non-conducting) boundary will dissipate. To sustain activity in such a system, we will therefore need to provide stimulation again and again. By contrast, the neurons in the brain, have a much more complicated arrangement by which they connect to each other. As neurons have elongated processes that can span large distances compared to the dimensions of the cell body, in principle, a single neuron can form connections with several thousands of other neurons in the brain. While a majority of these connections may be with other neurons whose cell bodies are within a short distance, connections with neurons that are located much farther are by no means rare. The deviation of the connection topology from that characterizing a regular lattice therefore provides yet another source of heterogeneity in excitable media, and we will look at its role on the collective dynamics of the system in this chapter.

8.2   A "small-world" excitable medium

Until quite recently, most theoretical studies on how the arrangement of connections between constituent elements affected system dynamics confined their attention to two broad categories: (i) *regular* lattices (mentioned above) where each element was coupled to all other elements that were located within a given neighborhood, whose geometry was defined by a pre-specified range of influence and the dimension in which the system was embedded, and (ii) *random* networks, where any pair of elements in the system are connected to each other with a probability independent of their physical location relative to each other. The contrasting connectivity structures in the two classes give rise to very distinct behavior for properties such as average path length and neighborhood clustering. Regular lattices are characterized by path lengths that scale with the total number of elements in the system, while in random networks this grows much more slowly, viz., logarithmically with system size. On the other hand, in sparse networks with random connections, elements would have very low clustering within their neighbors as the probability of connection between any pair of neighbors is just the probability that any two nodes in the network are connected, while regular lattices characterized by a relatively large *coordination number*, i.e., the number of neighbors of an element, will typically have highly clustered neighborhoods with many neighbors of a given element being also mutual neighbors. Therefore, it came as something of a surprise when Duncan Watts and Steve Strogatz proposed a new theoretical model for connection topologies that could yield networks exhibiting global behavior characteristic of random networks (viz., low average path length) but local properties reminiscent of regular networks (viz., highly clustered neighborhoods [Watts and Strogatz 1998]. The Watts–Strogatz model constructs such *small-world* networks by starting from a regular lattice and rewiring a fraction p of the existing links to randomly chosen elements of the system, thereby forming long-range shortcuts across large distances in the system. As the rewiring probability p increases from 0 (corresponding to a regular lattice) to 1 (a random network), the local and global properties of

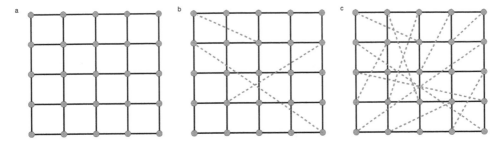

FIGURE 8.1 Constructing a small-world network on a two-dimensional square lattice substrate. Starting from a regular network (a) where each node is connected to its nearest neighbors, with probability p a link is added to a node that connects it to a randomly chosen node in the network. For small p, the resulting network (b) has a structure almost identical to that of the regular network with only a very few additional random long-range links. With increasing p (c), these random connections increase in number and the network begins to exhibit the global properties associated with random networks, such as low average path length. Qualitatively similar features are observed if instead of adding random links, a fraction of the links is randomly rewired keeping the total number of connections conserved.

the network gradually change, with the "small-world" regime being observed over a large, intermediate range of values of p. While the original Watts–Strogatz model had explicit conservation of the number of links, qualitatively similar behavior is observed in the absence of conservation, when the random connections are made in addition to the existing connections of the regular lattice (Fig. 8.1).

Here we look at excitable systems with similarly altered connection topology, i.e., one which has a regular topology with cells communicating only with nearest neighbors, but where there are a few random long-range connections, linking spatially distant cells. Such "small-world" topologies have been associated with self-sustained activity in a chain of model neurons [Roxin et al. 2004] as well as periodic epidemic patterns in disease spreading [Kuperman and Abramson 2001]. However, instead of looking at the role such long-range connections may play in maintaining pre-existing spiral waves [He et al. 2002], here we shall focus on the question of the emergence of, and transitions between, different types of self-sustained activity patterns as a result of such connection topology. The observation of spontaneous pattern formation in natural systems where sparse long-range connections coexist with fairly regular underlying connection topology, as in the brain [Harris-White et al. 1998], points toward intriguing possibilities for the functional role of these connections. In the following pages we shall investigate a generic model of excitable media that shows two qualitatively different regimes of self-sustained pattern formation obtained by gradually increasing the density of random long-range connections [Sinha et al. 2007]. The correspondence of the observed patterns with those observed in nature, e.g., epileptic bursts and seizures [Netoff et al. 2004], as well as their dependence on the topological structure of connections, makes these results highly relevant for natural systems.

The model for a small-world excitable medium consists of a two-dimensional array of $N \times N$ excitable cells coupled diffusively,

$$x_{t+1}^{i,j} = (1-D)f(x_t^{i,j}, y_t^{i,j}) + \frac{D}{4} \sum_{q=\pm 1} f(x_t^{i+q,j+q}, y_t^{i+q,j+q}).$$

Here D is the diffusion coefficient and the dynamics of individual cells are described by a pair of variables x_t, y_t, evolving according to a discrete-time model of generic excitable media [Chialvo 1995]:

$$x_{t+1} = f(x_t, y_t) = x_t^2 e^{(y_t - x_t)} + k,$$
$$y_{t+1} = g(x_t, y_t) = a y_t - b x_t + c,$$

where the parameter values used are $a = 0.89, b = 0.6, c = 0.28$, and $k = 0.02$. When this system is excited with a supra-threshold stimulation, the fast variable x shows an abrupt increase. This triggers changes in the slow variable y, such that the state of the cell is gradually brought down to that of the resting state. Once excited, the cell remains impervious to stimulation up to a refractory period, the duration of which is governed by the parameter a. Neighboring cells communicate excitation to each other with a strength proportional to the diffusion constant D, chosen to be $D = 0.2$ for the results described below. In addition to the diffusive coupling, long-range connections are introduced such that each cell receives a connection from a randomly chosen cell with probability p. The strength of this connection is chosen to be the same as that of the nearest neighbors, i.e., $D/4$. These random long-range links can be either quenched (i.e., chosen initially and kept fixed for the duration of the simulation), or annealed (i.e., randomly created at each time step). While the results reported below are for annealed random links, no qualitative difference is observed between these two cases. Both periodic and absorbing boundary conditions for the system have been used, and no significant differences in the results are observed. The results shown here are for absorbing boundary conditions.

8.3   Transitions between distinct dynamical regimes

The spatiotemporal patterns in the small-world excitable system observed on varying the value of the long-range shortcut density p can, generally speaking, be divided into three categories. First, below the lower critical probability, i.e., $0 < p < p_c^l$, the state of the system after an initial transient period is characterized by self-sustaining single or multiple spiral waves covering the entire system. Second, at $p = p_c^l$, the spiral wave mode is suppressed and the system undergoes a transition to a regime in which a large fraction of the system gets simultaneously active, and then refractory, in a periodic manner. Third, when the value of p is increased above a system-size-dependent upper critical probability p_c^u, the self-sustained activity ceases and the system falls into the absorbing state where $x^{i,j} = 0, \forall i, j$.

In the beginning of each simulation run the system is initialized such that $x = 1$ for a small number of cells, and $x = 0$ for the rest. For small shortcut probabilities

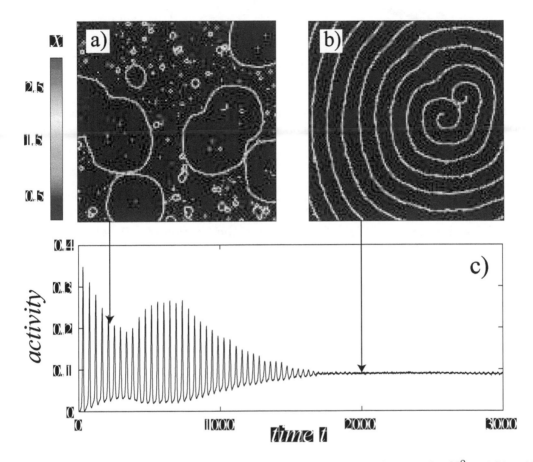

FIGURE 8.2 Emergence of spiral waves, recorded in a simulation of a $N^2 = 128 \times 128$ system, with shortcut probability $p = 0.25$. After initialization, the dynamics of the system can be characterized by circular waves. At around $t \sim 1,500$ time steps, a spiral wave is spontaneously created and subsequently takes over the dynamics. Panels (a) and (b) show the state of the system at different times in terms of the fast variable x. Grayscale bar indicates the excitation level of the cells. (c) Time series of the average activity, i.e., the fraction of cells with $x > 0.9$.

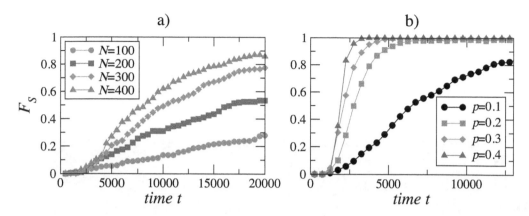

FIGURE 8.3 Fraction F_S of systems with spiral waves in 400 runs as function of time t: (a) fixed shortcut probability $p = 0.05$ with varying system size, (b) fixed system size $N^2 = 300 \times 300$ with varying shortcut probability p.

$p \ll p_c^l$, upon starting the simulation, multiple coexisting circular excitation waves are seen to emerge (see Fig. 8.2a), to be later taken over by spiral waves (Fig. 8.2b). The activity time series in Fig. 8.2c, displaying the fraction of cells where $x > 0.9$, shows a high frequency periodicity corresponding to the refractory period; then, the periodicity disappears as the system settles into the spiral wave mode. For the time series preceding the onset of the spiral wave, typically slow modulations of the envelope of the periodic oscillations are observed, which arise from the interaction between the waves as well as from the shortcut-induced excitations. Power spectral densities of such series are observed to show a power-law like decay, indicating the presence of $1/f$-noise.

The spiral waves are observed to be primarily created by a shortcut-induced excitation occurring in the refractory "shadow" of a circular wavefront, sparking a semi-circular wave whose transmission is partially hindered by the shadow. By explicitly using externally applied signal to stimulate an appropriate point in this region and then observing the resultant patterns, it has been verified that spiral waves can indeed be triggered in this fashion. Once a spiral wave is created, it will eventually take over the dynamics of the system, as the successive excitation wavefronts occur with the highest frequency compared to all other excitations which will then be swept away. Evidently, the probability of a spiral wave creation per unit time increases with the system size N^2 and the shortcut probability p. Thus for long times and large system sizes, excitation via random shortcuts will eventually always result in the formation of spiral waves. This is corroborated by Fig. 8.3, where the fraction of spiral wave configurations in 400 simulation runs is shown as a function of time t, for varying system sizes at $p = 0.05$ (panel a) and for fixed system size but increasing values of p (panel b). The spatial structure of the spiral waves, i.e., an excitation front followed by a refractory shadow, makes them very robust against perturbations. For example, if the state of the system is frozen, and the state of large areas (say, up to a quarter of the total area) or every second cell is set to $x = 0$, the spiral wave mode is quickly recovered once the simulation is restarted.

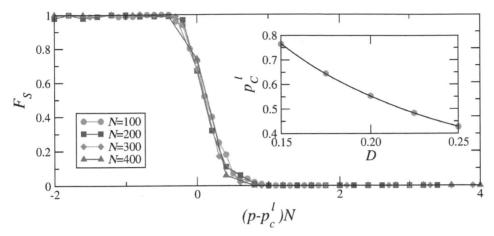

FIGURE 8.4 The fraction of spiral configurations F_S as a function of $\left(p - p_c^l\right)$ normalized by system size N, for shortcut probabilities p around $p_c^l \approx 0.553$. F_S is calculated at $t = 20,000$ time steps, averaged over 400 runs. Inset: Dependence of the critical value p_c^l on diffusion constant D. The circles are simulation results for $N = 200$ while the curve shows fitting with $p_c^l \sim D^{-1.14}$.

However, at high enough values of p, the shortcut-induced excitations become too numerous for sustaining the spiral wave dynamics. As almost every point is liable to be excited with a frequency proportional to its refractory period, the spirals become unstable and a transition to a new regime is seen at $p = p_c^l \approx 0.553$ (Fig. 8.4). This value of p_c is found to be independent of the system size N^2. Here, the spatial pattern becomes more homogeneous, as a large fraction of the system becomes simultaneously active and subsequently decays to a refractory state. However, some cells not participating in this wave of excitation carry on the activity to the next cycle, where it again spreads through almost the entire system. This results in a remarkably periodic behavior of the system in time, with a large fraction of cells being recurrently active with a period close to the refractory period of the cells (see Fig. 8.5 (e)-(f)). Often, small spiral-like waves are also observed (Fig. 8.5 (a)-(d)), but these are short-lived and spatial correlations are not maintained.

Finally, when p is increased still further, the very large number of shortcut connections guarantees almost simultaneous spread of excitation to nearly all cells. As a result, the dynamics of the system tends to "burn out" such that after a transient, almost all cells become refractory and not enough susceptible cells are left to sustain the excitation. Once a system-size dependent value $p = p_c^u(N)$ is approached, the probability of reaching the absorbing state $\left(x^{i,j} = 0, \forall i, j\right)$ grows rapidly (see Fig. 8.6) such that at high enough values of p, the system is never seen to sustain its activity. The limiting behavior for $N \to \infty$ is investigated by plotting p_c^u against $1/N$ and extrapolating its value for $1/N \to 0$. Fitting a quadratic function yielded $p_c^u \to 0.86$, hence it appears that the regime where activity is sustained has for all system sizes an upper limit for the shortcut density p. Note that this can be viewed as an approximation only, because there is no *a priori* reason to assume any particular form for the dependence of p_c^u on N.

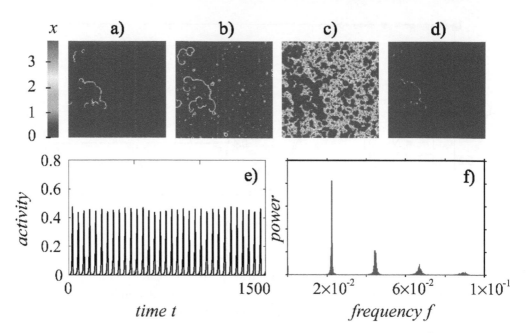

FIGURE 8.5 The periodic regime, recorded in a simulation of a $N^2 = 128 \times 128$ system, with shortcut probability $p = 0.6$. a)-d): Snapshots of the state of the system taken at intervals of $\Delta t = 10$ steps. e): Time series of the activity (see Fig. 1) and (f): the corresponding power spectrum. The main peak is at $f_0 \approx 0.022$, corresponding to a wavelength of ≈ 43 time steps; other peaks are harmonics at integer multiples of f_0.

As a biological system often exhibits significant variation in the intrinsic properties associated with different elements constituting the system, simulations with disorder in the parameters describing the properties of individual cells have also been carried out. For instance, the parameter a that controls the refractory period has been made a quenched random variable ranging over a small interval. One finds that there is a tendency for greater fragmentation of waves with disorder, and corresponding increase in the number of coexisting spiral waves, but otherwise no remarkable changes. The robustness of the results in the presence of disorder in the individual cellular properties underlines the relevance of this study to real-world systems, where cells are unlikely to have uniform properties. In addition, the role of the diffusion constant D has also been investigated. For higher D the wave propagates much faster, so that, in a fixed amount of time, the system will initiate excitation through many more long-range connections than the system with lower D. Therefore, the higher D system will be equivalent to a lower D system with a larger number of long-range connections, i.e., higher p (Fig. 8.4, inset).

8.4 Implications

The above results carry potential relevance to all natural systems that are excitable and have sparse, long-range connections. Experiments carried out in the excitable BZ reaction system with nonlocal coupling [Tinsley et al. 2005] have exhibited some of the features described above. For example, the coexistence of transient activity and sustained synchronized oscillations that are observed close to p_c^u has been reported in the above system, and can now be understood in terms of the model introduced here. Recently, there has also been a lot of research activity on the role of non-trivial network topology on brain function (e.g., Ref. [Eguiluz et al. 2005]). In particular, studies show a connection between the existence of small-world topology and epilepsy [Netoff et al. 2004; Percha et al. 2005]. This is one of the areas where the results described here can have a possible explanatory role. In the brain, glial cells form a matrix of regular topology with cells communicating between their nearest neighbors through calcium waves. Neurons are embedded on this regular structure and are capable of creating long-range links between spatially distant regions. It is now known that neurons and glial cells can communicate with each other through calcium waves [Charles 1994]. Therefore, the aggregate system of neurons and glial cells can be seen as a small-world network of excitable cells. Here the neurons, which are outnumbered by glial cells approximately by one order of magnitude, form the sparse, long-range connections. Recently, the role of neural-glial communication in epilepsy has been investigated (e.g., Ref. [Nadkarni and Jung 2003]), but the observation of spiral intercellular calcium waves in the hippocampus [Harris-White et al. 1998], and the similarity of the patterns seen in the small-world excitable media model with the observed features of epileptic seizures and bursts, makes it especially appealing to postulate the role of small-world topology in the generation of epilepsy. Experimental verification of this suggested scenario can be performed through calcium imaging in glial-neuronal co-culture systems.

The relationship between the fraction of long-range connections and normal brain function is indicated by experimental evidence not just for epilepsy but for various other mental disorders as well. A recent study has found that the brains of patients with clinically diagnosed schizophrenia, depression, or bipolar disease have lower glia to neuron ratio compared to normal subjects [Brauch et al. 2006]. Therefore, understanding the dynamical ramifications of increasing shortcuts in an excitable media can motivate experiments that have the potential significance of aiding clinical breakthroughs in treating a whole class of mental disorders. This is also connected with the question of the evolutionary significance for increasing glia to neuron ratio with brain size [Reichenbach 1989]. It is known that excitable media of larger dimensions are more likely to exhibit spiral waves [Sinha et al. 2001]. Therefore, decreasing the fraction of neurons (and therefore, long-range shortcuts) could be Nature's way of ensuring dynamical stability for neural activity.

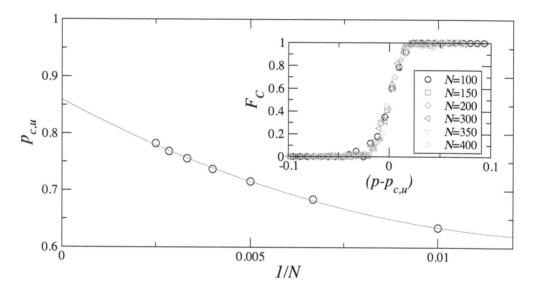

FIGURE 8.6   Inset: Fraction of configurations F_C where activity has ceased at $t = 20,000$ time steps in 100 simulation runs as function of $(p - p_c^u)$. Main: The upper critical shortcut probability p_c^u as function of inverse system dimension $1/N$ ($\circ$). The solid line displays a fitted quadratic function, where $p_c^u \, (N \to \infty) \approx 0.86$.

III

Controlling Dynamics of Excitable Media

III

9

Control and dynamical systems

9.1 Introduction

In this book we have so far looked at the intrinsic dynamics of excitable media, both homogeneous as well as in the presence of different types of heterogeneities. However, given the important functional role played by the patterns seen in such media (e.g., the connection between spiral waves and cardiac arrhythmias), it is important to ask whether the insights provided by the above studies can be used to devise methods to control these patterns. In the field of cardiac arrhythmia, empirical studies of how externally imposed electrical stimulation (such as in pacing or defibrillation discussed later) can be used to alter or suppress patterns such as pinned rotating waves or how spatio-temporal chaos has been used to establish heuristic principles. However, a deeper understanding of the nonlinear dynamics of the interaction between the dynamics patterns and external stimulation, may help in developing more efficient and safer techniques for clinical applications. In this and subsequent chapters, we focus on using insights from dynamical systems theory to come up with novel techniques of controlling the spatiotemporal patterns in excitable media using low-amplitude signals.

9.2 Control from the perspective of dynamical systems theory

To control a dynamical system, one needs a model of the system (either exact or approximate) which can, e.g., be expressed in terms of a system of differential equations [Fradkov and Pogromsky 2007]. Thus, a system evolving in continuous time with N-dimensional state space variables $\mathbf{x} = x_1, \ldots x_N$ and L parameters $\mathbf{r} = r_1, \ldots x_L$ can be described by $d\mathbf{x}/dt = \mathbf{F}(\mathbf{x}, \mathbf{r})$, where ideally $\mathbf{F}$ is a continuous function differentiable everywhere in the state space. One also needs to specify the set of observables or output for the system $\mathbf{O} = g(\mathbf{x})$. In the simplest case where all the state variables $x_1, \ldots x_N$ are accessible for measurement, we have $\mathbf{O} = \mathbf{x}$.

Once the system model has been chosen, the objective of the control needs to be stated. Usually, this involves stabilizing a desired state $\mathbf{x}^*$ of the system, which can be a fixed point, a periodic orbit or even an aperiodic trajectory. Starting from an initial state x_0, the aim of controlling the system will be to subject it to perturbations, in one or more of the state variables or the parameters, such that it approaches the desired state with time and eventually converges to it. If the control function does not depend on any of the state variables it is termed *open loop feedback*. However, if it uses information about the state the system is in or how the system responds to the control signal, it is known as *closed loop feedback* control.

9.3   Chaos control

Controlling chaos sounds like an oxymoron, because chaotic dynamics, where even the minutest of uncertainties in initial conditions rapidly results in complete loss of predictability, seems like the very antithesis of something that can be guided to conform to a pre-defined state.[†] Indeed until the pioneering papers on chaos control were published in 1990s [Ott et al. 1990], chaos was considered to be an unavoidable property of many non-linear systems, but something that was clearly not desirable. The possibility of controlling chaos hinged specifically on two aspects of such systems: (i) a chaotic dynamical attractor could be thought of as comprising an infinite number of unstable periodic orbits (UPOs) and (ii) a trajectory visits the neighborhood of every point in the chaotic attractor, so that it was guaranteed to come arbitrarily close to the UPO eventually. Thus the state of the system could at any given time approach arbitrarily close to a UPO if it found itself on the cycle's stable manifold, but then quickly move away along its unstable manifold.

Therefore, chaos can be controlled if one could somehow identify when the system comes close to a particular UPO and then perturb the system, viz., by manipulating a parameter or introducing a feedback perturbation to one of the dynamical variables, such that the state remains indefinitely on the stable manifold. An advantage of controlling chaotic systems is that a very small perturbation (small in comparison to the range of values that the dynamical variables span), can be sufficient to drastically alter the nature of the system state. Thus, one can choose a specifically desired state out of an infinite number of possible UPOs that is desired for a particular purpose or application and can nudge the system toward it by applying relatively weak signals. Such weak signals would have been insufficient to alter the state of an equivalent non-chaotic system, where one requires perturbation at least as large as the order of the deviation in the response that is desired [Boccaletti et al. 2000]. There is, in addition the possibility of rapidly switching from one desired state to a completely different state by simply switching between two

---

[†]Such sensitive dependence of initial conditions is quantitatively characterized by the notion of the Lyapunov exponent, which is the rate of exponential growth or decay of an initial deviation between two trajectories of the system in state space. If the Lyapunov exponent > 0, it implies chaos.

different control regimes. Extending this concept, one can produce almost any desired periodic or aperiodic dynamical trajectory by suitably controlling a chaotic system.

9.4 Controlling a low-dimensional dynamical system

To observe how control of chaos works in practice let us look at a simple example involving a low-dimensional dynamical system, viz., the logistic map. It describes the discrete-time evolution of a population as

$$x_{n+1} = F(x_n, r) = rx_n(1 - x_n), \tag{9.1}$$

where the dynamical variable $x_n \in [0, 1]$ represents the relative population size (expressed in terms of a maximum possible size that can be sustained with the resources available in the environment), time instant n, and the parameter r is the growth rate constrained to lie between the interval $[0, 4]$. It is easy to see that the map has two fixed points, one at $x_1^* = 0$ and the other at $x_2^* = 1 - (1/r)$. The fixed point at the origin, x_1^* is stable in the parameter range $0 \le r < 1$. At $r = 1$, the origin loses stability through a transcritical bifurcation and x_2^* becomes the new attractor for the system. However, this point also loses its stability at $r = 3$ through a pitchfork bifurcation and a period-2 cycle is born. In this regime the system switches repeatedly between the points $x_1^{2p} = [(1 + r) + \sqrt{(1+r)^2 - 4(1+r)}]/2r$ and $x_2^{2p} = [(1+r) + \sqrt{(1+r)^2 - 4(1+r)}]/2r$. This 2-cycle becomes unstable at $r = 1 + \sqrt{6} \simeq 3.45$, when a stable period-4 cycle is created. With increasing r, a sequence of several period-doublings is seen (Fig. 9.1), in each case a previously stable period-n cycle becomes unstable through a pitchfork bifurcation and a new period-$2n$ cycle becomes the new attractor, with the interval in parameter space between successive bifurcations diminishing rapidly. The length of the period characterizing the attractor diverges to infinity at the limit point $r^* \simeq 3.57$ where onset of chaotic behavior occurs. As r is increased above this value, more complicated behavior is observed, including a series of intervals where periodic cycles are again observed. Each of these periodic windows again shows period-doubling bifurcations to chaos. Finally, at $r = 4$, the chaotic attractor covers the entire unit interval and the Lyapunov exponent of the system reaches its maximum value $\lambda = \log_e 2 \simeq 0.693$.

As indicated earlier, if the system is in a chaotic regime when the parameter $r = r_0$ (say), it has an infinite number of unstable periodic cycles (including the unstable fixed points x_1^* and x_2^*), and we may want to stabilize any of them. Let us assume that the target of our control is an unstable period-M cycle ξ_i ($i = 1, \ldots, M$) such that $\xi_{i+1} = F(\xi_i, r)$ and $\xi_{M+1} = \xi_1$. If at any time t, the system state x_t is within a pre-specified neighborhood ϵ of ξ_i (i.e., $|x_t - \xi_i| < \epsilon$) we seek to apply a perturbation Δr_t to the parameter r so that, at the next iteration, the system state, viz., x_{t+1}, comes closer to the corresponding target state ξ_{i+1}. Thus, we seek to minimize the difference:

$$x_{t+1} - \xi_{i+1} = \frac{\partial F}{\partial x}(x_t - \xi_i) + \frac{\partial F}{\partial r}\Delta r_t. \tag{9.2}$$

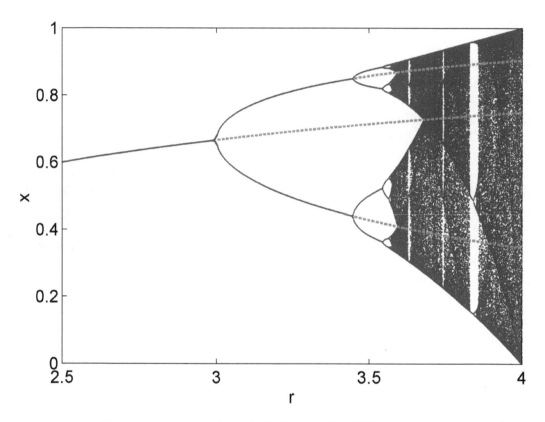

FIGURE 9.1 The attractor set of the logistic map for different values of growth rate $(2.5 \leq r \leq 4)$. The broken lines show the unstable nonzero fixed point x_2^* and the unstable 2-cycle $x_{1,2}^{2p}$.

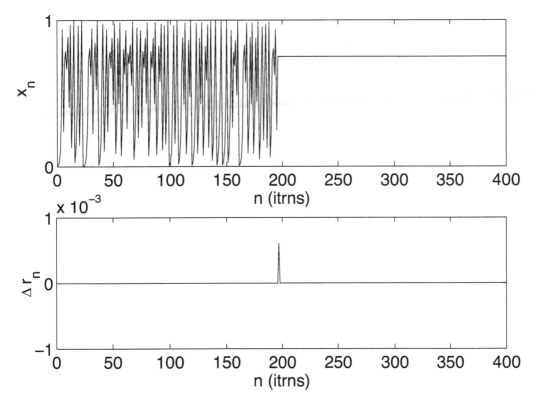

FIGURE 9.2 Demonstration of chaos control in the logistic map ($r = 3.999$) where the unstable fixed point $x_2^* = 1 - (1/r)(\simeq 0.75)$ is stabilized by parameter perturbation. The top panel shows the time-evolution of the state variable x with time (measured in terms of iterations of the map, n). Control is switched on when x enters a small neighborhood around the unstable fixed point. The bottom panel shows that a low-amplitude perturbation ($< 10^{-3}$) to the growth rate r applied for a single iteration is sufficient to stabilize the system at x_2^* (top panel).

Evaluating the partial derivatives of $F(\ )$ at $x = \xi_i$ and $r = r_0$ yields $x_{t+1} - \xi_{i+1} = r_0(1 - 2\xi_i)(x_t - \xi_i) + \xi_i(1 - \xi_i)\Delta r_t$. Ideally, we would like to make the difference disappear which suggests the perturbation we should apply:

$$\Delta r_n = r_0 \frac{(2\xi_i - 1)(x_t - \xi_i)}{\xi_i(1 - \xi_i)}. \tag{9.3}$$

Figure 9.2 shows the result of applying such parameter perturbation in order to stabilize the unstable fixed point (which can be thought of as a period-1 cycle) $x_2^* = 1 - (1/r)$ of the logistic map. The control is switched on only when the difference of the state variable x_n is less than $\epsilon = 0.01$ from x_2^*. A simple argument shows that the average transient period required before the system enters within the ϵ-neighborhood of a UPO is of the order of $1/\epsilon$. This is indeed what is observed in Fig. 9.2, where a small amplitude perturbation is seen to be triggered after almost 200 iterations, which is enough to stabilize the unstable fixed point.

Instead of applying perturbations to the parameter, one can also control the system by applying feedback signals to the state variable [Pyragas 1992]. In this method, one gives as an additive feedback to x a function of the state one desires to stabilize and information about the current state of the system, viz., $\Phi(\xi_i, x_t)$. For instance, one can add a term proportional to the deviation of the current state of the system from its state τ instant earlier, i.e., $K(x_{t-\tau} - x_t)$, where K is the strength of the feedback. This is known as *delayed feedback control* with delay period τ. If the UPO desired to be stabilized has a period τ, then the perturbation will disappear once control has been achieved. The actual implementation of the feedback method in experimental systems is generally simpler to achieve in comparison to the parameter perturbation method described earlier.

9.5   Control in spatially extended system

Control in spatially extended systems is a far more complicated problem than that of controlling dynamics in a low-dimensional system. Typically, the spectrum of Lyapunov exponents calculated for these systems will have more than one positive component. Applying perturbations at a single point of a spatially extended system may often be ineffective even in controlling the point itself, as the fluctuations resulting from the dynamical evolution of its neighboring elements will act as a source of noise that will interfere with the attempt by the control perturbations to stabilize a UPO. Thus, for control to be effective, the perturbations applied have to be felt by all points in the medium. A trivial way to achieve this is to apply control perturbations over the entire spatial extent of the medium. However, this may not be feasible — nor is it efficient in terms of the external signal applied to move the state of the system to a desired objective. One would prefer to impose control over the entire system by applying signals only to a finite set of points in the medium. For example, an early attempt to achieve this was envisaged by using control points regularly spaced apart over a medium (referred to as *pinning* control. Depending on how the signal is applied over a system, control schemes can be classified broadly into three categories [Sinha and Sridhar 2008]: local, spatially extended, and global, which will be discussed in later sections.

9.6   Why is control in excitable system different?

Controlling spatiotemporal chaos in excitable media has certain special features. Unlike other chaotic systems, response to a control signal is not proportional to the signal strength because of the existence of a threshold and a characteristic action potential with a specific maximum amplitude. As a result, an excitable system shows discontinuous response to control as regions, which have not yet recovered from a previous excitation or where the applied signal is below the threshold, will not be affected by the control algorithm at all. Also, the focus of control in excitable media is to eliminate all activity rather than to stabilize unstable periodic behavior. This is because the problem of chaos termination has tremendous practical importance in the clinical context, as the spatiotemporally chaotic state has been associated with the cardiac problem of ventricular fibrillation (VF). VF involves incoherent activation of the heart that stops it from pumping blood to the rest of the body, and is fatal within minutes in the absence of external intervention. At present, the only effective treatment is electrical defibrillation, which involves applying very strong electrical shocks across the heart muscles, either externally using a defibrillator or internally through implanted devices. The principle of operation for such devices is to overwhelm the natural cardiac dynamics, so as to drive all the different regions of the heart to rest simultaneously, at which time the cardiac pacemaker can take over once again. Although the exact mechanism by which this is achieved is still not completely understood, the danger of using such large amplitude control (involving ~ 1 A cm^{-2} externally and ~ 20 mA cm^{-2} internally [Winfree 1990]) is that, not only is it excruciatingly painful to the patient, but by causing damage to portions of cardiac tissue which subsequently result in scars, it can potentially increase the likelihood of future dynamical disorders of the heart. Therefore, devising a low-power control method for spatiotemporal chaos in excitable media promises a safer treatment for people at risk from potentially fatal cardiac arrhythmias.

9.7   Classifying chaos control schemes

In this section, we discuss most of the recent control methods that have been proposed for terminating spatiotemporal chaos in excitable media.[†] These methods are also often applicable to the related class of systems known as oscillatory media, described by the complex Landau–Ginzburg equation [Aranson and Kramer 2002], which also exhibit spiral waves and spatiotemporal chaos through spiral breakup. We have broadly classified all control schemes into three types, depending on the

---

[†]An earlier review, discussing methods proposed until 2002, can be found in Ref. [Gauthier et al. 2002].

nature of application of the control signal. If every region of the media is subjected to the signal (which, in general, can differ from region to region) it is termed as *global control*. On the other hand, if the control signal is applied only at a small, localized region from which its effects spread throughout the media, this is referred to as *local control*. Between these two extremes lie control schemes where perturbations are applied simultaneously to a number of spatially distant regions. We have termed these methods as *non-global, spatially extended control*. While global control may be the easiest to understand, involving as it does the principle of synchronizing the activity of all regions, it is also the most difficult to implement in any practical situation. On the other hand, local control (as it can be implemented using a single control point) will be the easiest to implement but hardest to achieve.

9.7.1   Comparing chaos control schemes

Most of the methods proposed for controlling spatiotemporal chaos in excitable media involve applying perturbations either globally or over a spatially extended system of control points covering a significant proportion of the entire system. However, in most practical situations this may not be a feasible option, either because of issues concerned with their implementation, or because of the large values of power involved. Moreover, if one is using such methods in the clinical context, e.g., terminating fibrillation, a local control scheme has the advantage that it can be readily implemented with existing hardware of the Implantable Cardioverter Defibrillator (ICD). This is a device implanted into patients at high risk from fibrillation that monitors the heart rhythm and applies electrical treatment when necessary through electrodes placed on the heart wall. A low-energy control method involving ICDs should therefore aim toward achieving control of spatiotemporal chaos by applying small perturbations from a few local sources.

However, the problem with most local control schemes proposed so far is that they use very high-frequency waves to overdrive chaos. Such waves are themselves unstable and may break up during propagation, resulting in re-initiation of spiral waves after the original chaotic activity has been terminated. The problem is compounded by the existence of inhomogeneities in real excitable media. Recently, Shajahan et al. [Shajahan et al. 2007] have found complicated dependence of spatiotemporal chaos on the presence of non-conducting regions and other types of inhomogeneities in an excitable system. Such inhomogeneities make the proposed local control schemes more vulnerable to failure, as it is known that high-frequency pacing interacting with, e.g., non-conducting obstacles, results in wavebreaks and subsequent genesis of spatiotemporal chaos [Panfilov and Keener 1993].

10

Controlling multiperiodic behavior

10.1 Alternans: A feature of non-linear properties of excitable system

An interesting property observed in certain biological excitable media, specifically the heart muscles, is the phenomenon of alternans. Alternans is the beat-to-beat variation in the response of the cardiac cells. At the cellular level, this beat-to-beat variation is observed in the duration of the action potential and the peak calcium concentration [Kockskämper and Blatter 2002]. The former is referred to as APD alternans, and the latter as transient calcium alternans. This beat-to-beat variation is manifested as a T-wave alternans in ECG [Pastore et al. 1999]. The T-wave in the ECG corresponds to the membrane repolarization at the level of the cell. Even micro-volt level variations in the T-wave amplitude could be related to very large alternations of the cellular repolarizations [Wilson and Rosenbaum 2007]. Detecting T-wave alternans is crucial and has implications for the prediction of the onset of fibrillation [Wilson and Rosenbaum 2007]. More generally as the transmembrane voltage and intracellular calcium concentration are two signals that determine the contraction of cardiac myocytes, any variation in the APD and/or peak calcium concentration can cause serious disruption to the normal rhythm of the heart. In fact it is now recognized that T-wave alternans (and the underlying cellular alternans) are pathological conditions that are closely linked to the onset of ventricular arrhythmias [Euler 1999; Pastore et al. 1999; Rosenbaum et al. 1994; Weiss et al. 2006].

Before going further, let us understand the phenomenon of alternans a little better. The typical response of the cardiac myocyte to an external supra-threshold stimulus is characterized in terms of its APD. Now for a sequence of such stimuli (also referred to as beat) under normal physiological conditions it has been observed that the response to the stimulus follows a simple 1:1 dynamics, i.e., the successive APDs are almost equal and do not differ from one beat to another. But conditions like rapid stimulation can lead to the disruption of this 1:1 response via a *period-doubling bifurcation* and result in more complicated dynamical behavior such as

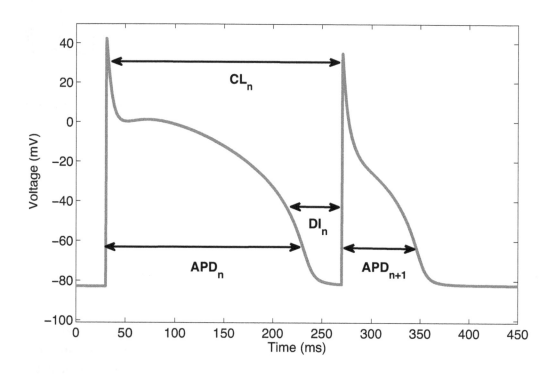

FIGURE 10.1 The time-series of the transmembrane potential V for a cell subjected to rapid periodic stimulation of period T demonstrates alternans of the action potential duration (APD). The nth action potential is shown to have a relatively long duration compared to the period between successive stimulations (denoted as the cycle length, CL_n). As the diastolic interval (recovery period) of the cell (denoted as DI_n) is short before the onset of the next excitation, the resulting $(n+1)$th action potential is shorter. Notice that CL_n, APD_n, and DI_n are the same as T_n, a_n, and d_n, respectively, in the text.

alternans and conduction block. For instance, as the period of stimulation is decreased, APD and intracellular Ca transient begin to alternate from beat to beat. This results in a sequence of long-short-long-short APDs or large-small-large-small peak calcium concentration. Since the transmembrane voltage and the intracellular calcium concentration are coupled to each other, alternans in one influences the other and vice versa.

10.1.1   Dynamical origins of alternans

The dynamical origin of the emergence of alternans can be best understood in terms of the cardiac restitution curve. For a cardiac system subjected to stimulation with an interval T_n between the generation of the nth and $(n + 1)$th action potentials, the duration of the latter depends on the former according to

$$a_{n+1} = f(d_n) = f(T_n - a_n), \tag{10.1}$$

where a_n and d_n are the *APD* and *DI* of the nth beat, respectively. The return map f relating successive APDs is the restitution function for the duration of an action potential that depends on the preceding diastolic interval DI. For a constant slow pacing period T, (i.e., fixed $T_n = T$), the APDs are constant corresponding to a stable fixed point of the restitution return map: a^* $[= f(T - a^*)]$. However, for fast enough pacing (i.e., small T), this fixed point becomes unstable giving rise to a period-2 attractor leading to a steady sequence of alternating long and short APDs. This mechanism for the origins of alternans in terms of the restitution relation was proposed in 1968 by Nolasco and Dahlen [Nolasco and Dahlen 1968; Weiss et al. 2006]. For a given pacing period or cycle length (CL), (consisting of an action potential duration (APD), followed by a recovery period or diastolic interval (DI) (see Fig. 10.1)), they showed that the slope of the APD restitution curve determines the occurrence of alternans. The intersection of the two curves, $CL_n = APD_n + DI_n$ and $APD_{n+1} = f(DI_n)$, is the solution that satisfies both equations. The solution is stable if the slope of the restitution curve is < 1 and unstable otherwise. When the slope is < 1, the solution is a period-1 behavior in APDs. For the case where the equilibrium point is unstable (i.e., slope of the restitution curve is > 1, the APD alternates successively between two values, and the resulting period-2 behavior corresponds to the alternans state.

The APD restitution curve thus provides a simple and elegant way to analyze a complicated non-linear phenomenon like the alternans. Unfortunately this simple analysis has several limitations and is not sufficient to make predictions that are in complete agreement with experimental observations in real cardiac tissue. The fundamental limitation of this approach lies in the assumption that APD is a function of the previous DI alone. Cardiac cells have short-term "memory" and the properties of the medium such as the APD and depolarization threshold, actually depend on the history of pacing [Fenton et al. 1999; Fox et al. 2002a; Watanabe and Koller 2002]. Another important factor influencing alternans is the bidirectional coupling between intracellular calcium and membrane voltage. The coupling from the transmembrane voltage to calcium concentration is positive, i.e, the APD alternation will be followed by a period-2 behavior in the peak calcium dynamics following a similar phase relationship. This means that a longer APD corresponds to a longer peak calcium transient. On the other hand, the influence of calcium on voltage can be either positive or negative which can result in an out of phase relationship between APD and Ca concentration [Weiss et al. 2006]. Restrepo and Karma refer to the situation corresponding to in-phase relation between APD and peak calcium as electromechanical concordant alternans and the state where they are out of phase as electromechanical discordant alternans [Restrepo and Karma 2009]. Mounting evidence seems to suggest that although voltage and calcium concentrations can independently display alternans, their mutual coupling results in the observation of period-2 behavior in both variables [Shiferaw et al. 2005]. More recently, fibroblast-myocyte coupling has been suggested as another potential mechanism underlying the onset of alternans in cardiac tissue [Xie et al. 2009]. In this simulation study it was observed that the electrotonic effects of fibroblast-myocyte coupling can promote both APD and Ca alternans.

10.2   Concordant and discordant alternans

While the phenomenon of alternating APDs can be observed even at the level of a single cell, spatial effects such as dispersion of the conduction velocity in an extended cardiac tissue can result in alternans being observed even for values of cycle length that would be expected to produce 1:1 response. Alternans can undergo a distinct change in characteristic with further decrease in the pacing period. In the alternans regime, for large cycle length all cells in the medium are in the same phase during every beat. That is all of them either show long APD (large peak Ca) or short APD (small peak Ca). Such a beat-to-beat variation in the *APD* where all the cells in the medium show the same behavior after each stimulus, is called *concordant alternans*. But under certain conditions, such as very rapid pacing, it might be possible that different regions of the medium show alternating responses. That is, for a given beat some cells might respond with a large *APD*, while some others might display a smaller *APD*. This spatial distribution of the long and short *APD*s reverses during the next stimulus. That is cells which displayed a short *APD* in response to n^{th} stimulus respond with a long APD at $n + 1^{th}$ beat and vice versa. This state with is called *discordant alternans*. The transition from concordant (at $T = 100$ ms) to discordant alternans (by $T = 500$ ms) can be seen in Fig. 10.2.

10.3   Why study alternans?: Instability and connection with arrhythmias

As we discussed in Chapter 3, it has been experimentally observed that the phenomenon of alternans is connected with the occurrence of cardiac arrhythmias via the process of spiral breakup. In its original form the "restitution hypothesis" suggested that spiral breakup can occur when maximum slope of the restitution curve exceeds one [Karma 1993, 1994]. Steep slope of the restitution curve results in an increased sensitivity of APD to changes in DI. Small changes in DI are followed by drastic increase or decrease in APD. In spatially extended systems, APD essentially determines the width of excitation, i.e., the region between the wavefront and waveback. Alternans in APD is then accompanied by oscillations in the width that give rise to wave instabilities causing breakup. But the criteria of the restitution slope < 1 giving rising to spiral breakup was later shown to be an over simplification [Cytrynbaum and Keener 2002]. Several other factors such as conduction velocity dispersion [Qu et al. 1999], dynamic restitution [Koller et al. 1998], and spatial dispersion of APD restitution [Banville and Gray 2002], have now been identified as potential mechanisms for spiral breakup. It has been observed that stable spirals exist even if the maximum slope of the restitution curve is actually greater than one [Nash et al. 2004, 2005; Tolkacheva et al. 2003]. In spite of these limitations, the "restitution hypothesis" is a useful paradigm to study wavebreaks in the cardiac tissue, especially in light of the fact that drug-induced flattening of the restitution curve increases the stability of waves in the heart and prevents spiral breakup [Garfinkel et al. 2000; Riccio et al. 1999].

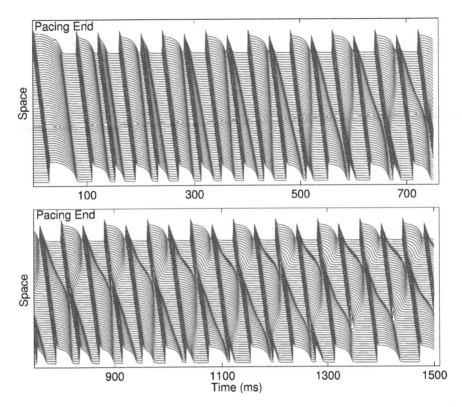

FIGURE 10.2 Development of spatially discordant alternans resulting in conduction block due to very rapid pacing. Superimposed traces of time series of transmembrane voltage along the length of a one-dimensional fiber of simulated cardiac cells. The "Pacing End" corresponds to the end at which the high frequency stimulus ($T = 160$ ms) is applied. The fiber consists of $L = 550$ cells. While initially the pacing and the distal ends display concordant alternans, by the time $T = 500$ ms the two ends have become out of phase with each other resulting in a spatially discordant alternans state. Following the beat at $T = 1300$ ms, a conduction block occurs and thereafter only every second stimulus propagates to the other end.

Rapid stimulation (pacing) of cardiac tissue is a commonly used technique [Fisher et al. 1978] to generate wave trains for removing sources of irregular activity responsible for arrhythmias [Karma and Gilmour 2007]. However, although generating waves by using very high-frequency stimulation should increase the probability of suppressing the abnormal activity [Pumir et al. 2010], extremely rapid pacing can also generate a time-varying response of the cardiac tissue [Cao et al. 1999] to the applied stimuli because of the non-linear recovery properties of excitable cells [Breuer and Sinha 2005]. Such a beat-to-beat variation in the duration of the cellular activity (action potentials), successively elicited by the applied stimuli, can themselves produce further sources of arrhythmia thereby defeating the very purpose of pacing.

10.4   Controlling alternans: Overview of feedback control

The fact that alternans is often the substrate for arrhythmogenesis such as VF has motivated a lot of attempts to develop external intervention schemes that can control or at least reduce the magnitude of alternans. While pharmaceutical interventions have been tried in this context, the results have been far from satisfactory. In fact doctors go to the extent to suggest that anti-arrhythmia drugs should not be prescribed to patients unless it is unavoidable [Fogoros 2008]. The skepticism of doctors is not entirely unfounded. The results from several studies on the effect of anti-arrhythmic drugs have not been encouraging. Further as the CAST trials showed, some of these drugs could have pro-arrhythmic effects and actually kill more people than they save [CAST; Coromilas et al. 1995]. Not surprisingly, there has been more interest in developing non-pharmacological treatment for arrhythmias. One of the most commonly used treatments these days involves the application of electrical signals from devices implanted on the heart. Implantable cardioverter defibrillator (ICD) are designed to monitor the rhythm of the heart continuously and detect the onset of abnormal rhythms. On detection of arrhythmia they are programmed to apply large defibrillatory shocks with typical strength being of the order of hundreds of volts [Jordan and Christini 2005].

One of the main drawbacks of ICD is the amplitude of the shock that can damage the cardiac tissue. Also such large shocks can be rather painful. From the late 1990s, several feedback control methods have been suggested toward the improvement of the efficacy of the ICD. Many of these schemes were based on methods developed for controlling the dynamics of non-linear physical systems. An example where such schemes have been shown to work is the termination of period-2 alternans in APD. The focus of these techniques has been to stabilize the unstable period-1 state corresponding to alternans. Most of these methods are inspired from the OGY chaos control scheme developed in 1990 [Ott et al. 1990]. The working principle of the OGY scheme is that by the application of small perturbations to one of the accessible parameters the system dynamics can be nudged toward the stable manifold of one of the unstable periodic orbits (UPOs) [Christini and Collins 1997]. In the context of cardiac alternans the accessible system parameter is the time of next stimulation.

The suppression of the period-2 dynamics is achieved by modifying the pacing period continuously using a feedback factor [Christini et al. 2006; Echebarria and Karma 2002a,b; Hall and Gauthier 2002; Hall et al. 1997]. In this method, the pacing period (CL_n) of the n-th step is controlled as:

$$CL_n = T + \frac{\gamma}{2}(APD_n - APD_{n-1}) \qquad (10.2)$$

where γ is a tunable parameter that defines that feedback gain. It can be easily shown that the originally unstable trivial fixed point can be made stable if γ is tuned to lie in the range $1 - 1/f' < \gamma < 2/f'$, where $f' \equiv f'(T - APD^*)$ [Echebarria and Karma 2002b]. Notice that the stable range of γ becomes small if f' is steep (the range vanishes as $f' = 3$). Furthermore, the pacing period of the proportional feedback method varies over many different values depending upon the difference between the last two APDs recorded. The range of variation can be quite large, especially at the initial stages soon after switching on the control and is a possible source of additional complications in potential clinical applications. Furthermore, the proportional feedback gain method has had only limited success in controlling alternans in spatially extended systems [Christini et al. 2006].

10.5 Generate stable yet rapid stimulation: Alternate period pacing

In this section, we will describe a recently proposed technique to significantly reduce the magnitude of alternans in an excitable cardiac system [Sridhar et al. 2013b]. In this method, suppression of period-2 alternans state is achieved by using just two values of pacing period. The applied pacing period is made to alternate between two values $T \pm \Delta T$ ($\Delta T \ll T$ being a predetermined control parameter) that are just above and below a period T which is known to produce a period-2 alternans state. This makes the algorithm more akin to an open loop control, unlike the proportional feedback method which is a closed loop scheme. As open loop control is easier to implement in general, this is an appealing feature for practical implementation. Finally, we will also describe a successful experimental implementation of this scheme in an isolated whole heart system.

10.5.1 Alternating period scheme

We will first describe the alternating period pacing algorithm that markedly reduces the variation between successive action potentials even when the system is driven by rapid periodic stimulation that would, in the absence of control, lead to significant APD alternans. As suggested by the name "alternating period," the method relies on switching continually between two different periods of successive stimulation of the system, measured by the cycle length CL_n for the period between the nth and $(n + 1)$th action potentials (Fig. 10.1). As alternans involves the variation in the APDs, the control method has to ensure that the APDs resulting from successive stimuli remain almost the same. In order to achieve this, the control algorithm decides the time of next stimulation by choosing between the two possible values of pacing period, $T_+ \equiv T + \Delta T$ and $T_- \equiv T - \Delta T$. Here ΔT is a free parameter chosen before the application of the control. The choice of the pacing period is based on the difference between the current and previous APDs,

$$\text{If } a_n > a_{n-1}, \text{ then } T_n = T_+, \text{ otherwise } T_n = T_-. \tag{10.3}$$

This ensures that if APD_n is short, the following APD will also be short (as the cycle length is short); in contrast, if APD_n is long, the use of a longer cycle length will ensure that the next APD will also be long. For most of the period during which the alternating period pacing is applied, the cycle lengths follow either the sequence $T_+T_-T_+T_- \ldots$ or the (effectively equivalent) sequence $T_-T_+T_-T_+ \ldots$, switching from one to the other, whenever, an APD is shorter (or longer) than both its preceding and following APDs. The choice of $T_\pm$ ensures that the mean value for the cycle length is T, the period of rapid stimulation for which we seek a steady response of the system. For the results reported here, the value of T has been chosen such that while for $\Delta T = 0$ we observe large alternans, the magnitude of the period-2 response is significantly reduced for a suitable choice of $0 < \Delta T \ll T$ (Fig. 10.3).

10.5.2 A numerical investigation of the alternating period control

In this section, we will describe the efficacy and mechanism of the alternans suppression scheme described above. We will first discuss the application of this control scheme on a single cell and then follow it up with the application on a spatially extended system of coupled excitable cells. The model system used to study the control scheme has the generic form,

$$\frac{\partial V_t}{\partial t} = \frac{-I_{ion}(V, g_i) + I_{noise}(x, t) + I_{ext}(x, t)}{C_m} + D\nabla^2 V, \tag{10.4}$$

where V (mV) is the potential difference across a cellular membrane, $C_m(= 1$ μFcm^{-2}) is the transmembrane capacitance, I_{ion} (μAcm^{-2}) is the total current density through ion channels on the cellular membrane (Luo–Rudy I model described in Chapter 2), and g_i describes the dynamics of gating variables of different ion channels. The stochastic current density term I_{noise} represents an additive thermal or channel noise [Clayton et al. 2003] that is randomly fluctuating in both time and space within a limited range (uniformly distributed between 0 and 0.5 μAcm^{-2} here). The space- and time-dependent pacing current density I_{ext} (μAcm^{-2}) represents the external stimuli applied in a local region in order to generate wave trains. The last term on the right corresponds to spatial coupling in a multicellular array, with an effective diffusion constant D= 0.001 cm^{2}s^{-1}. For all the results shown here, the maximum K^+ channel conductance G_K is increased to 0.705 mS cm^{-2} in order to reduce the APD [ten Tusscher and Panfilov 2003a]. The maximum Ca^{2+} channel conductance is set to $G_{si} = 0.09$ mS cm^{-2} in order to generate alternans in the single cell simulations. The results described here are robust and are not sensitively dependent on small variations in the model parameters. The equations are

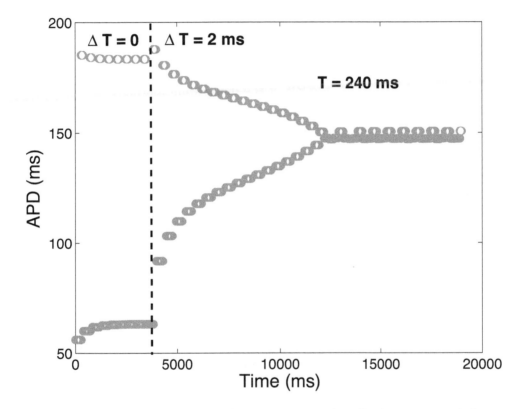

FIGURE 10.3 Suppression of APD alternans in a single cell subject to rapid alternating period pacing. When a cell is initially stimulated with constant period pacing ($T = 240$ ms), the APD shows a marked alternans behavior. After the alternating period pacing is switched on (indicated by the dashed line) where the stimulation period switches between $T_+ = T + \Delta T$ and $T_- = T - \Delta T$ with $\Delta T = 2$ ms, the APD variation becomes negligible within a period of 10 s.

solved using a forward-Euler scheme with a time-step $dt = 0.01$ ms. The spatially extended system is discretized on an array of size L with space step $dx = 0.0225$ cm and a standard three-point stencil for the Laplacian describing the spatial coupling between the units. For most results reported here L is between 30 and 40 cells, although we have used L up to 60. No-flux boundary conditions are implemented at the edges. Pacing stimuli are implemented by applying an external current I_{ext} of magnitude 100 μAcm^{-2} for a duration of 1 ms. For a one-dimensional cable, it is applied from one end of the system ($x = 0$) over a region of finite width (0.225 cm). Action potential duration is measured as the period between the successive instants when V crossed -60 mV from below and from above.

Figure 10.3 shows the result of applying the alternating period pacing method on a single cell to suppress APD alternans when stimulating at a mean period of T=240 ms. As seen from the variation of the APD before the alternating method is switched on (i.e., the region to the left of the dashed line), stimulation at constant period T results in a significant degree of alternans, with successive action potentials having durations of 183 and 63 ms, respectively. However, when alternating period stimuli with $\Delta T = 2$ ms are applied, the alternans is suppressed within 10 s to a relatively negligible magnitude, with the action potentials at the steady state having durations ranging between 147 and 151 ms. This finding is consistent with the experimental observations (that will be described later) that the alternans can be suppressed within 50 beats after the start of the control.

Notice that the switching condition between T_+ and T_- based on the difference between successive APDs [Eq. (10.3)] is crucial for successful control. A simple two-period pacing of the cell, using either the sequence $T_+T_-T_+T_- \ldots$ or $T_-T_+T_-T_+ \ldots$ exclusively without switching between them would result, after a short initial transient of converging APDs, in successively diverging APDs.

While the suppression of alternans in a single cell is encouraging, for the method to be applicable in practical situations the alternating period pacing should be successful in significantly reducing alternans in spatially extended systems. Figure 10.4 shows the results of applying the alternating period pacing method on a one-dimensional fiber of excitable cells with $L = 30$. When the system is stimulated (the pacing end is at $x = 0$) at a constant period $T = 250$ ms, different points on the fiber show alternans having magnitude that increases as one proceeds along the fiber away from the pacing end. For example, at $x = L/3$, the successive APDs alternate between 82 and 192 ms, respectively, while at the farthest end ($x = L$) APDs switch between 62 and 195 ms, respectively, for successive pacing stimuli (see the time-series to the left of the dashed line in Fig. 10.4). Within a few seconds of switching on the alternating period pacing, the APD alternans is significantly reduced and in the steady state the APDs fluctuate over a relatively small range between approximately 141 and 166 ms that does not vary significantly with the location on the fiber. While the method is effective for even longer fibers, the efficacy however rapidly diminishes with system size with the decrease in APD alternans as a result of alternating period pacing becoming less marked for larger values of L.

It is instructive to compare the alternating period pacing method with previously proposed algorithms for reducing alternans [Echebarria and Karma 2002a,b]. These

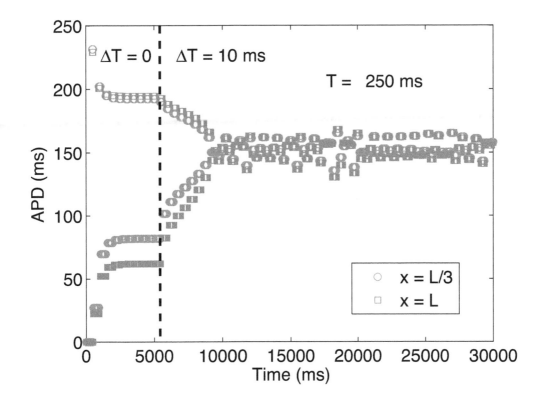

FIGURE 10.4 Reduction of APD alternans in a one-dimensional fiber of excitable cells by alternating period pacing. An array of excitable cells with $L = 30$ subjected to pacing at the $x = 0$ end shows variation in the duration of successive action potentials when it is subjected to a constant periodic stimulation ($T = 250$ ms), with the magnitude of alternans increasing as one moves further away from the pacing end along the fiber. Once the alternating period pacing is switched on (indicated by the broken line) with the stimulation at the pacing end switching between $T_+ = T + \Delta T$ and $T_- = T - \Delta T$ with $\Delta T = 10$ ms, the APD variation is significantly reduced within 5 s. Notice that different points in the fiber show a similar degree of reduction in the magnitude of alternans.

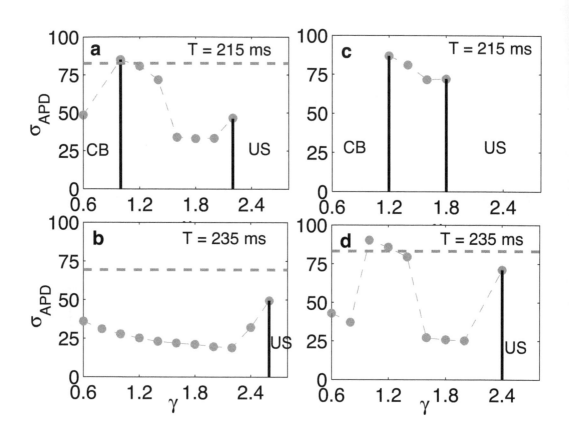

FIGURE 10.5   Performance of the proportional feedback control method in suppressing alternans in a one-dimensional fiber ($L = 40$) of excitable cells in the absence and presence of noise. The efficiency of the method is obtained from the standard deviation of the durations of a long sequence of action potentials measured at the center of the system ($x = L/2$) as a function of the feedback gain parameter γ. The system is subjected to two different imposed pacing periods (a,c) $T = 215$ ms and (b,d) 235 ms in the absence (a-b) and presence of noise (c-d). The thick broken line shows the standard deviation of the APD for constant pacing with period T in the absence of feedback control ($\gamma = 0$). Applying control can sometimes result in conduction block (CB) at a location downstream of the pacing site ($x = 0$). For certain cases, no response is elicited even at the pacing end; the control is said to be unstable (US) in such cases.

feedback control methods seek to stabilize the unstable fixed point of the effective periodically perturbed dynamical system by adopting the cycle length according to the proportional control scheme given by Eq. (10.2). If the duration of the n-th action potential is larger than the $(n-1)$th one, Eq. (10.2) ensures that the next action potential has a longer duration than would have been the case without control thereby suppressing alternans behavior. Figures. 10.5(a) and 8(b) show the performance of the proportional gain method on a fiber with $L = 40$ for two different values of T. The success of alternans reduction is measured in terms of the standard deviation for the sequence of APDs. The dashed line shows the APD standard deviation in the absence of control ($\gamma = 0$). A comparison with the corresponding values in the presence of control for different values of the gain parameter show that while this feedback control can reduce alternans, it also sometimes results in conduction block (CB) of the stimulation away from the pacing site. Moreover, for higher values of the gain parameter, even the pacing site may be incapable of being activated — a situation referred to as unstable control (US).

As any experimental implementation of the control will always have to encounter the effect of noise, the efficacy of the method by switching on the stochastic fluctuation term I_{noise} in Eq. (10.4) is examined. As seen from Fig. 10.5 (c) and 10.5 (d), the region of parameter space over which the method is successful in reducing alternans is markedly reduced in the presence of noise, with conduction block becoming prominent at lower values of γ while at higher values of γ, the control tends to become unstable. Comparison of the performance of the alternating period pacing method with that of the proportional gain method reveals that the former performs favorably in terms of the reduced number of instances where conduction block and instabilities of control occur even though the degree of reduction of alternans in both the schemes is of a similar magnitude (Fig. 10.6). This can be observed particularly in the presence of noise [Fig. 10.6 (c) and (d)], where the alternating pacing period scheme can consistently perform well compared to the proportional feedback control method.

One potential reason for the better performance of the alternating period pacing scheme in reducing alternans in the presence of noise is that unlike the proportional feedback control method, it does not seek to alter the stability of the fixed point. Instead the focus is on trying to confine the state of the system over a small volume of phase space such that all variations in the duration of action potential occur within a narrow interval. The function f of Eq. (10.1) can be obtained by stimulating the cell at different DIs and measuring the corresponding APDs. (The restitution curve may alter slightly depending on whether it is measured in a fully or partially recovered system because of memory effects. Here, to include the role of such memory in the medium dynamics, the cell is first subjected to 15 successive stimuli at intervals of 300 ms before the APD is measured for a desired value of T). For long inter-stimuli interval T, the action potentials have an almost constant duration and the corresponding system dynamics has a stable fixed point at $APD^* [= f(T - APD^*)]$. When T is reduced, this fixed point becomes unstable and a period-2 attractor is generated where a long APD is followed by a short APD and vice versa. It is easy to see that such stable alternans will correspond to three fixed points in the

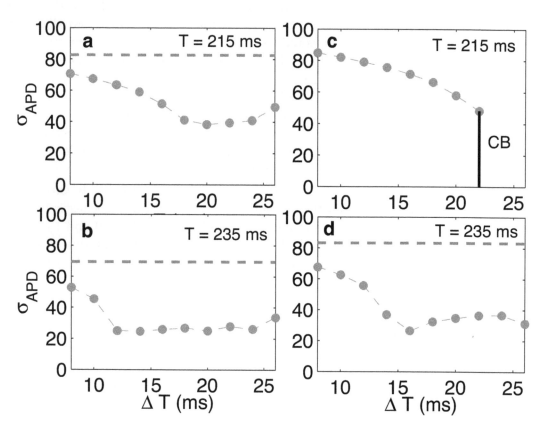

FIGURE 10.6 Performance of alternating period pacing in suppressing alternans in a one-dimensional fiber ($L = 40$) of excitable cells in the absence and presence of noise. As in Fig. 10.5, the efficiency in reducing alternans is obtained from the standard deviation of the durations of a long sequence of action potentials measured at the center of the system ($x = L/2$) as a function of the difference in the alternating periods ΔT. The system is subjected to two different pacing periods of (a,c) $T = 215$ ms and (b,d) 235 ms in the absence (a-b) and presence (c-d) of noise. The dashed blue line shows the standard deviation of APD for constant pacing with period T ($\Delta T = 0$). Alternating period pacing can occasionally result in conduction block (CB) at a location downstream of the pacing site ($x = 0$).

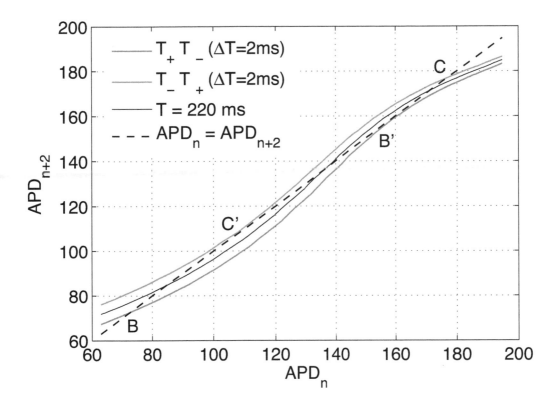

FIGURE 10.7 The alternating period pacing scheme reduces alternans by confining the system state in a small region in phase space. The curves correspond to the composite return map $f \circ f(APD)$ obtained from the restitution function numerically computed from a single cell. The middle curve corresponds to the situation where the cell is subjected to a constant period stimulation with $T = 220$ ms. The two other curves correspond to alternating period stimulation following the sequence $S : T_+ T_- T_+ T_- \ldots$ or $S' : T_- T_+ T_- T_+ \ldots$ with $T_\pm = T \pm \Delta T$ (here $\Delta T = 2$ ms). The intersection of the curves with the diagonal line $APD_{n+2} = APD_n$ shows the fixed points of the system dynamics. For alternating period pacing the system has a stable fixed point (B, C for S, S', respectively) and a constriction point (B', C' for S, S', respectively) where the system dynamics slows down considerably while it is transiting through this region of phase space. By switching appropriately between the sequences S, S' the system state can be maintained indefinitely within the interval $(APD_{B'}, APD_{C'})$ around the unstable fixed point of f which reduces the magnitude of the alternans (for constant T the system switches between APD_B and APD_C).

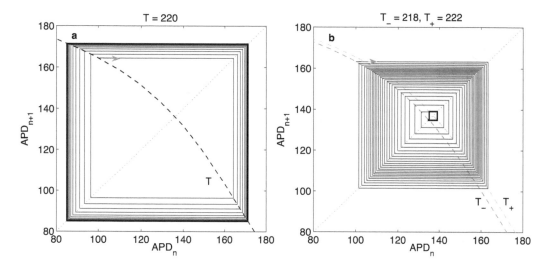

FIGURE 10.8 The trajectory of the system on the first return map Eq. (10.1) in the uncontrolled (a) and controlled (b) cases. (a) The cobweb diagram on the return map (dashed curve) for constant periodic pacing with $T = 220$ shows that the system converges to stable, large magnitude period-2 alternans. (b) The magnitude of the alternans reduces drastically when the pacing is alternated between the periods $T_- = 218$ and $T_+ = 222$ according to the proposed control scheme. The return maps corresponding to the two periods are shown as dashed curves and are indicated by T_- and T_+, respectively. The dynamical evolution of the return map in both cases begins from $APD = 100$ (the initial trajectory is indicated by an arrow). The dotted line corresponds to the $APD_n = APD_{n+1}$ line.

composite return map $f \circ f$ as shown in Fig. 10.7. In the figure, the fixed point in the middle is unstable while the other two are stable. When the alternating period pacing scheme is applied, the system follows one of the two possible maps depending on the exact sequence of the pacing periods $T_{\pm} = T \pm \Delta T$ being used, viz., $S : T_+T_-T_+T_- \ldots$ or $S' : T_-T_+T_-T_+ \ldots$. Both of these maps are characterized by the existence of a stable fixed point (close to one of the two APDs seen during alternans state induced by constant period pacing) and a constriction point near the unstable fixed point of the map where the curve comes very close to the diagonal line $APD_{n+2} = APD_n$. In a situation analogous to that observed during intermittency in deterministic systems [Pomeau and Manneville 1980], the system dynamics slows down significantly as it negotiates this region of phase space. By switching between the sequences according to Eq. (10.3), the system is maintained indefinitely in the narrow region whose limits are defined by the constriction points of the two maps corresponding to two S and S' (Fig. 10.7). One can observe this by comparing the system trajectory on the first return maps (Eq. 10.1) corresponding to the uncontrolled and controlled cases shown in Fig. 10.8. Figure 10.8 (a) shows the evolution from an initial $APD = 100$ ms to a stable period-2 behavior having a large alternans amplitude when only a single pacing period, T is used. Figure 10.8 (b) shows the result of using the alternate pacing period scheme, where instead of pacing at a constant T, T_- and T_+ are applied alternately. As seen in the figure, starting from the same initial APD as in the uncontrolled case, the magnitude of the alternans reduces remarkably [Fig. 10.8 (a)]. The final state in the presence of control corresponds to very low amplitude oscillations around the $a_n = a_{n+1}$ state.

This analysis of the control mechanism implies that there is an optimal range of values of ΔT for which the pacing scheme will be most effective. Increasing ΔT increases the space between the curves and the diagonal line at the constriction points that allows the trajectories to escape from the trapping region easily, thereby reducing the efficacy of the method. In contrast, reducing ΔT can decrease the robustness of the scheme as fluctuations can easily eject the system state from the very small trapping region between the constriction points. In practice, the optimal values of ΔT can be easily obtained through trial and error.

From the discussions above, it is clear that the T_+T_- control is capable of reducing the alternans in spatially extended models of the cardiac excitable media. Similar to the proportional scheme, the T_+T_- control is also not stable in spatially extended system if there is only one control point. It would be desirable to have a scheme that can have global stability even when there is only a single control point. On the other hand, the simulation results indicate that T_+T_- control has a smaller number of conduction blocks and control instabilities, especially in the presence of noise. It would be important to understand the physical mechanism of the T_+T_- control scheme; presumably investigations on the phase space dynamics of the control can provide deeper insights both theoretically and experimentally. Similar to the situation that alternans will occur only for fast enough pacing (less that a critical value of T), it can be anticipated that the T_+T_- control scheme will be able to suppress alternans magnitude for ΔT larger than some critical value.

10.6   Experimental implementation of the alternate pacing scheme

In this section, we will describe the experimental implementation of the alternating period control scheme. The experimental system comprises isolated hearts extracted from Wistar rats maintained in a Langendorff system. Briefly, the Langendorff system is used to maintain the physiological condition of an isolated heart and keeping it functional by providing perfusion with nutrient rich, oxygenated solution at a constant temperature. Usually, these preparations last for three to four hours. In the experiment, pseudo-electrocardiograms (ECGs) of the hearts at various locations are monitored by inserting the electrodes in the heart tissues. A pacing electrode placed on the septum between two ventricles is used to provide controlled stimulation to the isolated heart. The contraction pressure of the left ventricle (LVP) is monitored by a water-filled balloon (1 cm long, made of latex) inserted inside the left ventricle through a pressure transducer. Detailed descriptions of the setup and experimental conditions can be found in Ref. [Sridhar et al. 2013b]. The electric stimulation to the heart is delivered through an isolated stimulator. The form of a single electrical stimulation is a rectangular current pulse with duration of 1 ms and amplitude that is twice the diastolic threshold current. Instead of using the APDs in the isolated heart for control, the peak value of LVP (p_n for the nth beat) is used. As a longer APD leads to a stronger cardiac contraction [Hirth et al. 1983], there is a $1:1$ correspondence between APD_n and LVP_n. In fact, the first documented phenomenon of alternans was based on the pulse pressure [Karma and Gilmour 2007]. Thus, one can effectively replace APD_n with LVP_n. The experiments detailed below are all carried out at $20°C$ because it is much easier to generate alternans at lower temperature with slow pacing. Experiments carried out at other temperatures (viz., $37°C$) produce qualitatively similar results.

Before the start of an experiment, the excised heart was first maintained in the Langendorff system for at least 30 min to allow it to adapt to the Langendorff environment; beating with its own sinus rhythm period, T_{SR}. Then, external stimulations are used to produce alternans by slowly decreasing the pacing (beating) period from T_{SR} to T_0. Alternans can be seen as the occurrence of two values of measured LVP for a single pacing period. Fig. 10.9 shows the time course of the generation of the alternans. The pacing period in Fig. 10.9 is decreased in three steps to the desired T_0. It can be seen from Fig. 10.9 that while the stimulation has a single period, the response of the LVP undergoes a period-doubling transition. The alternans can be seen as the LVP alternating between two different values. The stability of this period-2 behavior depends on the state of the heart and the magnitude of the alternans. In our experiments, we have always checked that the alternating state is stable for at least 10 min before we start our experiments. To start a T_+T_- control experiment, procedures similar to those used in Fig. 10.9 are used to first produce the alternans and then the T_+T_- control is started 15-30 s after T_0 has been reached.

Figure 10.10 shows the time course of LVP in a typical T_+T_- control experiment.

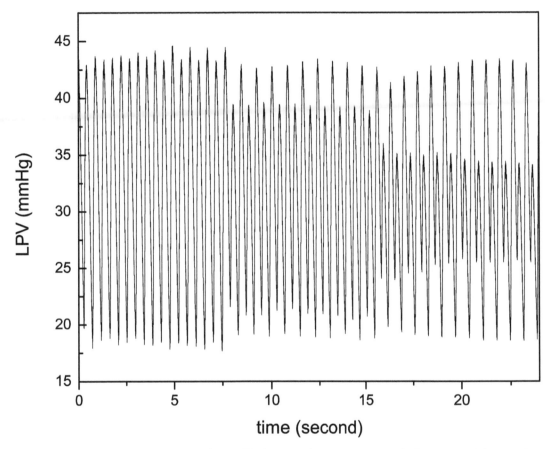

FIGURE 10.9 Time course of the LVP during the generation of alternans. The pacing period is decreased from 450 to 400 ms and then to $T_0 = 350$ ms. It can be seen that there is a transition from a single period to a period doubling response of the heart. Notice that if the rate of decrease of the pacing period is too fast or T_0 is too small, there might be inductions of tachycardia or fibrillations.

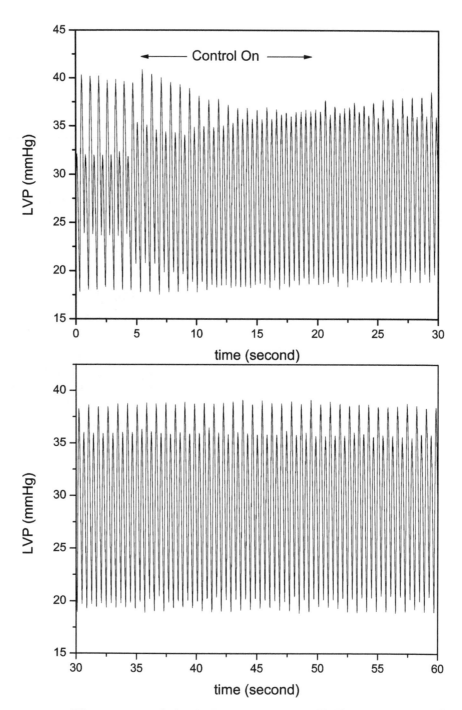

FIGURE 10.10   Time course of the LVP under 15 s of T_+T_- control with $\Delta T = 10$ ms and $T_0 = 400$ ms, (top) Control started at t=5 s and lasted until 20 s. (bottom) Continuation of (top) after the control has stopped. Notice that the magnitude of alternans after control is smaller than before control.

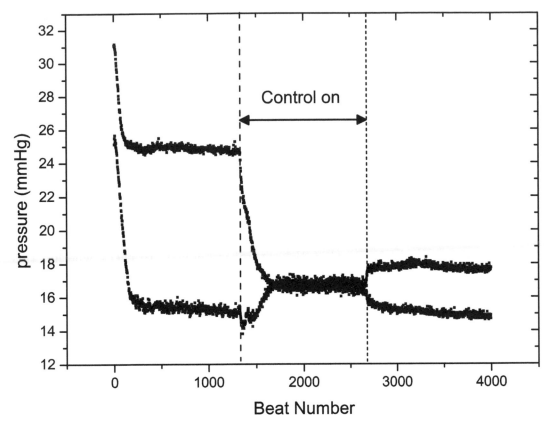

FIGURE 10.11 Time course of the peak value of LVP under 600 s of T_+T_- control with $\Delta T = 14$ ms and $T_0 = 500$ ms. Time is being shown as beat numbers. Control started at beat number 1350 and ended at 2650. It can be seen that the magnitude of alternans after control is smaller than before control.

The suppression of alternans can be seen as the decrease in the magnitude of the difference between two successive LPVs. In Fig. 10.10 (a), T_+T_- control is started at t= 5 s and kept on up to t = 20 s. It can be seen that the response of the heart changes immediately after the control is switched on and the magnitude of the alternans (difference between alternating peaks) decreases monotonically. From the figure it can be observed that the alternans is suppressed around t = 17 s or within 40 beats. Another remarkable feature seen in Fig. 10.10 is that the control itself has some systematic long-term effect on the alternans state. This effect can be seen in Fig. 10.10 (b) once the control is switched off. From the figure it is clear that the alternans state re-emerges at (t = 25 s) once control is switched off, and the new state has a smaller alternans magnitude compared to that before the control was applied. It seems that the T_+T_- control has changed the alternans response of the heart. There are some hysteresis effects of T_+T_- on the heart. The time for which the control is applied in Fig. 10.10 is kept short so as to demonstrate the efficacy of the scheme. To study the properties of the controlled state, we have also performed experiments with much longer control time as shown in Fig. 10.11. In this figure we show LVP_n (the peak values of the LVP at the nth beat) as a function of the beating number n. The initial alternans state is prepared between $n = 0$ and $n = 1350$ with $T_0 = 500$ ms. The alternans state can be seen as the existence of two main values of the LVP_n. Their difference is the magnitude of the alternans. Similar to Fig. 10.10, it can be seen from Fig. 10.11 that the suppression of alternans is fast once the control is switched on. Now, with a control time of 600 seconds, it can be seen clearly that the LVP_n under control is not constant but takes on a range of values more or less randomly. It would be desirable to control the system to have a very short range of LVP_n by using a smaller ΔT. Figure 10.12 shows such an attempt with $\Delta T = 2ms$. It can be seen that there is little suppression with this control. That is, ΔT cannot be too small and there seems to be a minimal value of ΔT for effective control. If we examine the re-emerged alternans state after the control is stopped in the experiments discussed above, it can be seen that the re-emerged state always has a smaller alternans magnitude. There seems to be a hysteresis effect on the heart when the control is applied. To check whether this observation is an artifact of the T_+T_- control scheme, we have also implemented the proportional control scheme of Eq. (10.2) in our experiment. In experiments with the proportional control scheme, we have also found similar effects (data not shown). However, this hysteresis is not a long-term effect due to a change in the physiology of the heart as we can always reproduce the initial alternans state by regenerating it from a non-alternans state with a much longer T_0 as depicted in Fig. 10.9. It seems that the control has put the heart in a different dynamical state even after the control has been stopped. Furthermore, to eliminate variabilities due to different samples, the above experiments have all been repeated for at least five different hearts and similar results were found.

For the whole heart experiments, it is still a puzzle why the alternans response of the heart would be different after alternans suppression control. Since the magnitude of the alternans response is governed by the restitution curve, a different response after the feedback control suggests that the restitution properties of the cardiac cells have been altered by the control. Experimentally, it is found that for both the proportional feedback and T_+T_- controls, the pacing interval changes from a constant value to an almost constant value plus a small random fluctuating part when compared to that of without control. This small fluctuating part is always only a few percent or less of the original constant value. Its large systematic effect on the alternans is quite surprising and understanding the mechanism by why such a small change in the pacing interval has large systematic effect on the restitution on the cardiac cells would be useful in further refining this scheme to improve its efficacy.

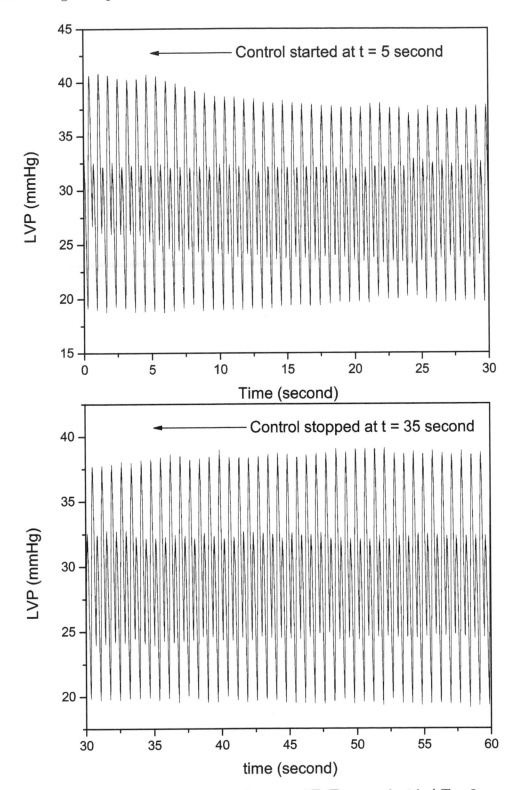

FIGURE 10.12 Time course of LVP under 30 s of T_+T_- control with $\Delta T = 2$ ms and $T_0 = 360$ ms. (a) Control started at t = 5 s. (b) Continuation of (a) and the control is stopped at t = 35 s. Notice that the magnitude of alternans after control is smaller than before control.

11

Control of reentrant waves and spirals

11.1 Introduction

Controlling spiral waves using low-amplitude external perturbation is not only a problem of fundamental interest in the study of dynamics of excitations in active media [Hörning et al. 2009; Isomura et al. 2008; Sinha and Sridhar 2008; Sinha et al. 2001; Sridhar and Sinha 2008; Takagi et al. 2004; Zhang et al. 2005a], but also has significant implications for the clinical treatment of cardiac arrhythmias [Fenton et al. 2009]. A potentially fatal arrhythmia occurring in the ventricles is tachycardia, or abnormally fast excitation, during which the heart can be activated as rapidly as 300 beats per minute. There are multiple mechanisms by which ventricular tachycardia (VT) may arise, but the most common one is due to the formation of a reentrant pathway, i.e., a closed path of excitation feedback. Reentry often has an anatomical substrate, with the excitation wave going round and round an existing inexcitable obstacle, e.g., a region of scar tissue as shown in Fig. 11.1 (left).

For people in chronic risk of VT, the most common treatment is implanting an ICD, a device capable of detecting the onset of VT and giving a periodic sequence of low-amplitude electrical stimuli (pacing) through an electrode, usually located in the ventricular apex, in order to restore the normal functioning of the heart [Josephson 1993]. The operating principle of this device is that, by pacing at a frequency higher than that of the VT,[†] the stimulated waves will eventually reach the reentrant circuit and terminate the reentry. However, the underlying mechanisms of the success and failure of pacing termination are not yet well-understood and, the algorithms currently used in such devices are often based on purely heuristic principles. As a result, occasionally, instead of terminating VT, pacing can accelerate it or can even promote its degeneration to lethal ventricular fibrillation (VF), leading to death within minutes if no immediate action is taken. Understanding the interaction

---

[†]The VT frequency is essentially the inverse of the reentry period ($T_{reentry}$), the time required by the excitation wave to go around the reentrant circuit once.

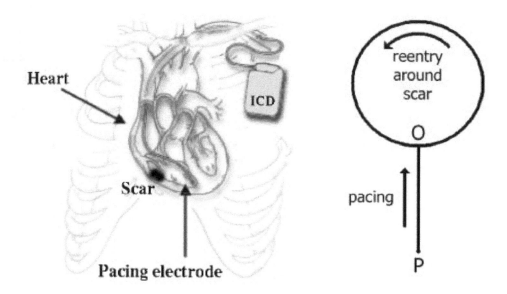

FIGURE 11.1 (left) Schematic diagram of antitachycardia pacing in the heart using an ICD (adapted from a figure courtesy of Guidant Corp.). Note the non-conducting scar tissue (in black) occupying a significant portion of the ventricle. Pacing is usually applied via an electrode placed at the ventricular apex (the lowermost point of the ventricle in the figure). (right) A simplified ring-and-sidebranch model of pacing. Reentrant activity occurring around a scar tissue is simplified into a wave going around a ring. The sidebranch joining the ring at O represents the external stimulation arriving from the pacing electrode located at P.

dynamics between pacing and reentrant waves is therefore essential for designing more effective and safer ICD pacing algorithms.

In this chapter we will discuss two studies that use simple models of heterogeneous excitable media in order to explain the conditions that allow the successful elimination of a reentrant spiral wave by generating rapid waves of excitation from a localized source [Breuer and Sinha 2005; Pumir et al. 2010].

11.2   Elimination of pinned spiral waves by rapid pacing

In Chapter 9, we broadly classified control schemes in excitable media into three categories. One of the categories involved the generation of high-frequency excitation waves by the local application of the control stimulus. Such periodic high-frequency stimulation (pacing) from a localized region in the excitable medium can generate wave-trains that interact with the spiral wave [Zipes and Jalife 2004]. If the frequency of external stimulation is higher than the rotation frequency of the spiral wave, the wave-train induces it to drift. In a finite medium, the vortex (spiral) is eventually driven to the boundary and thereby eliminated from the system [Agladze et al. 2007; Gottwald et al. 2001a; Krinsky and Agladze 1983].

However while the above argument can account for the removal of spirals in a homogeneous medium, it may not apply to the case of spirals in a medium with inexcitable obstacles. Spiral waves tend to anchor or pin themselves to heterogeneities. The anchoring can be considered as analogous to the pinning of vortices in disordered superconductors [Blatter et al. 1994; Pazo et al. 2004]. Attachment of the spiral to an obstacle prevents its removal by the use of a rapid external wave-train [Ginn and Steinbock 2004; Jimenez et al. 2009; Pertsov et al. 1984; Vinson et al. 1994]. In the heart, obstacles such as blood vessels or scar tissue, can play the role of pinning centers [Lim et al. 2006], leading to anatomical reentry, the sustained periodic excitation of the region around the obstacle. In this section, the conditions that can lead to the unpinning of an anchored spiral wave by pacing will be discussed. We will also discuss the relation between the size of the obstacle and the pacing period that can successfully detach the pinned vortex [Pumir et al. 2010]. We will start with a brief description of the the argument, based on a classical result of Wiener and Rosenblueth, that a pinned spiral cannot be detached using a wave-train.

11.2.1   Classical Wiener–Rosenblueth Theory

In a homogeneous active medium, a spiral wave can be controlled by a wave train, induced by periodic stimulation from a local source (pacing)[Zipes and Jalife 2004]. If the frequency of stimulation is higher than that of the spiral wave, the wave train induces the spiral to drift. In a finite medium, the vortex is eventually driven to the boundary and thereby eliminated from the system [Agladze et al. 2007; Gottwald et al. 2001a; Krinsky and Agladze 1983]. Inhomogeneities in the medium, such as inexcitable obstacles, can anchor the spiral wave preventing its removal by a stimulated wave-train [Pertsov et al. 1984]. This mechanism is analogous to pinning of vortices in disordered superconductors [Blatter et al. 1994; Pazo et al. 2004]. In the heart, obstacles such as blood vessels or scar tissue, can play the role of pinning centers [Lim et al. 2006], leading to anatomical reentry, the sustained periodic excitation of the region around the obstacle. In the immediate neighborhood of the obstacle, pinned vortices are qualitatively equivalent to waves circulating in a one-dimensional ring. They can be removed by external stimulation provided the electrode is located on the reentrant circuit, i.e., the closed path of the vortex around the obstacle, and the stimulus is delivered within a narrow time interval [Glass and Josephson 1995]. However, for the more general situation of pacing waves generated far away from the reentrant circuit, a classical result due to Wiener and Rosenblueth (WR) states that, all waves circulating around such obstacles are created or annihilated in pairs (see Ref. [Wiener and Rosenblueth 1946], in particular, pp. 216-224). This implies that it is impossible to unpin the spiral wave by a stimulated wave train.

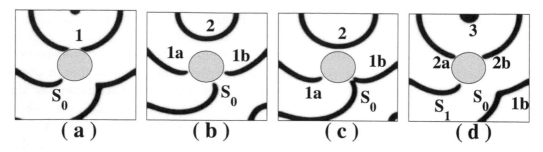

FIGURE 11.2   (a) Wave S_0, pinned to an obstacle (shaded), rotates counterclockwise; wave 1 is the first pacing wave. (b) Wave 1 hits the obstacle, and separates into a wave rotating counterclockwise ($1a$) and a wave rotating clockwise ($1b$). (c) Waves S_0 and $1b$ collide and merge leaving only one rotating wave $1a$ denoted S_1 hereafter. (d) The wave resulting from the merging of S_0 and $1b$ leaves the system. The interaction between the following pacing wave, 2 and S_1, is similar to that shown in (a-c). Thus, the pinned vortex persists. Numerical simulation of the Barkley model with parameters: $a = 0.9$, $b = 0.17$; the pacing period is $T_p = 6.7$ and the radius of the obstacle is $R = 6.5$.

11.2.2   Failure of the classical theory

However, as will see later in this chapter, the WR mechanism for the failure of pacing in unpinning spiral waves is valid only when the radius of the *free* spiral core (i.e., the closed trajectory of the spiral tip defined as a phase singularity [Winfree 1987]) is small compared to the size of the obstacle. There occurs a transition in behavior from the case of a free vortex to one attached to a large obstacle. This transition is due to the systematic reduction of the core radius of the free spiral, R_{FS}, relative to the obstacle size, R_{obst}, achieved by increasing the excitability of the medium. The Barkley model [Barkley et al. 1990] is used to illustrate the argument that an anchored rotating wave can be removed by a stimulated wave-train provided the condition $R_{FS} > R_{obst}$ holds.

The Barkley model, described earlier in Chapter 2, comprises an excitatory (u) and a recovery (v) variable, and two parameters a and b that describe the kinetics. In this study the relative timescale ϵ between the local dynamics of u and v is set to 0.02. The system is discretized on a square spatial grid of size $L \times L$, with a lattice spacing of $\Delta x = 0.25$ and time step of $\Delta t = 0.01$ in dimensionless units. For the simulations discussed here, $L = 200$. The model equation is solved using the forward Euler scheme with a standard nine-point stencil for the Laplacian. No-flux boundary conditions are implemented at the edges of the simulation domain. The obstacle is implemented by introducing a circular region of radius R_{obst} in the center of simulation domain, inside which diffusion is absent. Pacing is delivered by setting the value of u to $u_p = 0.9$ in a region of 6×3 points at the center of the upper boundary of the simulation domain. The maximum pacing frequency is limited by the *refractory period*, T_{ref}, the duration for which the stimulation of an excited region does not induce a response.

When the obstacle size is large relative to the core radius of the free spiral, R_{FS}, the failure of a wave-train in unpinning the vortex is illustrated in Fig. 11.2.

Initially, the spiral wave S_0 rotates counterclockwise around the obstacle. During the interaction with pacing waves, the number of waves attached to the obstacle can change due to two possible processes (see Ref. [Wiener and Rosenblueth 1946], p. 216 and 220). First, when the pacing wave reaches the obstacle, it splits into two oppositely rotating waves: one clockwise and the other counterclockwise. Second, collision between two rotating waves, as seen in Figure 11.2(c), results in the annihilation of a pair of counterclockwise and clockwise waves. In both cases, the number of waves rotating counterclockwise is always larger than the number rotating clockwise by 1. Thus, in addition to conservation of total topological charge (i.e., sum of the individual chiralities, $+1$ or -1) for *all* spiral waves in a medium [Glass 1977; Winfree 1987], topological charge *around* the obstacle also appears to be conserved. However, in the limiting case of infinitesimally small obstacle corresponding to a free vortex, a stimulated wave-train with frequency higher than that of the spiral wave will always succeed in displacing the latter, eventually removing it from a finite medium. Thus, there is a transition from failure to successful pacing as R_{obst} is reduced relative to R_{FS}.

The primary fact responsible for this transition is that the spiral wave is no longer in physical contact with an obstacle of size smaller than R_{FS} [Lim et al. 2006], contrary to the fundamental assumption of Ref. [Wiener and Rosenblueth 1946]. Figure 11.3 shows an explicit example of successful detachment of a pinned wave from the obstacle boundary, where the core radius of a free spiral in the medium is made larger than R_{obst} by diminishing the excitability of the system.

The possibility of unpinning the wave in Fig. 11.3 can be traced to the following fact: the collision between S_0 and the pacing wave-branch 1b occurs *a small distance away* from the obstacle boundary and does not result in complete annihilation of both waves. A small fragment 1c survives in the spatial interval between the collision point and the obstacle [Fig. 11.3(d)]. If the tip of S_0 is close to the obstacle, the fragment 1c is small, and rapidly shrinks and disappears. However, if the gap between the reentrant wavetip and the obstacle is large at the collision point, such that the size of 1c is larger than a critical value l_n, the fragment can survive. As 1c propagates further away, it collides with the pacing wave 1a and forms a new broken wave S_1 that is completely detached from the obstacle. Interaction with successive pacing waves progressively pushes the vortex further away from the obstacle, and eventually from a finite medium. The difference between the number of f spirals rotating counterclockwise and clockwise *around the obstacle* changes from 1 initially (Fig. 11.3, a), to 0 in Fig. 11.3(e), contrary to what happens for a larger obstacle (Fig. 11.2). The absence of topological charge conservation for waves rotating *around* a smaller obstacle underlines the breakdown of the fundamental assumption behind the WR argument for why pacing cannot detach pinned waves. The unpinned wave is subsequently driven outside the system boundaries by pacing (Fig. 11.3, f), thus eventually also reducing the total topological charge of the *finite* medium to 0. The relative size of the obstacle, compared to the free spiral core, is the key parameter that decides whether a pinned reentrant wave can be removed or not. Indeed, the radius of the free spiral core in the successful case, $R_{FS} = 9.05$ (Fig. 11.3) is significantly larger than in the unsuccessful one, $R_{FS} = 5.80$ (Fig. 11.2).

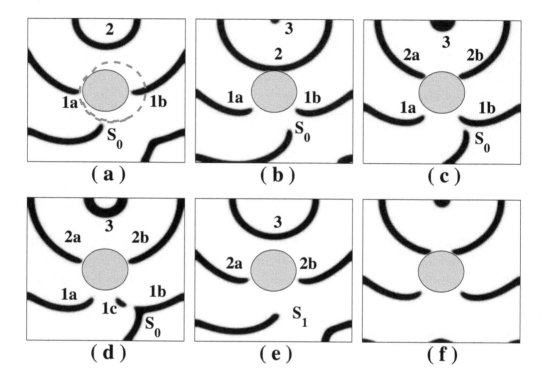

FIGURE 11.3   Lowering excitability results in successful detachment of pinned vortex by pacing. S_0 is a rotating wave whose core (dashed line) is larger than the pinning center (shaded). (a-c) are topologically as in Fig. 11.2. (d) A wavelet $1c$ is produced after collision of waves S_0 and $1b$, in contrast with Fig. 11.2(d). (e) The wavelet $1c$ collides with $1a$ and the resulting wave S_1 is displaced away from the obstacle. (f) Subsequent pacing induces drift of the spiral wave S_1 to the boundary, eventually removing it from the medium. The parameters are as in Fig. 11.2, except for $a = 0.895$ and $b = 0.1725$, resulting in increasing the vortex core size.

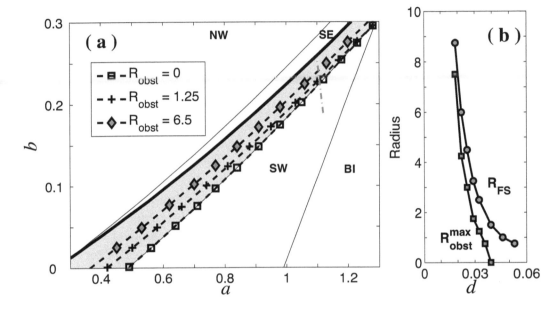

FIGURE 11.4 (a) Parameter space of the Barkley model. Unpinning is possible in the shaded portion of the SW region, which exhibits persistent spiral waves. The thick line indicates the boundary with the SE region, where spirals cannot form. The domain where unpinning is possible shrinks with increasing size of the pinning center, the three dashed lines corresponding to $R_{obst} = 0$, i.e., no obstacle (square), $R_{obst} = 1.25$ (plus), and $R_{obst} = 6.5$ (diamond). (b) Radius R_{FS} of the free spiral and the maximum obstacle radius R_{obst}^{max} from which wave trains can unpin vortices, as a function of the distance d from the SE-SW boundary, along the dot-dashed line indicated in (a). Note that $R_{FS} > R_{obst}^{max}$, and both increase with decreasing d. [In (a), NW (BI) indicates the parameters for which steady waves are absent (the medium is bistable).]

It is further confirmed by a detailed numerical study of the interaction between a pacing wave-train and a pinned spiral over the (a, b) parameter space of the Barkley model. As shown in Fig. 11.4(a), the rotating wave anchored to the obstacle can be removed by pacing only in the neighborhood of the sub-excitable (SE) region (using the terminology of Ref. [Alonso et al. 2003]), where R_{FS} diverges [Fig. 11.4(b)]. This is explained by noting that in the SE regime, the tangential velocity of a broken wavefront is negative, thus causing the front to shrink and not form a spiral. As the regime where spiral waves are persistent (SW) is approached, the tangential velocity of the wavebreak gradually increases to zero and becomes positive on crossing the SE-SW boundary, so that the broken wavefront can now evolve into a spiral. As R_{FS} increases with decreasing tangential velocity of the wavefront, the spiral core becomes large close to the SE region resulting in successful pacing-induced termination of pinned reentry.

It is observed that there is a maximum radius of the obstacle (R_{obst}^{max}) close to R_{FS} above which pacing is unsuccessful in detaching the anchored spiral wave [Fig. 11.4(b)]. Figure 11.5(a) shows that the pacing period for successful unpinning from the obstacle is bounded by the refractory period (T_{ref}) and a maximum value T_p^{max} that is independent of R_{obst} for small obstacles. On approaching R_{obst}^{max}, the upper bound sharply decreases, becoming equal to the refractory time at R_{obst}^{max}, which indicates that pacing will be unsuccessful in unpinning waves attached to obstacles of radii larger than R_{obst}^{max}. Thus, the results shown in Figs. 11.4(b) and 11.5(a) demonstrate the assertion made earlier that pacing induced removal of anchored waves is possible only when the obstacle is smaller than the core radius of the free spiral wave in the medium.

11.2.3   Relation between spiral period and obstacle size

The numerical results shown here indicate that the maximum pacing period necessary for detaching a pinned spiral wave is a decreasing function of the obstacle size [Fig. 11.5(a)]. This can be explained semi-quantitatively by the following geometric argument, valid when the size of the obstacle is small compared to the core size of the spiral, and supported by the simulations shown in Fig. 11.5(c-f). The tip of the spiral S moves along its circular trajectory, shown by the broken line in Fig. 11.5(b), and interacts with the pacing wave coming from the top, represented by a solid line. The part $1b$ of the pacing wave collides with S at the point C characterized by an angle θ that the spiral tip makes with the symmetry axis (i.e., the line joining the centers of the obstacle and spiral core); the resulting wave eventually leaves the system [Fig. 11.5(d)]. The remaining section of the pacing wave splits into two waves, $1a$ and $1c$, propagating along either side of the obstacle. The wave tip moves approximately in a straight line from C, so that the length of the wave $1c$ at the symmetry axis is $l = R_{FS}(1 + \cos\theta) - 2R_{obst}$. When the fragment $1c$ is larger than the nucleation size l_n, it expands into a wavefront that reconnects with wave $1a$. This results in a displacement of the wave $1a$ away from the obstacle, leading to unpinning (as in Fig. 11.3). For $l < l_n$, $1c$ shrinks and eventually disappears, resulting in unsuccessful pacing.

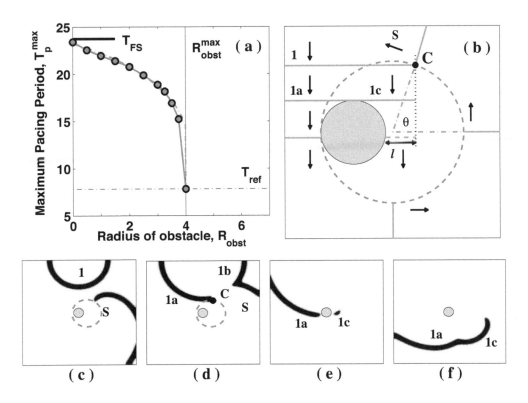

FIGURE 11.5 (a) The maximum pacing period T_p^{max} at which unpinning is possible as a function of the obstacle radius R_{obst}. For the parameters $a = 1.1323, b = 0.2459$ that have been used, the maximum radius of obstacle from which unpinning can occur is $R_{obst}^{max} = 4$. T_{FS} is period of a free spiral wave and T_{ref} is the refractory period. The dashed line indicates the prediction from Eq. (11.1). (b) The wavelet formation mechanism leading to the detachment of the pinned vortex (schematic). (c-f) Numerical simulation of the Barkley model. S collides with wave 1 at point C at an angle θ. The part $1b$ of the pacing wave merges with S, moving out of the system. The remaining part of the pacing wave collides with the obstacle (shaded) separating into $1a$ and a small wavelet $1c$. When the length l of wavelet $1c$ is larger than the critical nucleation length, $1c$ survives and collides with S. This results in unpinning of S.

Thus, the condition for detachment is $l \geq l_n$. The length l is a decreasing function of the angle θ, which in turn, is a decreasing function of the pacing period, T_p, as explained below. The relation between T_p and θ can be established by estimating the time interval for two successive collisions of the spiral with the pacing waves. From the point of collision C, the pacing wave reaches the obstacle after time $T_1 = (R_{FS} \sin\theta - R_{obst})/v$, and the symmetry axis after time $T_2 = T_1 + (R_{obst}T_{FS}/4R_{FS})$. From the symmetry axis, the new reentrant wave S moves by an angle $(\theta + \pi)$ to arrive at C at time $T_3 = T_2 + [T_{FS}(\theta + \pi)/2\pi]$, where it collides with the next pacing wave. Noting that $T_3 = T_p$ allows T_p to be expressed as a function of θ, and thereby, l. The maximum pacing period leading to detachment is obtained when $l = l_n$, as:

$$T_p^{max} = \frac{R_{FS}}{v}(\sin\theta_c - f_R) + \frac{f_R T_{FS}}{4} + \frac{T_{FS}(\theta_c + \pi)}{2\pi}, \qquad (11.1)$$

where, $\theta_c = \arccos(2f_R - 1 + [l_n/R_{FS}])$ and $f_R = R_{obst}/R_{FS}$. When $R_{obst} > R_{obst}^{max} = R_{FS} - (l_n/2)$, T_p^{max} has complex values, indicating that for larger obstacles the fragment is too small to survive. The nucleation length l_n can thus be estimated from R_{obst}^{max}, which allows us, in turn, to determine the dependence of T_p^{max} as a function of R_{obst} from Eq. (11.1). Figure 11.5(a) shows this to be in fair agreement with the numerical simulations.

The arguments used here are model independent, and are based only on the property that waves in excitable media annihilate on collision. Similar wave-train induced unpinning is also observed in a more detailed and realistic description of cardiac tissue, the Luo–Rudy I model [Luo and Rudy 1991], under conditions of reduced excitability. Meandering, which occurs in the Barkley model at low a, b values (Fig. 11.4, a), does not affect the physical effect discussed here. Note that the proposed unpinning mechanism is for the case of an obstacle smaller than the vortex core. It is possible under certain circumstances to unpin waves from obstacles larger than the core because of other effects such as the presence of slow conduction regions [Sinha and Christini 2002; Sinha et al. 2002] and non-linear wave propagation (alternans) [Breuer and Sinha 2005].

These results, thus predict that in cardiac tissue, the removal of spiral waves pinned to a small obstacle by high-frequency wave-trains is facilitated by decreasing the excitability of the medium. This is consistent with previous experimental results on cardiac preparations using Na^+-channel blockers [Lim et al. 2006] and the prediction could be directly tested in a similar experimental setup [Agladze et al. 2007; Lim et al. 2006].

In conclusion, this study shows that for a pinned vortex interacting with a pacing wave-train, unpinning is possible when the size of the obstacle is smaller than that of the spiral core. The minimum wave-train frequency necessary for unpinning in the presence of an inexcitable obstacle is higher than that for inducing drift in a free vortex toward the boundaries of a finite domain, and it increases with the size of the pinning center. These results suggest that lowering the excitability of the medium makes it easier to unpin vortices by pacing.

11.3 Terminating reentry by generating dynamical hetero-geneities

In this section, we will consider another kind of inhomogeneity, namely those generated dynamically due to the non-linear properties of wave propagation in an otherwise homogeneous cardiac medium. We will discuss the role of these dynamic heterogeneities in the successful termination of spiral waves by high-frequency stimulation, which could be a potential mechanism by which ICDs terminate VT [Breuer and Sinha 2005].

Although propagation of excitation in the heart occurs in a three-dimensional tissue, for ease of analysis and numerical computations most theoretical studies of pacing have focussed on reentry in a one-dimensional ring of cardiac cells, which is essentially the region immediately surrounding an anatomical obstacle [Comtois and Vinet 2002; Glass and Josephson 1995; Glass et al. 2002; Nomura and Glass 1996; Sinha and Christini 2002; Sinha et al. 2002]. The conventional view of reentry termination has been that each pacing wave splits into two branches in the reentry circuit, the retrograde branch traveling opposite to the reentrant wave and eventually colliding with it, annihilating each other. The other, anterograde, branch travels in the same direction as the reentrant wave, and depending on the timing of the pacing stimulation, either resets the reentry by becoming the new reentrant wave, or leads to termination, if it is blocked by a refractory region left behind in the wake of the preceding wave. If the pacing site is on the ring itself, continuity arguments can be used to show that there will always exist a range of stimulation times, such that the reentry will be terminated. However, this argument breaks down when we go beyond the 1-D ring geometry and consider a pacing site situated some distance away from the reentry circuit. As, in reality, the pacing site is fixed (usually in the ventricular apex) while the reentry can occur anywhere in the ventricles, this leaves the question open about how pacing terminates VT.

Previous investigations in the reentry in a quasi-1-D geometry consisting of a ring attached to a sidebranch whose other end is the pacing site [Fig. 11.1 (right)] have attempted to address this issue. This approach allows for the treatment of the issue of propagation of the pacing wave across a 2-D domain to the reentry site, even while retaining the elegant simplicity of the 1-D formulation of the problem. These studies showed that the existence of inhomogeneities in the reentry circuit is essential for successful termination of VT by pacing [Sinha and Christini 2002]. Further work in two dimensions upheld the qualitative results [Sinha et al. 2002]. However, these studies focussed exclusively on the role of static (structural) inhomogeneities, such as a zone of slow conduction. Inhomogeneities can be also generated through the non-linear dynamics of excitation wave propagation in otherwise homogeneous cardiac tissue. Specifically, non-linear effects like action potential restitution and conduction velocity dispersion have been shown to create disorder (inhomogeneities) in the properties of cardiac tissue [Echebarria and Karma 2002a,b; Qu et al. 2000a; Watanabe et al. 2001], sometimes leading to conduction block [Fox et al. 2002b,c].

The system considered has a quasi-1-D geometry consisting of a ring of model

cardiac cells, attached to a sidebranch. The propagation of excitation in this model
is described by the partial differential equation:

$$\partial V/\partial t = I_{ion}/C_m + D\nabla^2 V, \tag{11.2}$$

where V (mV) is the membrane potential, $C_m = 1$ μF cm^{-2} is the membrane
capacitance, D (cm^2 s^{-1}) is the diffusion constant and I_{ion} (μA cm^{-2}) is the cellular membrane ionic current density. Different models are characterized by the
equations used to describe I_{ion}. The excitation kinetics is captured by the Karma
model [Karma 1994], discussed in Chapter 2, which is one of the simplest sets of
equations describing cardiac excitation that incorporate both restitution and dispersion effects.

In the simulations, the parameters $D = 0.8115$ cm^2 s^{-1}, $\tau_V = 0.0022$ s, $\tau_n =$
0.22 s, $V^* = 1.5415$, $V_h = 3$, $V_n = 1$, $M = 4$, and $Re = 1.5$ (the last two values are
chosen to make both restitution and dispersion significant in the study we carried
out). Figure 11.6 (top) shows a representative time series of V at a single point of
a one-dimensional array of cells, one end of which is paced with a period of 0.232 s.
The sharp rise in V (from the resting state $V = 0$) followed by a plateau and then a
gradual decline back to the resting state is characteristic of an *action potential*. Its
duration (APD) is seen to be clearly a monotonic function of the preceding *diastolic
interval* (DI), the period between the onset of the current action potential and the
decline to the resting state for the previous one. The alternation between long and
short APD (called *alternans*) is a result of the high degree of restitution effect,
also seen in the steepness of the APD$_{N+1} = f$(DI$_N$) curve in Fig. 11.6 (bottom).
Note that, if the diastolic interval becomes less than a critical value DI$_c$, a wave
cannot be initiated at that point (because the tissue has not recovered sufficiently
from the previous excitation) and so the restitution curve does not show any points
corresponding to DI $<$ DI$_c$.

Dispersion effect is observed by noting the conduction velocity (CV) of the excitation wavefront and how it varies with the preceding DI. Analogous to restitution,
the degree of the dispersion effect can be measured by the steepness of the CV
vs, DI curve. In the absence of dispersion, a given wave has a constant APD as it
traverses the tissue (even though different waves may have widely different APDs);
however, dispersion causes the waves to slow down or speed up depending upon
the dispersion, and therefore the wavefront has different APDs at different regions.
The resultant modulation of the wave profile plays a key role in this termination
mechanism.

In the simulation study, a stable reentrant wave, whose propagation would not be
terminated by the spontaneous emergence of a conduction block, is first initiated
in the ring. The ring length is chosen to be long enough so that the reentrant
wave displays no spatial variation in the APD. Pacing is started the moment the
reentrant wave arrives at the point (P) at a certain time $t = 0$ (taken to be the
origin). The pacing is then started after waiting for a fixed fraction of the reentry
period (referred to as the *pacing phase*). Each pacing stimulus is of duration 0.00025
s and amplitude 40. The number of applied stimuli is a measure of the duration of
pacing, and is a parameter in the study, as is the pacing interval (T_{pacing}), the

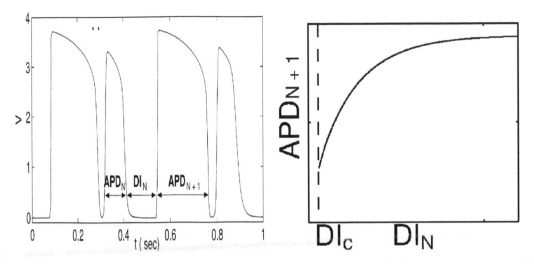

FIGURE 11.6 (left) Time evolution of membrane potential (V) in the Karma model at a single point in a one-dimensional array of cells which is paced at one end with a period of 0.232 s. The alternation between long and short APD indicates the presence of strong restitution effect as is seen from the steepness of the schematic restitution curve $\text{APD}_{N+1} = f(\text{DI}_N)$ (right). The dashed line indicates the critical value DI_c.

inverse of the frequency with which the stimuli are applied. The outcome of the pacing of reentry with different numbers of stimuli and pacing intervals (for a pacing phase of 80%) is summarized in Fig. 11.7. Similar figures are obtained for different values of the pacing phase, which is a free parameter in the pacing algorithm.[†] A general feature of these figures is that the boundary of the regions where termination is possible is given by a hyperbola-like curve, with the number of pacing stimuli necessary for termination increasing with pacing interval (and hence decreasing pacing frequency). This is related to the minimum number of stimuli necessary for entering the reentrant circuit. The higher the frequency of the pacing compared to that of the reentrant wave, the faster the pacing waves will be able to reenter the ring; as a result, fewer pacing stimuli are required to terminate reentry. In addition, after entering the ring, a number of stimuli are required to create or amplify APD modulations such that conduction block can occur at the inhomogeneities leading to reentry termination. Another feature one observes in the regions where termination does occur successfully is the presence of a few white horizontal bands indicating failure of termination. In most cases, these bands are due to early conduction block of the pacing wave in the sidebranch. The figure also shows regions where success and failure alternate with increasing number of pacing stimuli, as well as regions where termination occurs for a contiguous range of pacing stimuli. These two regions

---

[†]However, if the pacing phase is chosen to be too small, the first pacing pulse will be applied to the medium when it is still refractory at the pacing point. Thus no wave can be initiated and the pacing does not affect the reentry.

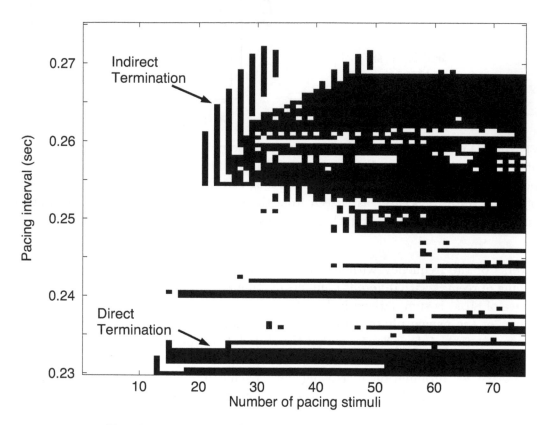

FIGURE 11.7 Termination success diagram showing the result of pacing as a function of the number of pacing stimuli (i.e., duration of the pacing) and the pacing interval (~ 1/pacing frequency). The system is a ring of perimeter length 6.85 cm and the pacing point is located 2.5 cm away from the ring. The reentry period is 0.3 s. Black represents successful termination of reentry while white represents failure. The figure shows parameter regions at which different termination mechanisms operate: direct and indirect (see text for details).

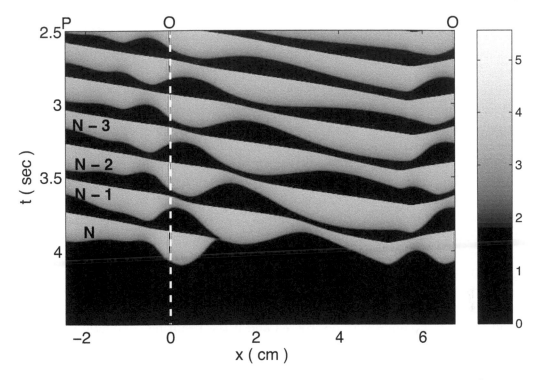

FIGURE 11.8 A space-time diagram showing direct termination of reentry for pacing interval of 0.232 s and using a total of 16 pacing stimuli (pacing phase 80%). The grayscale shows the magnitude of V at different points. The dashed line indicates the junction of the ring and sidebranch at O ($x = 0$ cm), while P ($x = -2.5$ cm) refers to the pacing point. Termination of reentry occurs at $x \simeq 1.1$ cm at $t \simeq 3.92$ s.

indicate the presence of two distinct mechanisms of reentry termination, which is referred to as indirect and direct termination, respectively.

11.3.1 Direct termination

Figure 11.8 shows an instance of successful termination of reentry by conduction block of the anterograde branch of the last (N-th) pacing wave entering the ring. Although the pacing is stopped at this point, it is obvious that additional pacing waves will not reinitiate the reentry. This mechanism, where the termination occurs with the pacing still switched on, is referred to as *direct termination*.

To understand the mechanism, one has to consider the combined effects of diffusion, restitution, and dispersion. While the steepness of a restitution curve decides whether alternation between long and short APDs will be observed, diffusion can significantly alter the curve and produce alternans in a tissue when such effect is absent in the single cell. However, stable alternans is not enough to produce reentry termination, as there will occur wave propagation at all regions or conduction block at all regions depending upon the pacing interval. The introduction of dispersion changes the situation, as there can now be different diastolic intervals at different regions of the tissue. The resulting unstable alternans may, for a sufficiently high pacing frequency, cause a particular point to have a diastolic interval $< \mathrm{DI}_c$. This will cause conduction block to occur for the next wave and successfully terminate reentry.

This phenomenon has been observed previously in a 1-D fiber of simulated cardiac tissue, where increasing the pacing frequency resulted in a transition to unstable alternans, where the waveform showed modulations as a consequence of having different APDs at different locations in the fiber [Echebarria and Karma 2002a,b; Qu et al. 2000a; Watanabe et al. 2001]. Increasing the pacing frequency further typically created conduction blocks [Fox et al. 2002b,c]. The difference in these results with the present case of a ring is that, in the latter, the reentry period is determined by the size of the reentrant circuit (in actual cardiac tissue, the circuit around the anatomical obstacle). If the reentry circuit is too small, then conduction blocks may occur spontaneously during VT as a result of too low reentry period. However, if the circuit is large enough to sustain stable reentry, by pacing at a high enough frequency one can decrease the effective period of reentry and thereby create or amplify existing modulations of the waveform. The APD of a wave, as it circulates around the ring, becomes very disordered and if the resulting conduction block occurs in the ring, reentry is terminated. For example, observe the situation shown in Fig. 11.9. If wave A starts in a region where the diastolic interval is small and enters afterward a region with large diastolic interval (due to modulation of the waveform), the propagation of the waveback slows down dramatically. In contrast to that, the propagation velocity of the wavefront is nearly constant (i.e., the dispersion effect is much smaller). Thus the wavefront of another wave B that follows wave A can collide with the back of the latter, leading to conduction block.

Note that conduction block may also occur as the pacing wave is approaching the reentrant circuit (i.e., at the sidebranch) and this will result in the failure of pacing to terminate reentry. In order to create conduction block in the ring before sufficiently large modulations can occur in the sidebranch, one has to avoid using too high pacing frequencies, although superficially this may seem to be the ideal course of action.

11.3.2   Indirect termination

The other mechanism of reentry termination is referred to as indirect termination as the conduction block occurs after the pacing has been switched off. As seen in Fig. 11.10, the last pacing stimulus produces a wave N, which circulates through the ring and then reenters both the ring and the sidebranch $(N + 1)$. However, at the end of the $(N + 1)$-th circulation around the ring, the wave finds that the point O has not fully recovered and is blocked, resulting in reentry termination. It is a characteristic of this mechanism that the conduction block always occurs at the junction of the ring and the sidebranch, i.e., the point of first contact between the reentrant wave and the pacing wave.

The mechanism can be understood by looking at the sequence of the waves leading to termination. At the junction point O, if one notes the APD and DI of

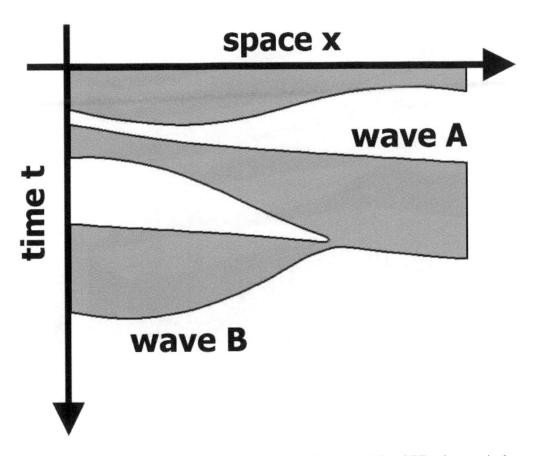

FIGURE 11.9 Mechanism of direct termination of reentry. The APD of wave A shows large modulations as a result of the preceding DI gradually increasing along x. The following wave B collides with the waveback of A, leading to conduction block of B.

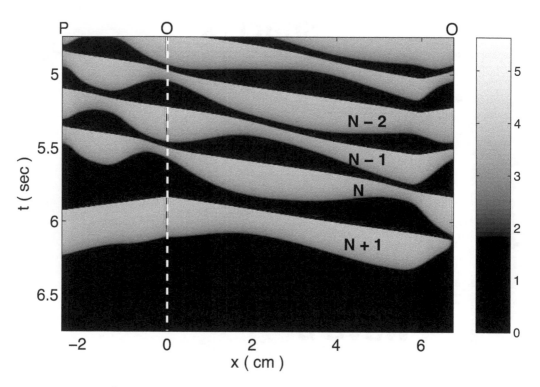

FIGURE 11.10   A space-time diagram showing indirect termination of reentry for pacing interval of 0.255 s and using a total of 21 pacing stimuli (pacing phase 80%). The grayscale shows the magnitude of V at different points. The dashed line indicates the junction of the ring and sidebranch at O ($x = 0$ cm), while P ($x = -2.5$ cm) refers to the pacing point. Termination of reentry occurs at O at $t \simeq 6.13$ s.

successive waves, then the following sequence is observed: APD_{N-1} is short, which implies a long DI_{N-1} (since APD+DI = pacing interval, is constant). However, in addition, since the pacing has been switched off after the N-th wave, the period between successive waves has suddenly increased from the pacing interval to the original reentry period. This additional time interval $(T_{reentry} - T_{pacing})$ is now added to DI_{N-1}, making it extraordinarily long. As a result, it is observed that a very long APD_N, leads to a DI_N that is too short to support conduction (i.e., $DI_N < DI_c$) and this results in block of the $(N+1)$-th wave.

It is obvious from this argument that stopping the pacing after the $(N-1)$-th wave would not have resulted in termination, as in that case the circulating wave would have encountered a recovered region at the junction point O (since DI_{N-1} is long). This implies that, for this mechanism, increasing the number of pacing stimuli will result in alternating success and failure of termination, with $N, N+2, N+4,...$ resulting in success and $N-1, N+1, N+3,...$ resulting in failure. This is quite evident in the termination success diagram (Fig. 11.7).

In this section, we have discussed interesting mechanisms of reentry termination in the Karma model based on the non-linear dynamics underlying wave propagation in excitable media. The generation of waveform modulations through dynamical instabilities as a result of restitution and dispersion effects in cardiac tissue leads to formation of conduction inhomogeneities in the reentry circuit. This disorder in turn leads to conduction block during rapid pacing and therefore results in successful reentry termination.

Preliminary results on the Luo–Rudy I model [Luo and Rudy 1991], which incorporates details of ion channel currents of the cardiac cell, show that dynamical inhomogeneities can successfully terminate reentrant wave propagation in more realistic models of heart tissue [Sinha and Breuer 2004]. This confirms the generality of the results and points to the application of these findings to the design of better pacing algorithms for ICDs.

Based on the simulation results, it can be concluded that the various pacing parameters have optimal values for successful reentry termination. For example, the pacing frequency has to be carefully chosen. While the pacing interval has to be shorter than the reentry period to be able to enter the reentrant circuit, it cannot be too short, as the propagation of high-frequency waves causes instability and wave breakup, leading to formation of spiral waves around transiently inactive cores ("functional" reentry). This may be the mechanism responsible for rapid pacing occasionally giving rise to faster arrhythmias. Wave instability can initiate further breakup of the spiral wave leading to the spatiotemporal chaos of VF. In addition, the number of pacing stimuli also has an optimal value. It has to be high enough to be able to enter the reentrant circuit, but not so high that it causes additional conduction blocks, and therefore, restarts the reentry. This shows that designing an optimal pacing algorithm is essentially a complex optimization problem.

The ultimate goal of antitachycardia pacing is to terminate reentrant activity with stimuli of smallest magnitude in the shortest possible time with the lowest probability of giving rise to faster arrhythmias or VF. The constant frequency pacing described here is only a partial solution to this end, and a more efficient algorithm might have to adjust the pacing intervals on a beat-to-beat basis. The results discussed in this chapter are aimed toward answering how such an optimized pacing scheme may be designed.

12

Controlling spatiotemporal chaos

12.1 Introduction

As mentioned earlier in this book, the dynamical spatial patterns in excitable media, such as spiral and scroll waves, can become unstable in certain conditions and give rise to a state characterized by spiral turbulence, corresponding to spatiotemporal chaos. As mentioned in Chapter 1, such spatiotemporally chaotic states have been implicated in clinically significant disturbances of the natural rhythm of the heart [Gray et al. 1998; Witkowski et al. 1998], e.g., fibrillation. Ventricular fibrillation in particular is lethal as it results in complete loss of coordination of activity between different regions in the heart [Winfree 1987]. The resulting cessation of the mechanical pumping action necessary for blood circulation leads to a drastic fall in blood pressure. If not treated immediately death follows within a few minutes. However the conventional methods of defibrillation require the application of large electrical shocks that are undesirable for a variety of reasons. Developing low-amplitude control schemes involving as few control electrodes as possible is an exciting challenge and has potential clinical relevance [Christini et al. 2001; Gauthier et al. 2002; Pumir et al. 2007]. As mentioned in Chapter 9, devising such control using low voltages or current has to take into account the special features of excitable media like the existence of a refractory period and activation threshold. The amplitude and timing of the control signal needs to be appropriately chosen so that it results in the desired response from the medium. Note that the excited state is meta-stable, and the cell eventually recovers to the *resting* state associated in different biological systems with a characteristic resting transmembrane potential ($\simeq -84$ mV for cardiac myocytes). Thus, the control of spatiotemporal chaos in excitable media may be viewed as essentially a problem of synchronizing the excitation phase of every cell, so that the entire system returns to the resting state, resulting in the termination of all activity.

12.2 Characterizing spatiotemporal chaos

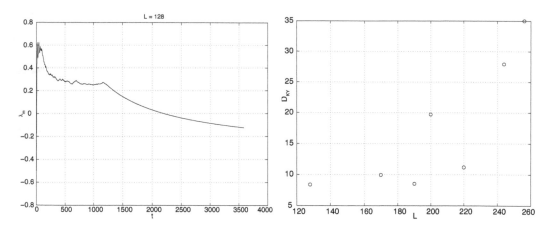

FIGURE 12.1 (left) Time-evolution of the maximum Lyapunov exponent λ_{max} for Panfilov model dynamics on a two-dimensional square domain (with 128×128 grid points). The exponent initially converges to a positive value ($\simeq 0.2$) indicating the presence of spatiotemporal chaos in the system, before eventually decaying to negative values at long times indicating that the chaotic behavior is a long-lived transient. The duration of the chaotic activity increases rapidly with system size. (right) The Kaplan–Yorke dimension (D_{KY}) measured during chaotic activity is seen to increase with the system size L. The observed fluctuations can be attributed to finite-size effect, as well as varying rates of convergence in the calculation of the Lyapunov spectrum as L is changed.

Mathematically spatiotemporal chaos is usually characterized by the spectrum of Lyapunov exponents and the Kaplan–Yorke dimension (Fig. 12.1). Lyapunov exponent is the mathematical measure of how fast two trajectories that are initially infinitesimally close, diverge. For example, in a one-dimensional system the Lyapunov exponent λ is given as

$$\lambda = \frac{1}{n} \log \frac{d_n}{d_0} \qquad (12.1)$$

where d_i is the distance at the ith time-step between a pair of dynamical states following trajectories that initially had a difference d_0 between them. The property of sensitive dependence on initial conditions associated with chaos suggests that, if the system is chaotic, then two initially adjacent trajectories would diverge exponentially fast with time, so that the Lyapunov exponent $\lambda > 0$. Higher dimensional systems will be characterized by a spectrum of such exponents $\lambda_1, \lambda_2, \ldots, \lambda_{max}$ (when sequentially placed in increasing order) equalling in number the phase space dimensions required to completely describe the system. The sign of the largest of the exponents, the maximal Lyapunov exponent (MLE), λ_{max}, governs the nature of the dynamics, with $\lambda_{max} > 0$ implying that it is chaotic. Typically, for spatiotemporal chaos, multiple elements of the Lyapunov exponent spectrum will be positive. Knowledge of the Lyapunov exponents can be used to determine the Kaplan–Yorke dimension D_{KY} as

$$D_{KY} = k + \sum_{k=1}^{k} \lambda_i / |\lambda_{k+1}| \qquad (12.2)$$

where k is the maximum integer for which $\sum_{i=1}^{k} \lambda_i$ is greater than zero.

12.3 Classifying chaos control schemes

12.3.1 Global control

Possibly the first attempt at controlling chaotic activity in excitable media dates back almost to the beginning of the field of chaos control itself, when proportional perturbation feedback (PPF) control was used to stabilize cardiac arrhythmias in a piece of tissue from rabbit heart [Garfinkel et al. 1992]. In this method, small electrical stimuli were applied at intervals calculated using a feedback protocol, to stabilize an unstable periodic rhythm. Unlike in the original proposal for controlling chaos [Ott et al. 1990], where the location of the stable manifold of the desired unstable periodic orbit (UPO) was moved using small perturbations, in the PPF method it is the state of the system that is moved onto the stable manifold. However, it has been later pointed out that PPF does not necessarily require the existence of UPOs (and, by extension, deterministic chaos) and can be used even in systems with stochastic dynamics [Christini and Collins 1995]. Later, the PPF method was used to control atrial fibrillation in human heart [Ditto et al. 2000]. However, the effectiveness of such control in suppressing spatiotemporal chaos, when applied only at a local region, has been questioned, especially as other experimental attempts in feedback control have not been able to terminate fibrillation by applying control stimuli at a single spatial location [Gauthier et al. 2002].

More successful, at least in numerical simulations, have been schemes where control stimuli are applied throughout the system. Such global control schemes either apply small perturbations to the dynamical variables (e.g., the transmembrane potential) or one of the parameters (usually the excitation threshold). The general scheme involves introducing an external control signal A into the model equations, e.g., in the Panfilov model: $\partial e / \partial t = \nabla^2 e - f(e) - g + A$, for a control duration τ. If A is a small, positive perturbation added to the fast variable, the result is an effective reduction of the threshold (Fig. 12.2), thereby making simultaneous excitation of different regions more likely.

In general, A can be periodic, consisting of a sequence of pulses. Figure 12.3 shows the results of applying a pulse of fixed amplitude but varying durations. While in general, increasing the amplitude or the duration increases the likelihood of suppressing spatiotemporal chaos it is not a simple, monotonic relationship. Depending on the initial state at which the control signal is applied, even a high amplitude (or long duration) control signal may not be able to uniformly excite all regions simultaneously. As a result, when the control signal is withdrawn, the inhomogeneous activation results in a few regions becoming active again and restarting

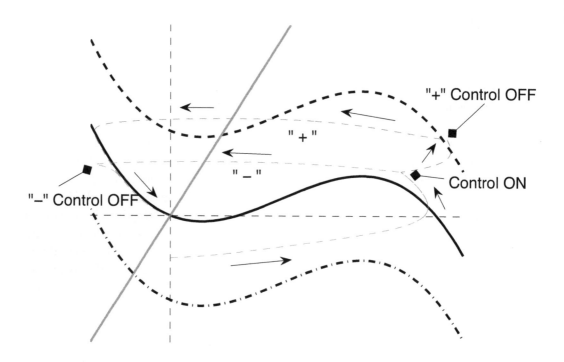

FIGURE 12.2   The result of applying a positive ("+") or negative ("−") additive perturbation of the same duration to the e variable in the Fitzhugh–Nagumo model: "+" control decreases the threshold and makes excitation more likely, while "−" control decreases the duration of the action potential and allows the system to recover faster. For the duration of the control signal, the e-nullcline shifts upward (downward) for positive (negative) perturbation as indicated by the dashed (dash-dotted) curve.

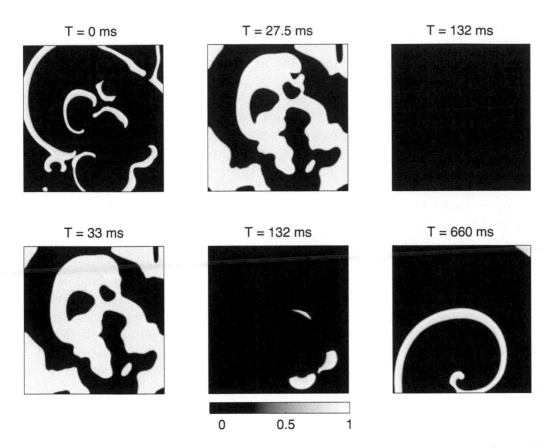

FIGURE 12.3 Global control of the two-dimensional Panfilov model with $L = 256$ starting from a spatiotemporally chaotic state (top left). Pseudo-grayscale plots of excitability e show the result of applying a pulse of amplitude $A = 0.833$ between $t = 11$ ms and 27.5 ms (top center) that eventually leads to elimination of all activity (top right). Applying the pulse between $t = 11$ ms and 33 ms (bottom left) results in some regions becoming active again after the control pulse ends (bottom center) eventually re-initiating spiral waves (bottom right).

the spatiotemporal chaotic behavior.

Most global control schemes are variations or modifications of the above scheme. Osipov and Collins [Osipov and Collins 1999] have shown that a low-amplitude signal used to change the value of the slow variable at the front and back of an excitation wave can result in different wavefront and waveback velocities which destabilizes the traveling wave, eventually terminating all activity, and, hence, spatiotemporal chaos. Gray [Gray 2002] has investigated the termination of spiral wave breakup by using both short- and long-duration pulses applied on the fast variable, in 2-D and 3-D systems. This study concluded that while short-duration pulses affected only the fast variable, long-duration pulses affected both fast and slow variables and that the latter is more efficient (using less power) in terminating spatiotemporal chaos. The external control signal can also be periodic [$A = Fsin(\omega t)$], in which case the critical amplitude F_c required for terminating activity has been found to be a function of the signal frequency ω [Sakaguchi and Fujimoto 2003].

Other schemes have proposed applying perturbations to the parameter controlling the excitation threshold, e.g., the parameter b in the Barkley model (see Chapter 2). Applying a control pulse on this parameter ($b = b_f$, during duration of control pulse; $b = b_0$, otherwise) has been shown to cause an excitation wave to split into a pair of forward and backward moving waves [Woltering and Markus 2002]. Splitting of a spiral wave causes the two newly created spirals to annihilate each other on collision. For a spatiotemporally chaotic state, a sequence of such pulses may cause termination of all excitation, there being an optimal time interval between pulses that results in fastest control. Another control scheme that also applies perturbation to the threshold parameter is the uniform periodic forcing method suggested by Alonso et al. [Alonso et al. 2003, 2006] for controlling scroll wave turbulence in three-dimensional excitable media. Such turbulence results from negative tension between scroll wave filaments, i.e., the line joining the phase singularities about which the scroll wave rotates. In this control method, the threshold is varied in a periodic manner [$b = b_0 + b_f cos(\omega t)$] and the result depends on the relation between the control frequency ω and the spiral rotation frequency. If the former is higher than the latter, sufficiently strong forcing is seen to eliminate turbulence; otherwise, turbulence suppression is not achieved. The mechanism underlying termination has been suggested to be the effective increase of filament tension due to rapid forcing, such that the originally negative tension between scroll wave filaments is changed to positive tension. This results in expanding scroll wave filaments to instead shrink and collapse, eliminating spatiotemporal chaotic activity. In a variant method, the threshold parameter has been perturbed by spatially uncorrelated Gaussian noise, rather than a periodic signal, which also results in suppression of scroll wave turbulence [Alonso et al. 2004].

As already mentioned, global control, although easy to understand, is difficult to achieve in experimental systems. A few cases in which such control could be implemented include the case of eliminating spiral wave patterns in populations of the *Dictyostelium* amoebae by spraying a fine mist of cAMP onto the agar surface over which the amoebae cells grow [Lee et al. 2001]. Another experimental system where global control has been implemented is the photosensitive Belousov–Zhabotinsky reaction, where a light pulse shining over the entire system is used as a control signal [Munuzuri et al. 1997]. Indeed, conventional defibrillation can be thought of as a kind of global control, where a large amplitude control signal is used to synchronize the phase of activity at all points by either exciting a previously unexcited region (advancing the phase) or slowing the recovery of an already excited region (delaying the phase) [Gray and Chattipakorn 2005].

12.3.2   Non-global spatially extended control

The control methods discussed so far apply control signal to all points in the system. As the chaotic activity is spatially extended, one may naively expect that any control scheme also has to be global. However, we will now discuss some schemes that, while being spatially extended, do not require the application of control to all points of the system.

The control method of Sinha et al. [Sinha et al. 2001] involving supra-threshold stimulation along a grid of points, is based on the observation that spatiotemporal chaos in excitable media is a long-lived transient that lasts long enough to establish a non-equilibrium statistical steady state displaying spiral turbulence. The lifetime of this transient, τ_L, increases rapidly with linear size of the system, L, e.g., increasing from 850 ms to 3200 ms as L increases from 100 to 128 in the two-dimensional Panfilov model. This accords with the well-known observation that small mammals do not get life-threatening VF spontaneously whereas large mammals do [Winfree 1987]. This has been experimentally verified by trying to initiate VF in swine ventricular tissue while gradually reducing its mass [Kim et al. 1997]. A related observation is that non-conducting boundaries tend to absorb spiral excitations, which results in spiral waves not lasting for appreciable periods in small systems.

The essential idea of the control scheme is that a domain can be divided into electrically disconnected regions by creating boundaries composed of recovering cells between them. These boundaries can be created by triggering excitation across a thin strip. For two-dimensional media, the simulation domain (of size $L \times L$) is divided into K^2 smaller blocks by a network of lines with the block size $(L/K \times L/K)$ small enough so that spiral waves cannot form. For control in a 3-D system, the mesh is used only on one of the faces of the simulation box. Control is achieved by applying a supra-threshold stimulation via the mesh for a duration τ. A network of excited and subsequently recovering cells then divides the simulation domain into square blocks whose length in each direction is fixed at a constant value L/K for the duration of control. The network effectively simulates non-conducting boundary conditions (for the block bounded by the mesh) for the duration of its recovery period, in so far as it absorbs spirals formed inside this block. Note that τ need not be large at all because the individual blocks into which the mesh divides the system (of linear size L/K) are so small that they do not sustain long spatiotemporally chaotic transients. Nor does K, which is related to the mesh density, have to be very large since the transient lifetime, τ_L, decreases rapidly with decreasing L. The method has been applied to multiple excitable models, including the Panfilov and Luo–Rudy models (Fig. 12.4).

An alternative method [Sakaguchi and Kido 2005] for controlling spiral turbulence that also uses a grid of control points has been demonstrated for the Aliev–Panfilov model. Two layers of excitable media are considered, where the first layer represents the two-dimensional excitable media exhibiting spatiotemporal chaos that is to be controlled, and the second layer is a grid structure also made up of excitable media. The two layers are coupled using the fast variable but with asymmetric coupling constants, with excitation pulses traveling $\sqrt{D}$ times faster in the second layer compared to the first. As the second layer consists only of grid lines, it is incapable of exhibiting chaotic behavior in the uncoupled state. If the coupling from the second layer to the first layer is sufficiently stronger than the other way round, the stable dynamics of the second layer (manifested as a single rotating spiral) overcomes the spiral chaos in the first layer, and drives it to an ordered state characterized by mutually synchronized spiral waves.

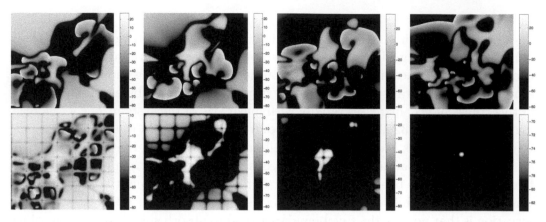

FIGURE 12.4 Spatiotemporal chaos (top row) and its control (bottom row) in the two-dimensional Luo–Rudy I model with $L = 90$ mm. Pseudo-grayscale plots of the transmembrane potential V show the evolution of spiral turbulence at times $T = 30$ ms, 90 ms, 150 ms, and 210 ms. Control is achieved by applying an external current density $I = 150\mu A/cm^2$ for $\tau = 2.5$ ms over a square mesh with each block of linear dimension $L/K = 1.35$ cm. Within 210 ms of applying control, most of the simulation domain has reached a transmembrane potential close to the resting state value; moreover, the entire domain is much below the excitation threshold. The corresponding uncontrolled case shows spatiotemporal chaos across the entire domain.

Another method of spatially extended control is to apply perturbations at a series of points arranged in a regular array. Rappel et al. [Rappel et al. 1999] had proposed using such an arrangement for applying a time-delayed feedback control scheme. However, their scheme only prevented a spiral wave from breaking up and did not suppress pre-existing spatiotemporal chaos.

12.3.3   Local control of spatiotemporal chaos

We now turn to the possibility of controlling spatiotemporal chaos by applying control at only a small localized region of the spatially extended system. Virtually all the proposed local control methods use *overdrive pacing*, generating a series of waves with frequency higher than any of the existing excitations in the spiral turbulent state. As low-frequency activity is progressively invaded by faster excitation, the waves generated by the control stimulation gradually sweep the chaotic activity to the system boundary where they are absorbed. Although we cannot speak of a single frequency source in the case of chaos, the relevant timescale is that of spiral waves which is limited by the recovery period of the medium. Control is manifested as a gradually growing region in which the waves generated by the control signal dominate until the region expands to encompass the entire system. The time required to achieve termination depends on the frequency difference between the control stimulation and that of the chaotic activity, with control being achieved faster, the greater the difference.

Stamp et al. [Stamp et al. 2002] have looked at the possibility of using low-amplitude, high-frequency pacing using a series of pulses to terminate spiral turbulence. However, using a train of pulses (having various waveform shapes) has met with only limited success in suppressing spatiotemporal chaos. By contrast, a periodic stimulation protocol [Zhang et al. 2003] has successfully controlled chaos in the 2-D Panfilov model, as well as in other models.[†] The key mechanism underlying such control is the periodic alternation between positive and negative stimulation. A more general control scheme proposed in Ref. [Breuer and Sinha 2004] uses *biphasic pacing*, i.e., applying a series of positive and negative pulses, that shortens the recovery period around the region of control stimulation, and thus allows the generation of higher frequency waves than would have been possible using positive stimulation alone. This method is discussed in detail later in this chapter.

Recently, another local control scheme has been proposed [Zhang et al. 2005a] that periodically perturbs the model parameter governing the threshold. In fact, it is the local control analog of the global control scheme proposed by Alonso et al. [Alonso et al. 2003] discussed earlier. As in the other methods discussed here, the local stimulation generates high-frequency waves that propagate into the medium and suppress spiral or scroll waves. Unlike in the global control scheme, $b_f >> b_0$, so that the threshold can be negative for a part of the time. This means that the regions in resting state can become spontaneously excited, which allows very high-frequency waves to be generated.

We will in the following sections describe in detail three low-amplitude control techniques that would illustrate the mechanisms, efficacy, advantages, and limitations of low amplitude control scheme. The first requires stimulation only at a local region, while the second is a non-global spatially extended scheme. Both involve the application of supra-threshold stimulus. In contrast, the last scheme we describe involves the global application of weak sub-threshold stimulus.

12.4 Controlling spatiotemporal chaos with local stimulation

In this section we outline a method for controlling chaos in an excitable media by locally applying low-amplitude biphasic stimuli, i.e., a sequence of alternating positive and negative pulses [Breuer and Sinha 2004]. In the context of applying control in clinical situations, such as for terminating ventricular fibrillation, such local control schemes have the advantage that they can be readily implemented with existing hardware of the implantable cardioverter-defibrillator (ICD). This is a device implanted into patients at high risk from VF that monitors the heart rhythm and applies electrical treatment, when necessary, through electrodes placed on the heart wall. A low-energy control method involving ICDs aims at achieving control of spatiotemporal chaos by applying small perturbations from a single source (or at

---

[†]A related case of this control scheme is that proposed in Ref. [Yuan et al. 2005], where the high-frequency periodic signal is applied from the boundaries.

most a few electrodes).

The excitable media model used for illustrative purpose here is the modified Fitzhugh–Nagumo equations proposed by Panfilov [Panfilov and Hogeweg 1993]. For simplicity an isotropic medium is considered; in this case the model is defined by the two equations governing the excitability e and recovery g variables,

$$
\begin{aligned}
\partial e/\partial t &= \nabla^2 e - f(e) - g + I_{control}, \\
\partial g/\partial t &= \epsilon(e,g)(ke - g).
\end{aligned}
\tag{12.3}
$$

To achieve control of spatiotemporal chaos, a perturbation, $I_{control} = AF(2\pi ft)$ of amplitude A and frequency f is applied at a local region of the medium. F represents any periodic function, e.g., a series of pulses having a fixed amplitude and duration, applied at periodic intervals defined by the stimulation frequency f. The control mechanism can be understood as a process of overdriving the chaos by a source of periodic excitation having a significantly higher frequency. As noted in Refs. [Lee 1997; Xie et al. 1999], in a competition between two sources of high-frequency stimulation, the outcome is independent of the nature of the wave generation at either source, and is decided solely on the basis of their relative frequencies. This follows from the property of an excitable medium that waves annihilate when they collide with each other [Krinsky and Agladze 1983]. The lower frequency source is eventually entrained by the other source and will no longer generate waves when the higher frequency source is withdrawn. Although we cannot speak of a single frequency source in the case of chaos, the relevant timescale is that of spiral waves which is limited by the refractory period of the medium, τ_{ref}, the time interval during which an excited cell cannot be stimulated until it has recovered its resting state properties. To achieve control, one must use a periodic source with frequency $f > \tau_{ref}^{-1}$. This is almost impossible with purely excitatory stimuli as reported in Ref. [Sinha et al. 2001]; the effect of locally applying such perturbations is essentially limited by refractoriness to the immediate neighborhood of the stimulation point.

To see why a negative rectangular pulse decreases the refractory period for the Panfilov model, we note that (in the absence of the diffusion term) the stimulation vertically displaces the e-nullcline of Eq. (12.3) and therefore, the maximum value of g that can be attained is reduced. Consequently, the system will recover faster from the refractory state. To illustrate this, let us assume that the stimulation is applied when $e > e_2$. Then, the dynamics reduces to $\dot{e} = -C_3(e-1) - g, \dot{g} = \epsilon_2(ke - g)$. In this region of the (e,g)-plane, for sufficiently high g, the trajectory will be along the e-nullcline, i.e., $\dot{e} \simeq 0$. If a pulse stimulation of amplitude A is initiated at $t = 0$ (say), when $e = e(0), g = g(0)$, at a subsequent time t, $e(t) = 1 + \frac{A-g(t)}{C_3}$, and $g(t) = \frac{a}{b} - [\frac{a}{b} - g(0)]exp(-bt)$, where, $a = \epsilon_2 k(1 + \frac{A}{C_3}), b = \epsilon_2[1 + (k/C_3)]$. The negative stimulation has to be kept on until the system crosses into the region where $\dot{e} < 0$, after which no further increase of g can occur, as dictated by the dynamics of Eq. (12.3). Now, the time required by the system to enter this region is $\frac{1}{b}\ln\frac{a/b-g(0)}{a/b-\phi}$, where $\phi = C_3(1 - e_2) + A$. Therefore, this time is reduced when $A < 0$ and contributes to the decrease of the refractory period. Note that the above discussion also indicates that a rectangular pulse will be more effective than

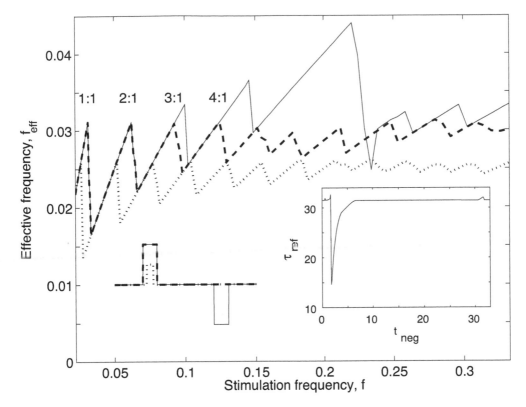

FIGURE 12.5 Stimulation response diagram for one-dimensional Panfilov model ($L = 40$) for different stimulation frequencies f. The dotted and broken curves represent purely excitatory pulses of amplitude $A = 5$, pulse duration $\tau = 0.05f^{-1}$, and $A = 10$, $\tau = 0.1f^{-1}$, respectively, while the solid curve represents biphasic pulses of amplitude $A = 10$ and pulse duration $\tau = 0.1f^{-1}$ (as shown in the bottom left corner). Note that, the highest effective frequencies f_{eff} for the three cases are very different. The ratio $f : f_{eff}$ is shown for the first four peaks. The inset shows the decrease in refractory period in absence of the diffusion term when a negative stimulation is applied at different times (t_{neg}) after the initial excitation.

a gradually increasing waveform, e.g., a sinusoidal wave (as used in [Zhang et al. 2003]), provided the energy of stimulation is the same in both cases, as the former allows a much smaller maximum value of g. Therefore, phase plane analysis of the response to negative stimulation allows us to design waveshapes for maximum efficiency in controlling spiral turbulence.

To understand how negative stimulation affects the response behavior of the spatially extended system, let us first look at a one-dimensional system. Figure 12.5 shows the relation between the stimulation frequency f and effective frequency f_{eff}, measured by applying a series of pulses at one site and then recording the number of pulses that reach another site located at a distance, without being blocked by any refractory region. Depending on the relative value of f and τ_{ref}, instances of $n : m$ response are observed, i.e., m responses are seen to be evoked by n stimuli. From the resulting effective frequencies f_{eff}, one sees that for purely excitatory stimu-

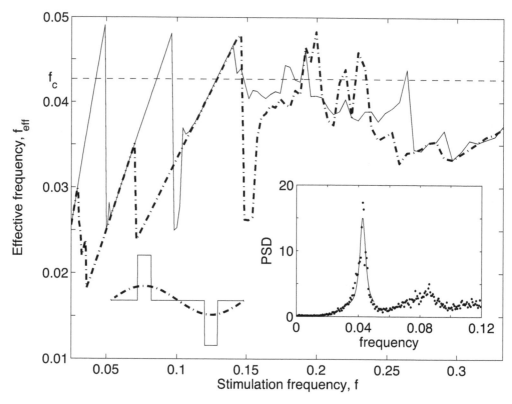

FIGURE 12.6 Stimulation response diagram for two-dimensional Panfilov model ($L = 26$) showing relative performance of different waveforms. The dash-dotted line represents a sinus wave ($A = 6$) and the solid curve represents a wave of biphasic rectangular pulses ($A = 18.9$), as shown in the bottom left corner, such that they have the same total energy. The inset shows the power spectra of spatiotemporal chaos in the 2-D Panfilov model ($L = 500$). The e variable was recorded for 3300 time units and the resulting power spectral density was averaged over 32 points. The peak occurs at the characteristic frequency $f_c \simeq 0.0427$ which is indicated in the main figure by the broken line.

lation, the relative refractory period can be reduced by increasing the amplitude A. However, this reduction is far more pronounced when a negative stimulation is applied between every pair of positive pulses. The inset in Fig. 12.5 shows that there is an optimal time interval between applying the positive and negative pulses that decreases the refractory period by as much as 50%. The highest effective frequencies correspond to a stimulation frequency in the range $0.1 - 0.25$, agreeing with the optimal time period of 2-5 time units between positive and negative stimulation.

A response diagram similar to the one-dimensional case is also seen for stimulation in a two-dimensional medium (Fig. 12.6). A small region consisting of $n \times n$ points at the center of the simulation domain is chosen as the stimulation point. For the simulations reported here $n = 6$; for a smaller n, one requires a perturbation of larger amplitude to achieve a similar response. To understand control in two dimensions, the characteristic timescale of spatiotemporal chaotic activity is obtained

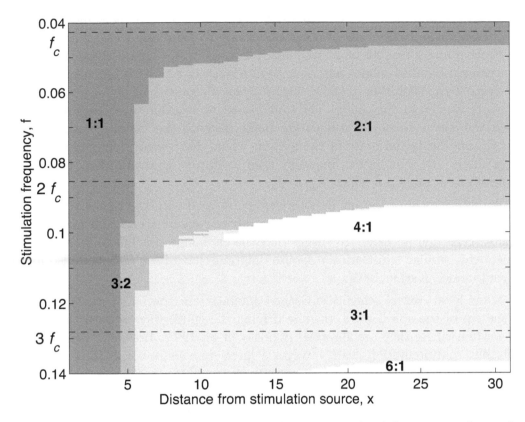

FIGURE 12.7   Distance dependence of stimulus response for different stimulation frequencies f in the two-dimensional Panfilov model. Biphasic rectangular pulses ($A = 18.9$) having duration $\tau = 0.1f^{-1}$ are applied, which elicit a response having an effective frequency f_{eff} at a particular location. The first three cells ($x = 1, 2, 3$) are within the region subject to direct stimulation. The shaded regions represent different response ratios $f : f_{eff}$. Integral multiples of the characteristic frequency f_c are indicated on the f-axis.

from its power spectral density (Fig. 12.6, inset). A peak is observed at a frequency $f_c \simeq 0.0427$. As seen in Fig. 12.6, there are ranges of stimulation frequencies that give rise to effective frequencies higher than this value. As a result, the periodic waves emerging from the stimulation point will gradually impose control over the regions exhibiting chaos. If f is only slightly higher than f_c control takes very long; if it is too high the waves suffer conduction block at inhomogeneities produced by chaotic activity that reduces the effective frequency, and control fails. Note that at lower frequencies the range of stimulation frequencies for which $f_{eff} > f_c$, is smaller than at higher frequencies. The performance of sinusoidal waves can be now compared with that of rectangular pulses by adjusting the amplitudes so that they have the same energy. The former is much less effective than the latter at lower stimulation frequencies, which is the preferred operating region for the control method.

The effectiveness of overdrive control is limited by the size of the system sought to be controlled. As shown in Fig. 12.7, away from the control site, the generated waves are blocked by refractory regions, with the probability of block increasing as a function of distance from the site of stimulation.[†] To see whether the control method is effective in reasonably large systems, it is used to terminate chaos in the two-dimensional Panfilov model, with $L = 500$.[†] Figure 12.8 shows a sequence of images illustrating the evolution of chaos control when a sequence of biphasic rectangular pulses is applied at the center. The time necessary to achieve the controlled state, when the waves from the stimulation point pervade the entire system, depends slightly on the initial state of the system when the control is switched on. Not surprisingly, the stimulation frequency used to impose control in Fig. 12.8 is seen to belong to a range for which $f_{eff} > f_c$.

Although most of the simulations were performed with the Panfilov model, the arguments involving phase plane analysis apply in general to excitable media having a cubic-type non-linearity. To ensure that the explanation is not sensitively model dependent, similar stimulation response diagrams have been obtained also for the Karma model [Karma 1993].

Some local control schemes envisage stimulating at special locations, e.g., close to the tip of the spiral wave, thereby driving the spiral wave toward the edges of the system where they are absorbed [Krinsky et al. 1995]. However, aside from the fact that spatiotemporal chaos involves a large number of coexisting spirals, in a practical situation it may not be possible to have a choice regarding the location of the stimulation point. One should therefore look for a robust control method which is not critically sensitive to the position of the control point in the medium. There have been some proposals to use periodic stimulation for controlling spatiotemporal chaos. For example, recently Zhang et al. [Zhang et al. 2003] have controlled some excitable media models by applying sinusoidal stimulation at the center of the simulation domain. Looking in detail into the mechanism of this type of control, one can conclude that the key feature is the alternation between positive and negative stimulation, i.e., biphasic pacing, and it is, therefore, a special case of the general scheme presented here.

Previous explanations of why biphasic stimulation is better than purely excitatory stimulation (that uses only positive pulses) have concentrated on the response to very large amplitude electrical shocks typically used in conventional defibrillation [Anderson and Trayanova 2001; Keener and Lewis 1999] and have involved details of cardiac cell ion channels [Jones and Tovar 2000]. The study described above gives a simpler and more general picture for understanding the efficacy of the biphasic scheme using very low amplitude perturbation, as the results do not depend on the details of ion channels responsible for cellular excitation. There are some limitations

---

[†]The profile of the stimulated wave changes as it propagates along the medium, from biphasic at the stimulation source to gradually becoming indistinguishable from a purely excitatory stimulation. As a result, far away from the source of stimulation, the response cannot have a frequency higher than $\sim \tau_{ref}^{-1}$.

[†]The initial condition used for this purpose is a broken plane wave which is allowed to evolve for 5000 time units into a state displaying spiral turbulence.

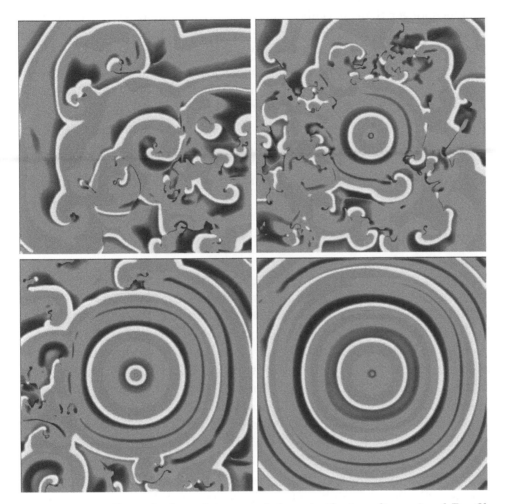

FIGURE 12.8   Control of spatiotemporal chaos in the two-dimensional Panfilov model ($L = 500$) by applying biphasic pulses with amplitude $A = 18.9$ and frequency $f = 0.13$ at the center of the simulation domain. The pulse shape is rectangular, having a duration of ~ 0.77 time units. Snapshots are shown for (top left) $t = 0$, (top right) $t = 1000$, (bottom left) $t = 2700$ and (bottom right) $t = 3800$ time units. The excitation wavefronts are shown in white, black marks the recovered regions ready to be excited, while the shaded regions indicate different stages of refractoriness.

to achieving control over a large spatial domain in an excitable medium by pacing at a particular point. Under some parameter regimes, the circular waves propagating from this point may themselves become unstable and undergo conduction block at a distance from the origin (similar to the process outlined in Ref. [Fox et al. 2002b]). In addition, the control requires a slightly higher amplitude and has to be kept on for periods much longer than spatially extended control methods [Sinha et al. 2001]. However, these drawbacks may be overcome if one uses multiple stimulation points arranged so that their regions of influence cover the entire simulation domain.

Next, we describe a novel method of controlling spatiotemporal chaos in excitable media, using an array of control points. The points are stimulated in a sequence so as to generate a traveling wave of activity across the medium, which interacts with the chaotic excitations and eliminates them. The proposed method is robust even in the presence of significant conduction heterogeneities in the medium which have often been an impediment to the success of other control schemes. We will describe the effect of this this control scheme on 2-dimensional and 3-dimensional excitable media.

12.5　Control using an array of point electrodes

Non-global control methods use less power and also are relatively easier to implement practically, needing fewer control points. However, strictly local control methods almost always involve very high-frequency stimulation [Zhang et al. 2005a], that can by itself lead to reentrant waves in the presence of inhomogeneities [Panfilov and Keener 1993; Shajahan et al. 2007]. Moreover, the effect of local stimulation at a point can affect the rest of the system only through diffusion. As wavefronts annihilate on collision, control-induced waves are restricted to the local neighborhood of the stimulation point during spiral turbulence, with the existing excited fragments closer to the control point shielding chaotic activity further away. By using a spatially extended but non-global scheme [Sinha et al. 2001] one can potentially avoid these drawbacks. In this section, we describe a study that seeks to terminate chaos in excitable media by applying spatiotemporally varying stimulation along an array of control points [Sridhar and Sinha 2008]. The control signal appears to propagate along the array, triggering an excitation wavefront in the underlying medium, that is regenerated after each collision with chaotic fragments and in the process eliminating all existing activity. Stimulating each point once (or at most, twice) is seen to successfully control chaos in almost all instances. Although an array of control points has been used earlier to prevent the breakup of a single spiral [Rappel et al. 1999], this study is possibly the first to show control of fully developed spatiotemporal chaos (Fig. 12.9) in excitable media using only a finite number of control points without repeated stimulations at high frequency.

As shown earlier in the book the spatiotemporal dynamics of activation in excitable systems can be described by partial differential equations of the form:

$$\partial V/\partial t = \frac{-I_{ion} + I_{ext}(x, y, t)}{C_m} + D\nabla^2 V, \tag{12.4}$$

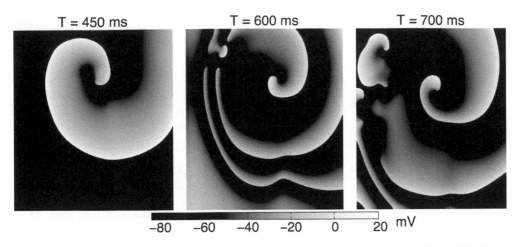

T = 450 ms T = 600 ms T = 700 ms

−80 −60 −40 −20 0 20 mV

FIGURE 12.9 Pseudo-gray-scale plot of the transmembrane potential V for the two-dimensional LRI model showing the time-evolution from single spiral (left) through breakup (center) to fully developed chaos (right).

where V (mV) is the transmembrane potential, $C_m = 1$ μF cm^{-2} is the transmembrane capacitance, D (cm^{2}s^{-1}) is the diffusion constant, I_{ion} (μA cm^{-2}) is the transmembrane ionic current density, and $I_{ext}(x, y, t)$ is the space- and time-dependent external stimulus current density that is applied for the purpose of control on a two-dimensional surface. For the specific functional form of I_{ion}, the Luo–Rudy I (LRI) action potential model [Luo and Rudy 1991] has been used. In order to verify the model independence of the results and for undertaking three-dimensional simulations, a simpler description of the action potential as given by Panfilov (PV) [Panfilov 1998; Panfilov and Hogeweg 1993] has also been used. The simulation domain is a square lattice of $L \times L$ points in two dimensions or a cuboid with $L \times L \times L_z$ points in three dimensions. For the LR I simulations, $L = 400$, while for PV, $L = 256$ (for 2-D) and $L = 128, L_z = 8$ (for 3-D). The standard five-point and seven-point difference stencils are used for the Laplacian in two and three dimensions, respectively. The time step for integration is chosen to be $\delta t = 0.01$ ms (for LR I) and $= 0.11$ ms (for PV). No-flux boundary conditions are implemented at the edges of the simulation domain.

Let us now focus on the control term $I_{ext}(x, y, t)$. For a 2-D domain of size $L \times L$, one can express

$$I_{ext} = I(x, y, t)\delta_{x,md} \, \delta_{y,nd} \, , \tag{12.5}$$

where the Kronecker delta function is defined as $\delta_{i,j} = 1$ if $i = j$, and $= 0$, otherwise, d is the spatial interval between points in an array where the control signal is applied and m, n are integers in the interval $[0, L/d]$. The current density $I(x, y, t) = I_0$ for $t \in [\sqrt{x^2 + y^2}/v, (\sqrt{x^2 + y^2}/v) + \tau]$, and $= 0$ otherwise, corresponds to a rectangular control pulse of amplitude I_0 of duration τ that is traveling with velocity v. At the onset of control ($t = 0$), the point at $(0,0)$ is stimulated, followed a short duration later by the points at $(0, d)$ and $(d, 0)$, and this process continues as the control pulse proceeds like a traveling wave across the array. At each control

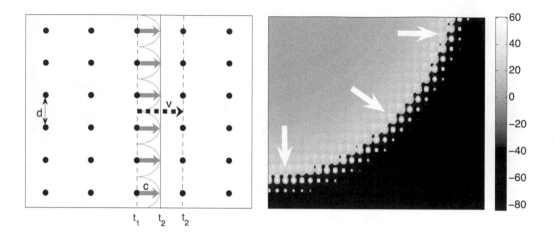

FIGURE 12.10   (left) Schematic diagram and (right) pseudo-grayscale plot of the transmembrane potential V for the two-dimensional LRI model, showing the propagation of the control-induced excitation wavefront. The control signal traveling with velocity v stimulates at time t_1 a column of points (spaced d apart) from which secondary excitations traveling with velocity c are generated. At time t_2 the control signal stimulates the next column of points, which the induced excitation wave may or may not have reached depending on the relative values of v and c [$v/c = 4$ in (right)].

point, the stimulation may excite the underlying region depending on its recovery phase. An excited region can in turn spread the effect of the stimulation to the surrounding regions through diffusion. Figure 12.10 shows this process of secondary wave generation at the stimulated control points, creating a sustained excitation wavefront. Note that any portion of this stimulated wavefront in the medium that is broken through collision with chaotic fragments can be regenerated by subsequent control points. Hence, there is effectively an unbroken wavefront that travels through the medium, sweeping away the spatiotemporal chaos and leaving the system in a recovering state. As each control point needs to impose order over a region of size $\sim d^2$ to eliminate spatiotemporal chaos, this highlights the critical role of d. Indeed, for large v, the method approaches global control as $d \to 0$, while for $d \sim L$ the activity resulting from the control stimulation is confined to a single, localized region. This implies that as d increases, terminating chaos becomes increasingly difficult. For example, in LRI model, chaos control fails for $d \geq 13$.

Before reporting the simulation results, we consider the role played by the traveling wave nature of the control pulse. The success of the proposed method depends on the control signal velocity, v, relative to the velocity of the excitation wavefront propagating via diffusion, c ($= 60.75$ cm s^{-1} for LR-1, $= 50.9$ cm s^{-1} for PV, for the parameters used here). When $v \to 0$, the control method reduces to a local scheme regardless of d, as the effect of the external stimulation can only propagate in the system via diffusion. On the other hand, when v is very large, all the control points are stimulated almost simultaneously. While the traveling wave nature of the control pulse allows propagation of stimulation independent of diffusion through the

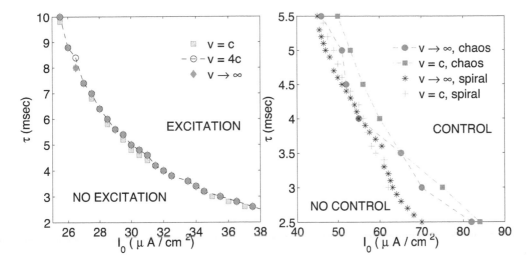

FIGURE 12.11 Strength-duration curves for LRI model (in a two-dimensional domain with $L = 400$) when control stimulation is applied on (left) quiescent homogeneous medium and (right) medium with existing excitation activity, either a single spiral or spatiotemporal chaos. In both figures, control is applied over a grid of points which are spaced apart by $d = 10$. Different curves correspond to different control signal velocities v, relative to c, the excitation wavefront velocity in the medium. For values of I_0 and τ above the curves, the external stimulation results in (left) excitation of the domain or (right) control of existing activity.

excitable medium, for $v \simeq c$ the excitation propagating by diffusion reinforces the external stimulation at each control point. Further, for a control signal propagating with a finite velocity (thus engaging only a few points at any given time), the energy applied per unit time to the medium is much lower than that for simultaneous stimulation of all the control points (i.e., $v \to \infty$).

For a system undergoing chaotic activity, the medium will at any time be at an extremely heterogeneous state, with certain regions excited and other regions partially or fully recovered. The vital condition for successful termination of chaos is that after the passage of the control-stimulated wave there should not remain any unexcited region which is partially recovered and which can be subsequently activated by diffusion from a decaying excitation front. This places a lower bound on the control signal parameters, i.e., the signal amplitude I_0 and its duration τ. If either is decreased below this bound, the external stimulation is unable to excite certain partially recovered regions. If these regions have neighboring chaotic fragments, whose activity is slowly decaying after collision with the control-induced wave, then, there will be a diffusion current from the latter. Depending on the phase of recovery, this may be sufficient to stimulate activity in the partially recovered regions, thereby re-initiating spatiotemporal chaos after the control signal has passed through.

To understand in detail the lower bound on the external stimulation parameters

I_0 and τ, let us first look at the condition for exciting a completely homogeneous medium in the resting state. The stimulation at each point must exceed the local threshold in order to generate an action potential. This could be achieved either directly through an external current I_{ext} or indirectly through diffusion from a neighboring excited region. Figure 12.11 (left) shows the result of applying control signals with different I_0 and τ at points which are spaced a distance d apart. The resulting *strength-duration curve* [Geddes et al. 1970; Gold and Shorofsky 1997] indicates that the response of the system is not sensitively dependent on the propagation velocity v of the control signal along the grid. As d decreases, excitation is possible at lower values of I_0 and τ, the minimum being for the case when all points are subject to direct external stimulation ($d \rightarrow 0$). This is because the entire applied current I_0 at any point is used to raise its state above the threshold, no part being lost to neighboring regions through diffusion.

For systems with existing activity, such as self-sustaining spiral waves or spatiotemporal chaos, the regions in the relative recovery period can be excited by stimuli larger than those needed for a fully recovered medium. Hence, the strength-duration curve for control of such a system will shift toward higher values of I_0 and τ (Fig. 12.11, right). The minimum external stimulus required for control does not vary significantly if the medium is undergoing fully developed spatiotemporal chaos as opposed to having a single spiral wave. Further, the velocity of the control signal along the array is not critical to the success in eliminating existing activity, provided v is not significantly smaller than c. Note that, if v is sufficiently small or d increases beyond a critical value, the control fails, as the effect of the control signal is confined to the region immediately surrounding the stimulated point.

Figure 12.12 shows the successful termination of spatiotemporal chaos in the LRI model using a control signal that travels across the two-dimensional domain while exciting points in the medium that are spaced apart by $d = 10$. The control scheme is verified to be not sensitively model dependent by using it to eliminate spatiotemporal chaotic activity in the PV model. As most systems in reality have thickness, it is crucial to verify that the method is successful in controlling chaos in a three-dimensional domain, even when the external stimulus is applied *only on one surface*. This latter restriction follows from the fact that, in most practical situations it may not be possible (or desirable) to penetrate the medium physically in order to apply control signals inside the bulk. The method is also seen to work in thin slices of excitable media of size $L \times L \times L_z$ ($L_z \ll L$), when the array of control points is placed on one of the $L \times L$ surfaces (Fig. 12.13). Even in cases where a single control-stimulated wave across the medium is unable to terminate all activity, it results in driving the chaotic activity further toward the boundaries and away from the origin of control stimulation. Thus, using multiple waves through application of control signals at intervals which are larger than the recovery period of the medium, the chaos in the bulk of three-dimensional systems is successfully terminated.

It has been verified that small distortions in the regular array of control points do not result in the failure of the proposed method. Similarly, starting the control signal at different points of origin (and indeed, using a planar wave rather than a curved wave) does not affect the efficacy of the scheme. Further, the method is

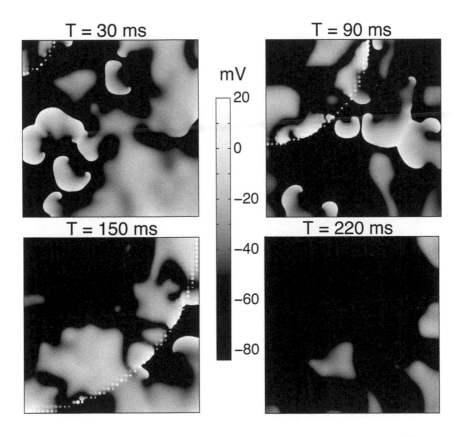

FIGURE 12.12 Pseudo-grayscale plots of the transmembrane potential V for the two-dimensional LRI model ($L = 400$) showing the elimination of all chaotic activity within 300 ms after initiation of control. A single wave of control stimulus ($I_0 = 75$ μA/cm^2, $\tau = 3$ ms) begins at the top left corner ($T = 0$) and travels across the domain with velocity $v = c$. This results in stimulating the region around the control points (spaced apart by $d = 10$) in a sequential manner, creating a stimulated wavefront seen as an arc consisting of excited points in the panels above.

FIGURE 12.13 The effect of applying control on a single surface of a three-dimensional domain using the PV model ($L = 128$ and $L_z = 8$). Two waves of control signal ($I_0 = 10$, $\tau = 16.5$ ms, $v = c$) are applied 165 ms apart resulting in termination of spatiotemporal chaos. The interval between control points, $d \to 0$. The panels show isosurface plots at T=178 ms (top left), 242 ms (top right), 308 ms (bottom left), and 341 ms (bottom right).

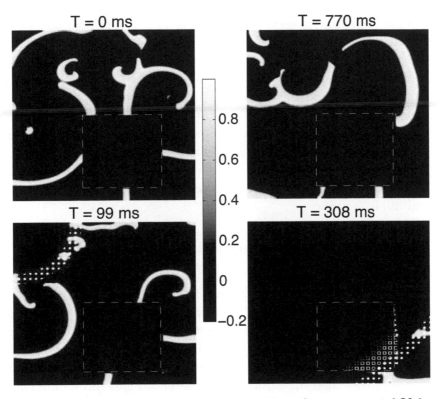

FIGURE 12.14 Pseudo-grayscale plots of the transmembrane potential V for the two-dimensional PV model ($L = 256$) with a conduction inhomogeneity (of size 110×110, indicated by the broken lines), showing the time evolution of the system from a chaotic state at $T = 0$ ms in the absence (top) and presence (bottom) of control. Inside this region, $D_{inhomogen} = 0.01D$, D being the diffusion constant of the rest of the medium. A single pulse of control stimulus ($I_0 = 8$, $\tau = 17$ ms, $v = c$) is applied at $T = 0$ ms over an array of control points spaced apart by $d = 8$, resulting in termination of all activity by $T = 350$ ms.

robust in the presence of conduction inhomogeneities (such as inexcitable obstacles) that tend to destabilize local control schemes. Figure 12.14 shows the occurrence of spatiotemporal chaos in a medium containing a large region of slow conduction, i.e., an extremely small value of D compared to the rest of the medium, which is successfully controlled by the proposed scheme.

To conclude, in this section, a control scheme involving external stimulation applied over an array of points has been described, which is successful in terminating spatiotemporal chaos in both simplified as well as realistic models of biological excitable media. The control signal amplitude is varied both spatially and temporally, such that it appears as a propagating wave along the array of control points. This results in a stimulated wavefront in the excitable medium that, depending on the propagation velocity of the control signal and the space interval between control points, eliminates all existing activity. The method requires very low-amplitude control currents applied for short durations at a finite number of points, each point being stimulated once (or at most, twice) in most situations. Further, it is successful in terminating chaos in the bulk of a three-dimensional medium even when applied only on one surface. The use of a significantly lower number of control points than that necessary for global control methods makes the proposed scheme more suitable for practical implementation.

12.6   Controlling chaos using sub-threshold stimuli

The excitation threshold is the key parameter governing the dynamics of spatial patterns in excitable systems. A stimulus that drives the system above this excitation threshold leads to a transition from quiescent to an active state, thereby generating an action potential (AP). Such a signal is called a supra-threshold signal. On the other hand, if an external signal is not able to initiate an action potential in a medium or cell at rest then the external signal is called sub-threshold. The demonstration of stochastic resonance (SR) [Jung and Mayer-Kress 1995b] and coherence resonance (CR) [Pikovsky and Kurths 1997] in excitable media suggests that weak sub-threshold signals could have a significant effect on the dynamics of these media [Alonso et al. 2001; Muratov et al. 2007]. Further, SR-like response resulting from chaotic dynamics in simple systems [Sinha 1999; Sinha and Chakrabarti 1998] raises the intriguing possibility that spatially heterogeneous activity may enhance the response of an excitable medium to sub-threshold signals.

In this section we show that these sub-threshold perturbations produce surprising changes in the dynamics of spatiotemporally heterogeneous activity in the medium [Sridhar and Sinha 2010]. Stimuli that cannot initiate action potentials in a resting medium can significantly alter the time-evolution of spatially heterogeneous activity by modifying the recovery dynamics of the medium. The application of a sub-threshold stimulus leads to a significant increase of the action potential duration and reduction in waveback velocity. This in turn causes a differential slowing down of recovery of the cells in the medium, leading to the development of spatial coherence, terminating all activity in the medium including spatiotemporal chaos.

The results described here not only suggest alternate low-amplitude mechanism of chaos control in excitable medium, but can also potentially explain how signals that are extremely attenuated during passage through intervening biological media, can still affect the excitation dynamics of organs such as the heart.

Prevailing methods of spatiotemporal chaos control in excitable systems are almost exclusively dependent on using supra-threshold signals, either through a local high-frequency source [Pumir et al. 2010; Zhang et al. 2005a] or using a spatially extended array [Shajahan et al. 2009; Sinha et al. 2001; Sridhar and Sinha 2008]. Controlling spatial patterns with sub-threshold stimulation would not only utilize new physical principles, but also avoid many of the drawbacks in previously proposed schemes. Throughout this section *sub-threshold* refers to stimuli that are insufficient to drive the resting tissue above the excitation threshold. In the context of cardiac control, occasionally threshold may also be used to refer to the minimum stimulus amplitude required for successful defibrillation. However, if such a stimulus causes an AP in resting tissue (i.e., the medium crosses the excitation threshold), it is referred to as supra-threshold here.

We show here that sub-threshold stimulation, while having no significant effect on a quiescent medium, can induce a remarkable degree of coherence when applied on a system with spatially heterogeneous activity. Synchronizing the state of activation of all excited regions ensures that they return to rest almost simultaneously, in the process completely terminating activity in the medium. Thus, control of spatially extended chaos is achieved efficiently using a very low-amplitude signal. The mechanism of this enhanced coherence can be explained in terms of the role played by sub-threshold stimulus in increasing the recovery period of the medium. It significantly reduces the propagation velocity of the *recovery* front, thereby increasing the extent of the inexcitable region in the medium. A semi-analytical derivation is presented of the relation between strength and duration of the globally applied sub-threshold signal necessary for complete elimination of spatially heterogeneous activity in excitable media.

A time-dependent external current is the spatially uniform signal, applied at all points of the simulation domain and is represented by the time-dependent current density, $I(\mu \text{Acm}^{-2})$. For the results reported here, the Luo–Rudy I (LRI) model that describes the ionic currents in a mammalian ventricular cell [Luo and Rudy 1991] has been used. The maximum K^+ channel conductance G_K has been increased to 0.705 mS cm^{-2} to reduce the duration of the action potential (APD) [ten Tusscher and Panfilov 2003a]. To study the effect of sub-threshold stimulus on a stable spiral and on spatiotemporal chaos, values of maximum Ca^{2+} channel conductance $G_{si} = 0.04$ and 0.05 mS cm^{-2} have been used, respectively. The model-independence of the results have been verified explicitly by observing similar effects in other realistic channel-based descriptions of the ionic current, such as the TNNP model [Sridhar et al. 2010; ten Tusscher et al. 2004].

Let us consider in turn the response of a single cell, a one-dimensional cable and a two-dimensional sheet of excitable units to a sub-threshold current I. The spatially extended systems are discretized on a grid of size L (for 1-D) and $L \times L$ (for 2-D). For most results reported here $L = 400$, although L up to 1200 has been used. The space step used for all simulations is $\delta x = 0.0225$ cm, while the time-step $\delta t = 0.05$ ms (for 1-D) or 0.01 ms (for 2-D). The equations are solved using a forward Euler scheme with a standard 3-point (for 1-D) or 5-point (for 2-D) stencil for the Laplacian describing the spatial coupling between the units. No-flux boundary conditions are implemented at the edges. The external current is applied globally, i.e., $I_{ext}(x, y, t) = I$ in Eq. (12.4) at all points in the system for the duration of stimulation, τ. A stimulus $\{I, \tau\}$ is sub-threshold if it does not generate an action potential when applied on a quiescent medium. The initial spiral wave state is obtained by generating a broken wavefront which then dynamically evolves into a curved rotating wavefront. Over a large range of parameter values, this spiral wave is a persistent dynamical state of the excitable medium. Under certain conditions (e.g., for some parameter values or external perturbations), the spiral wave can become unstable, eventually breaking up into multiple wavelets leading to a spatiotemporally chaotic state. An alternative method for obtaining the spatiotemporally chaotic state is to randomly apply supra-threshold stimuli at different points in the medium over a small duration. In biological systems, spiral waves and chaos often appear spontaneously as a result of existing heterogeneities or stochastic fluctuations. In the cardiac medium, such dynamical phenomena can be initiated experimentally by applying cross-field stimulation [Jalife et al. 1999]. The spatially extended chaotic state is a long-lived transient whose lifetime increases exponentially with the system size [Sinha et al. 2001]. For biologically realistic simulation domain sizes, such as those used here, the chaotic state persists longer than any reasonable duration of simulation.

Figure 12.15 shows that when sub-threshold stimulation is applied to an excitable medium with spatially heterogeneous activity, viz., either a single spiral wave (a-c) or spatiotemporal chaos (d-f), there is a striking change in the subsequent dynamics of the system. Within a short duration (comparable to the APD) there is complete suppression of all activity in the medium, although in absence of this intervention, the existing dynamical state would continue to persist for an extremely long time. This result is surprising as the weak sub-threshold signal appears to be incapable of significantly altering the dynamics of an excitable system. For example, if sub-threshold stimulation is applied on a large domain with a *single* wave propagating across it, there would have been no significant change in the time-evolution of the system with the wave continuing to move across the medium. This differentiates sub-threshold from other weak (i.e., low-amplitude) but suprathreshold stimuli which, on being applied globally to the above system, would have excited almost the entire medium resulting in eventual cessation of all activity.

To understand this apparent paradox, let us first note that the sub-threshold stimulation rapidly decreases the number of cells that can be excited by existing activity in the medium (Fig. 12.16,a). Indeed, global suppression of activity results when, by the end of the stimulation, the number of cells susceptible to excitation is insufficient to sustain the activity. This decrease in their number is because cells tend to remain in the recovering state for a longer period in the presence of a subthreshold stimulus. This can be clearly seen in the response of a single *excited* cell to a subsequent sub-threshold current I applied for a fixed duration τ (Fig. 12.16,b).

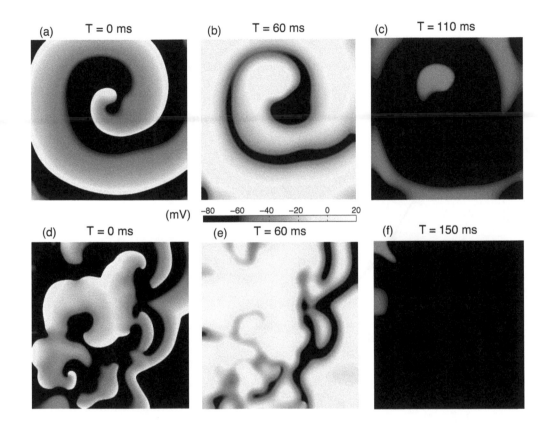

FIGURE 12.15 Pseudocolor plots of transmembrane potential V for the two-dimensional LRI model ($L = 400$) showing the elimination of all activity on applying a sub-threshold current. The current I is switched on at $T = 0$ for a duration $\tau = 60$ ms on (a-c) a single spiral, with $I = 1.6\mu$A cm^{-2}, and (d-f) a spatiotemporally chaotic state, with $I = 1.8\mu$A cm^{-2}. By $T = 150$ ms, excitation has been effectively terminated throughout the simulation domain.

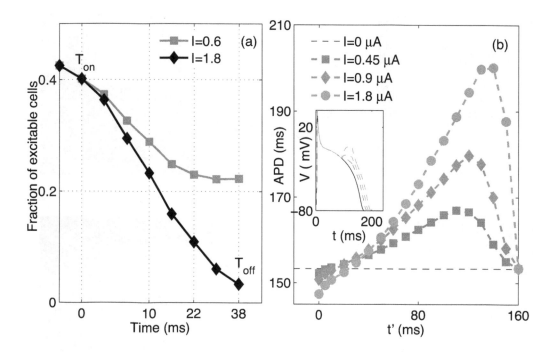

FIGURE 12.16 (a) The fraction of cells that can be potentially excited (i.e., with $V \leq -60mV$) decreases with time during stimulation by sub-threshold current I in a two-dimensional system with spatiotemporal chaos. The total duration of the external signal is $\tau = T_{on} - T_{off} = 38$ ms. Results shown correspond to failed (squares) and successful termination (diamonds) of activity in the medium. (b) Effect of sub-threshold current on APD of a single cell. The current I is varied keeping the duration τ fixed ($= 38$ ms). In all cases, the APD is shown as a function of the time interval t' between the initiation of AP and T_{on}. The inset shows the corresponding effect on the AP profile.

Increasing I significantly alters the recovery period resulting in a change in the APD. The time t' (measured from the initiation of the AP) at which the sub-threshold stimulation begins, also affects the response of the cell to the signal. These results clearly indicate that the dominant effect of a sub-threshold stimulus is to increase the time period that a cell spends in recovering from prior excitation.

In a spatially extended system, this enhanced recovery period of the cells results in altering the propagation characteristics of the traveling waves. Figure 12.17 (a) shows a spiral wave propagating in a two-dimensional medium, where each turn of the wave is a region of excited cells, with the successive turns separated by recovering regions. As the state of the cells evolves with time, it is manifested in space as movement of excitation and recovery fronts. Their propagation speeds are referred to as wavefront (c_f) and waveback (c_b) velocities, respectively. In the absence of any external stimulation, $c_f \simeq c_b$, ensuring that the width of the excited region remains approximately constant as the waves travel through the medium. However, on applying a sub-threshold stimulus, the waveback velocity becomes significantly lower than that of the wavefront which is almost unchanged. Figure 12.17 (b) shows that, once stimulation begins, c_b quickly decreases to a minimum value dependent on I. It then gradually rises to eventually become equal to c_f again. For a large sub-threshold stimulus I, the waveback velocity rapidly falls to its lowest value and changes very slowly thereafter. Under these conditions, one can ignore the time-variation of c_b for small τ and use the time-averaged value $\tilde{c}_b(I)$. Increasing I leads to an increased difference in the velocities of the excitation and recovery fronts, $c_f - \tilde{c}_b(I)$ (Fig. 12.17, c). For short stimulus durations, this difference is almost independent of τ. A significantly lower waveback velocity results in the inexcitable region between the excitation and recovery fronts of a wave becoming extended through the course of the stimulation (compare the profiles of APs in a 1-D cable shown in Fig. 12.18). This increases the overall area of the medium that cannot be excited, thereby making it progressively unlikely for the system to sustain recurrent activity. This is explicitly shown for a one-dimensional cable in Fig. 12.18. When two successive waves propagate along the cable, globally applying the sub-threshold stimulus reduces the excitable gap between the recovery front of the leading wave (whose velocity c_b has decreased) and the excitation front of the following wave (whose velocity c_f is unchanged). For a high sub-threshold I applied for a long enough duration, the waveback of the first wave slows down sufficiently to collide with the succeeding wavefront. This collision results in termination of the excitation front for the second wave which subsequently disappears from the medium. The above physical picture is fundamentally unchanged for a rotating spiral wave with multiple turns as shown in Fig. 12.17 (a). A simple semi-analytical theory for the mechanism by which the sub-threshold stimulus suppresses spatially heterogeneous activity can now be proposed.

In the absence of any external stimulus, the width of the excited region of a wave lying between its excitation and recovery fronts is $l = c_f \tau_r$, where τ_r is the period for which the active cells remain excited. This time period is operationally measured as the duration for which the transmembrane potential of a cell (V) remains above its excitation threshold. On applying a sub-threshold external current

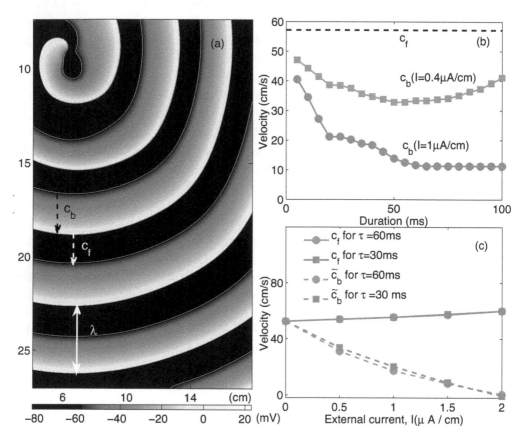

FIGURE 12.17 (a) Pseudocolor plot of a spiral wave indicating the wavelength λ, the wavefront velocity c_f, and the instantaneous waveback velocity $c_b (= c_f)$ in the absence of external stimulation. (b) Time-evolution of c_b during external stimulation of duration $\tau = 100$ ms using two different I corresponding to failed (squares) and successful termination (circles) of activity in the medium. During the course of stimulation, c_f is unchanged. (c) The average waveback velocity $\tilde{c}_b$ (broken lines) reduces with I in contrast to c_f (solid lines)

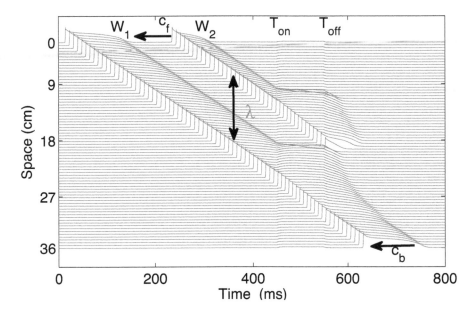

FIGURE 12.18 Spatiotemporal evolution of two successive waves propagating along a one-dimensional cable of excitable cells when a sub-threshold stimulus is applied over the time interval $[T_{on}, T_{off}]$. The propagation of the wave W_2 is blocked by the recovery front of W_1 due to reduction of its velocity, c_b. The wavefront velocities c_f of both waves are almost unchanged. The spatial interval between the two successive waves at a time instant is indicated by λ.

I, c_f is almost unchanged but the resultant waveback velocity, $c_b(I,t)$, which varies with time over the duration of the stimulus τ, is seen to decrease with increasing I. If I is large or τ is small, the time-variation of c_b can be neglected and it is reasonably well-approximated by the time-independent average value $\tilde{c}_b(I)$ over the stimulus duration. Thus, the width of the excited region of the wave increases to $l(I) = l + [c_f - \tilde{c}_b(I)]\tau$. If λ is the distance between excitation fronts of two successive waves in the medium, then collision between the recovery front of the leading wave and the excitation front of the following wave takes place when $l(I) \geq \lambda$. Thus, for a sub-threshold stimulus I, the shortest stimulus duration τ_{min} necessary to eliminate a source of recurrent activity such as a spiral wave is,

$$\tau_{min} = \frac{\lambda - c_f \tau_r}{c_f - \tilde{c}_b(I)}. \qquad (12.6)$$

Eq. (12.6) provides us with an analytical relation between the stimulus magnitude and its minimum duration necessary for terminating activity in the medium in terms of measurable dynamical characteristics of the system. Figure 12.19 shows that this theoretical strength-duration curve for the external stimulation necessary to terminate activity matches very well with the empirical data obtained from numerical simulations for both single spiral wave as well as spatiotemporal chaos. In general, the weaker the sub-threshold current, the longer it has to be applied in order to alter the dynamical behavior of the system. However, there is a lower bound for I below which there is no discernible effect of the sub-threshold stimulation regardless of its duration. Note that, for values of I just above this lower bound, the required τ_{min} is extremely long and the temporal variation of c_b over the duration of the stimulation can no longer be neglected. By explicitly considering the time-dependence of c_b in Eq. (12.6), one can theoretically estimate the value of I where the strength-duration curve becomes independent of τ.

The mechanism of the sub-threshold response of excitable media proposed here depends only on the recovery dynamics of the system. In detailed ionic models of biological excitable cells, this dependence is manifested as a decrease in the ion channel conductance responsible for the slow, outward K^+ current during the sub-threshold stimulation.[†] Thus, simplistic models of excitable media which do not incorporate the effect of external current stimulation on the recovery dynamics are inadequate to reproduce this enhanced sub-threshold response reported here. The above results provide a framework for explaining earlier experimental observations that, in the human heart, sub-threshold stimulation can prevent subsequent activation [Skale et al. 1983].

The study described here may have potential significance for understanding the spatiotemporal dynamics of excitable media in several practical situations such

---

[†]The decrease in waveback velocity compared to the wavefront velocity under certain conditions has been associated with the existence of a steep slope in the APD restitution curve for the excitable medium, which relates the APD to the time-interval between the recovery and subsequent re-excitation of a local region [Courtmanche 1996].

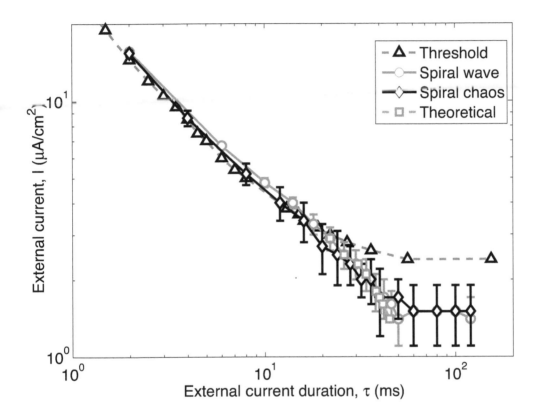

FIGURE 12.19 Strength-duration curves for a two-dimensional medium with the external current applied at all points of a quiescent medium (triangles) or a medium with existing excitation activity, either a single spiral (circles) or spatiotemporal chaos (diamond). The theoretical prediction given in Eq. (12.6) is also shown (square). Each (τ, I) point is averaged over 10 initial conditions.

as during clinical treatment of life-threatening arrhythmias. Current methods are primarily aimed toward synchronizing the activity of all regions by using supra-threshold stimuli having relatively larger amplitudes.[†] However, in such an approach, regions that have been rendered temporarily inexcitable due to prior activity remain unaffected. Thus, these regions can subsequently be re-activated by any remaining excitation after the stimulus is removed, leading to failure of control. By contrast, a sub-threshold stimulus slows the recovery of excited regions, thereby reducing the pool of cells available for excitation by existing activity in the medium. This suggests an alternative mechanism for the efficient termination of spatially extended chaos in excitable systems. It may provide a key toward understanding how spatially irregular activity in biological systems (e.g., fibrillation) is significantly affected by signals that have been strongly attenuated during passage through the intervening medium [Keener and Panfilov 1996; Krinsky and Pumir 1998].

[†]Proposed methods for controlling spatiotemporal chaos in simple models of excitable media have occasionally applied stimuli to an aggregated *recovery* variable [Osipov and Collins 1999] which has no direct correspondence in biologically realistic models such as LRI. However, such methods rely on driving the medium to a bistable regime which requires the application of a supra-threshold external current in a realistic model of biological excitable media. Hence, such control schemes are also supra-threshold as per the definition used here.

13

Sample programs

13.1 FHN single cell

```matlab
1   % %The following code simulates FitzHUgh Nagumo dynamics
2   % %for a single cell
3   % %The variables are u,v are the fast and slow variables
4   % %I_stim is the stimulus for time 'dur' for getting an excitation
5   % %a, k, epsilon are the parameters
6   % %The output values for u and v at specified time steps
7   % % are stored in rec_u and rec_v
8
9   T=10000; %Number of time steps
10  dt = 0.01; % Time step for integration
11  %Initial values of u and v corresponding to stable resting state
12  u=0;v=0;
13  %Parameters
14  a=0.139;
15  k=0.6;
16  epsilon=0.02;
17  dur=20;
18  t_init=100;
19  %Initialise empty vector rec_u and rec_v to store
20  %the values at different points in time%
21  rec_u=zeros(T,1);rec_v=zeros(T,1);
22  %The time loop%
23  for i1=1:T,
24      % Applying the stimulus for 'dur' iterations
25      if((i1>(t_init) && (i1<=(t_init+dur)))
26          I_stim=10;
27      else I_stim=0;
28      end;
29      %Reaction terms for excitation and relaxation
30      del_u = u*(1-u)*(u-a)-v+I_stim;
31      del_v = epsilon*(k*u-v);
32      %Integration using the Euler scheme
33      unew = u+del_u*dt;
34      vnew = v+del_v*dt;
```

```
35      %Updating u and v with unew and vnew
36         u=unew;v=vnew;
37      %Recording data at every time step
38         rec_u(i1,1)=u;rec_v(i1,1)=v;
39  end;
40  %Plotting the time series of the fast and the slow variables
41  figure,  plot(rec_u,'--b'), %fast
42  hold on, plot(rec_v,'r')  %slow
43  xlabel('Iterations'), ylabel('Instantaneous value of u and v)')
```

13.2 Oscillations arising from FHN coupled to a passive cell

```
1   % % The following code simulates the onset of oscillations when
2   % % a FitzHugh-Nagumo excitable cell is coupled with a
3   % % passive element to produce oscillations.
4   % % Variables u,v are the fast and slow variables
5   % % up is the value of the passive cell
6   % % I_stim is applied stimulus for getting an excitation
7   % % a, K,epsilon, are the parameters for the FHN system
8   % % up_r is the resting potential for the passive cell
9   % % cr is the strength of the interaction between passive
10  % % and excitbale cells.
11  % % Output values for u, v, up at specified time steps are
12  % % stored in rec_u, rec_v and rec_up
13
14  %Number of iterations%
15  T=1e6;
16  %Time step of integration%
17  dt = 0.001;
18  %Resting state values of all variables%
19  u=0;
20  v=0;
21  up=0;
22  %Parameters of the FHN model%
23  a = 3;epsilon=0.08;alpha=0.2;K=0.25;
24  %Passive cell characteristics%
25  np=1; %Number of passive cells
26  cr=1.9;%Strength of the coupling between excitable and passive cell
27  up_r=1.5;% Resting potential for the passive cell
28  %Initialising the vector for recording the data%
29  rec_u=zeros(T,1);rec_v=zeros(T,1);rec_up=zeros(T,1);
30  %Time loop%
31  for i1=1:T,
32      %Terms for excitation and relaxation and passive cell %
33      del_u = a*u*(1-u)*(u-alpha) -v + np*cr*(up-u);
34      del_v = epsilon*(u-v);
35      del_up = K*(up_r-up)-cr*(up-u);
36      %Euler method for integration
37      unew = u+del_u*dt;
38      vnew = v+del_v*dt;
39      upnew = up+del_up*dt;
```

```
40      %Updating
41      u=unew;v=vnew;up=upnew;
42      %Recording the data of u,v, and up%
43      rec_u(i1,:)=u;rec_v(i1,:)=v;rec_up(i1,:)=up;
44   end;
45   %Plotting the data%
46   figure, plot(rec_u,'--b'), %fast
47   hold on, plot(rec_v,'r') %slow
48   xlabel('Iterations'), ylabel('Instantaneous value of u and v)')
```

13.3 Rapid pacing of Luo–Rudy single cell

```
1    % % The following code simulates Luo-Rudy dynamics for a single cell
2    % % v is the transmembrane potential
3    % % m, h, j, d,f, x, xi are the different gates
4    % % External signal I_stim is applied for a duration 'dur' to generate a
5    % % sequence of action potentials
6    % % g_K is changed from default 0.282 (in original LR model) to 0.705
7    % % to reproduces APD close to that observed in human ventricles
8
9    %Total number of  iterations
10   T = 100000;
11   %Time step for integration
12   del_t=0.01;
13   %Pacing period and duration of a stimulus
14   pac_period=25000;
15   dur=100; % duration of 1 ms
16
17   %MODEL PARAMETERS
18   g_Na=23.0;     %Max Na conductance (units: mO^-1.cm^-2)
19   g_Ca=0.09;     %Max Ca conductance
20   g_K=0.705;     %Max K (time dep.) conductance
21   g_K1=0.6047;   %Max K (time indep.) conductance
22   E_Na=54.4;     %Reversal potential of Na (units: mV)
23   E_K=-77;       %Reversal potential of K (time dep.)
24   E_K1=-87.26;   %Reversal potential of K (time-indep.)
25   E_Kp=-87.26;   %Reversal potential of K (plateau)
26
27   %Initial value of membrane potential
28   v=-84.0;       %mV
29   %voltage gating parameters
30   alpha_m=0.32*(v+47.13)./(1-exp(-0.1*(v+47.13)));
31   beta_m=0.08*exp(-v/11.0);
32   alpha_h=0.5*(1-sign(v+40+eps))*0.135.*exp(-1*(80+v)/6.8);
33   beta_h=(3.56*exp(0.079*v)+3.1*1e5*exp(0.35*v))*0.5.*(1-sign(v+40+eps))...
34   +(1./(0.13*(1+exp(-1*(v+10.66)/11.1))))*0.5.*(1+sign(v+40+eps));
35   alpha_j=0.5*(1-sign(v+40+eps)).*(-1.2714*1e5*exp(0.2444*v)-
36   3.474*1e-5*exp(-0.04391*v)).*(v+37.78)./(1+exp(0.311*(v+79.23)));
37   beta_j=0.5*(1-sign(v+40+eps))...
38   .*0.1212.*exp(-0.01052*v)./(1+exp(-0.1378*(v+40.14)))
30   +0.5*(1+sign(v+40+eps))*0.3.*exp(-2.535*1e-7*v)./(1+exp(-0.1*(v+32)));
```

```
40  alpha_d=0.095*exp(-0.01*(v-5))./(1+exp(-0.072*(v-5)));
41  beta_d=0.07*exp(-0.017*(v+44))./(1+exp(0.05*(v+44)));
42  alpha_f=0.012*exp(-0.008*(v+28))./(1+exp(0.15*(v+28)));
43  beta_f=0.0065*exp(-0.02*(v+30))./(1+exp(-0.2*(v+30)));
44  alpha_x=0.0005*exp(0.083*(v+50))./(1+exp(0.057*(v+50)));
45  beta_x=0.0013*exp(-0.06*(v+20))./(1+exp(-0.04*(v+20)));
46  alpha_k1=1.02./(1+exp(0.2385*(v-E_K1-59.2915)));
47  beta_k1=(0.49124*exp(0.08032*(v-E_K1+5.476))
48  +exp(0.06175*(v-E_K1-594.31)))./(1+exp(-0.5143*(v-E_K1+4.753)));
49
50  %Initial conditions: gating variables & Ca concentration
51  m=alpha_m./(alpha_m+beta_m);        %Steady state values
52  h=alpha_h./(alpha_h+beta_h);
53  j=alpha_j./(alpha_j+beta_j);
54  d=alpha_d./(alpha_d+beta_d);
55  f=alpha_f./(alpha_f+beta_f);
56  x=alpha_x./(alpha_x+beta_x);
57  Ca =2e-4; %mmol/L - resting intracellular Ca concentration
58
59  for i1=1:T,    % 1 iteration = 0.01 ms
60   %Updating v_old
61    v_old=v;
62   %Stimulation current
63    I_stim =0;
64
65  % External current applied at t = 20 ms for a duration of 1 ms
66  if(mod(i1,pac_period)>0&&mod(i1,pac_period)<=dur)
67          I_stim = 50.0; %microamp/cm^2: stimulus current
68  end;
69  E_Ca = 7.7 - 13.0287*log(Ca);    % Calcium reversal potential
70  x_i=0.5*(1+sign(v+100-eps))*2.837...
71  .*(exp(0.04*(v+77))-1)./((v+77).*exp(0.04*(v+35)))+...
72  0.5*(1-sign(v+100-eps));
73
74  %Ionic currents
75  %Fast sodium current (inward)
76   I_Na = g_Na*(m.^3)*h*j*(v-E_Na);
77  %Slow calcium current (inward)
78   I_Ca = g_Ca*d*f*(v-E_Ca);
79  %Time-dependent potassium current (outward)
80   I_K = g_K*x*x_i*(v-E_K);
81  %Time-independent K current(outward)
82   I_K1 = g_K1*(alpha_k1/(alpha_k1+beta_k1))*(v-E_K1);
83  %Plateau potassium current (outward)
84   I_Kp = 0.0183*(v-E_Kp)/(1+exp((7.488-v)/5.98));
85  %Background current (outward)
86   I_b = 0.03921*(v+59.87);
87  %Total ionic current
88   I_ion = I_Na + I_Ca + I_K + I_K1 + I_Kp + I_b;
89
90  %TAU_X AND X_INFINITY (variables for the gate ODEs)
91  tau_m=1./(alpha_m+beta_m);m_inf=alpha_m./(alpha_m+beta_m);
92  tau_h=1./(alpha_h+beta_h);h_inf=alpha_h./(alpha_h+beta_h);
93  tau_j=1./(alpha_j+beta_j);j_inf=alpha_j./(alpha_j+beta_j);
94  tau_d=1./(alpha_d+beta_d);d_inf=alpha_d./(alpha_d+beta_d);
```

```
95    tau_f=1./(alpha_f+beta_f);f_inf=alpha_f./(alpha_f+beta_f);
96    tau_x=1./(alpha_x+beta_x);x_inf=alpha_x./(alpha_x+beta_x);
97
98    del_v = -I_ion + I_stim;
99    del_m = (m_inf-m)./tau_m;%alpha_m*(1-m)-beta_m*m;
00    del_h = (h_inf-h)./tau_h;%alpha_h*(1-h)-beta_h*h;
01    del_j = (j_inf-j)./tau_j;%alpha_j*(1-j)-beta_j*j;
02    del_d = (d_inf-d)./tau_d;%alpha_d*(1-d)-beta_d*d;
03    del_f = (f_inf-f)./tau_f;%alpha_f*(1-f)-beta_f*f;
04    del_x = (x_inf-x)./tau_x;%alpha_x*(1-x)-beta_x*x;
05    del_Ca = -1e-4*I_Ca+0.07*(1e-4-Ca);
06
07    %UPDATE (forward euler)
08    v=v+del_t*del_v;m=m+del_t*del_m;h=h+del_t*del_h;j=j+del_t*del_j;
09    d=d+del_t*del_d;f=f+del_t*del_f;x=x+del_t*del_x;Ca=Ca+del_t*del_Ca;
10
11    %updating voltage gating variables
12    alpha_m = 0.32*(v+47.13)./(1-exp(-0.1*(v+47.13)));
13    beta_m = 0.08*exp(-v/11.0);
14    alpha_h = 0.5*(1-sign(v+40+eps))*0.135.*exp(-1.0*(80+v)/6.8);
15    beta_h=(3.56*exp(0.079*v)+3.1*1e5*exp(0.35*v))*0.5.*(1-sign(v+40+eps))...
16    +(1./(0.13*(1+exp(-1*(v+10.66)/11.1))))*0.5.*(1+sign(v+40+eps));
17    alpha_j = 0.5*(1-sign(v+40+eps)).*(-1.2714*1e5*exp(0.2444*v)...
18    -3.474*1e-5*exp(-0.04391*v)).*(v+37.78)./(1+exp(0.311*(v+79.23)));
19    beta_j = 0.5*(1-sign(v+40+eps)).*0.1212...
20    .*exp(-0.01052*v)./(1+exp(-0.1378*(v+40.14)))...
21    + 0.5*(1+sign(v+40+eps))*0.3.*exp(-2.535*1e-7*v)./(1+exp(-0.1*(v+32)));
22    alpha_d = 0.095*exp(-0.01*(v-5))./(1+exp(-0.072*(v-5)));
23    beta_d = 0.07*exp(-0.017*(v+44))./(1+exp(0.05*(v+44)));
24    alpha_f = 0.012*exp(-0.008*(v+28))./(1+exp(0.15*(v+28)));
25    beta_f = 0.0065*exp(-0.02*(v+30))./(1+exp(-0.2*(v+30)));
26    alpha_x = 0.0005*exp(0.083*(v+50))./(1+exp(0.057*(v+50)));
27    beta_x = 0.0013*exp(-0.06*(v+20))./(1+exp(-0.04*(v+20)));
28    alpha_k1 = 1.02./(1+exp(0.2385*(v-E_K1-59.2915)));
29    beta_k1 = (0.49124*exp(0.08032*(v-E_K1+5.476))...
30    + exp(0.06175*(v-E_K1-594.31)))./(1+exp(-0.5143*(v-E_K1+4.753)));
31
32    %Measuring the time when the voltage crosses a threshold value
33    %In this case we set it to be -60 mV
34    if((v>(-60) && v_old<=(-60))||(v<=-60 && v_old>(-60)))
35        rec_v_counter(i1)=i1;
36    end;
37
38    %Recording the time series of the voltage and currents
39        rec_I_Na(i1)=I_Na; %Sodium current
40        rec_I_Ca(i1)=I_Ca; %Slow inward calcium current
41        rec_I_K(i1)=I_K;   %Time dependent potassium current
42        rec_I_K1(i1)=I_K1; % Time independent potassium current
43        rec_I_Kp(i1)=I_Kp; % Plateau current
44        rec_I_b(i1)=I_b; % Background current
45        rec_I_ion(i1)=I_ion; % Total ionic current
46        rec_Ca(i1)=Ca; % Intracellular calcium concentration
47        rec_v(i1)=v;
48    end; % end of time loop
49    %Plotting profile of the action potential
```

```
150   figure, plot(rec_v,'b')
151   %Calculating the APD and DI from the time series
152   %the odd numbers correspond to action potential duration
153   % APD and the even correspond to diastolic interval (DI)
154   A_I=diff(find(rec_v_counter));
```

13.4 Spiral waves and spatiotemporal chaos in Panfilov model

```
1    % Program for creating a spiral wave and allowing it to breakup into
2    % a chaotic state using the Panfilov model
3     size1=256; % Size of the system
4     T=80000; %Number of iterations
5     %Space and time steps
6     dx=0.5;
7     dt=0.022;
8    %Boundary conditions - no flux
9    n=[1 1:(size1-1)];
10   s=[2:size1 size1];
11   e=n;
12   w=s;
13
14   %Diffusion coeffecient
15   diffcoff = 1;
16
17   %Model parameters
18   x1=0.0026;x2=0.837;
19   y1=1.8;
20   silon1=1/75;silon2=1;silon3= 0.3;%0.1<silon3<2
21   c1=20;c2=3;c3=15;
22   a=0.06;
23   k1=3.0;
24
25   %Initial conditions that sets up a plane wavefront
26    u=zeros(size1,size1);v=zeros(size1,size1);
27    u(1:size1,1:2)=1;
28
29   j1=1; % counter for the frame loop
30   for i1=1:T,
31       %Creating a broken wavefront
32       if(i1==800)
33          u(64:size1,1:size1)=0.0;
34       end;
35
36       % Spatial term
37       spat=diffcoff*(u(n,:)+u(s,:)+u(:,e)+u(:,w)-4*u)/(dx*dx);
38
39       %Reaction terms for the fast and slow variables
40        del_u_1=-c1*0.5*u.*(1-sign(u-x1));
41        del_u_2=0.5*(1-sign(u-x1).*sign(u-x2)).*(c2*u-a);
```

```
42    del_u_3=-0.5*c3*(1-sign(x2-u)).*(u-1);
43    del_u=del_u_1+del_u_2+del_u_3-v;
44
45    del_v_1=silon3*0.5*(1+sign(sign(x1-u)+sign(y1-v)-0.5)).*(k1*u-v);
46    del_v_2=silon1*0.25*(1-sign(del_v_1-0.0001)).*(1-sign(u-x2)).*(k1*u-v);
47    del_v_3=silon2*0.5*(1-sign(x2-u)).*(k1*u-v);
48    del_v=del_v_1+del_v_2+del_v_3;
49
50    %Forward Euler
51      unew=u+dt*(spat+del_u);
52      vnew=v+dt*del_v;
53
54    %Creating frames capturing the pseudocolor plots of fast variable
55      if(i1==100*(j1-1)+1)
56        imagesc(unew),caxis([-0.2 1.2])
57        urec(j1)=getframe;
58        j1=j1+1;
59      end;
60
61    %Update the variables
62      u=unew;
63      v=vnew;
64 end;
65 % saving the data for u and v variables
66    save('Spiral_chaos_t.mat','u','v');
67 %Run the movie created by using the command
68 figure, movie(urec)
```

13.5   Controling chaos in Panfilov model using a wave of control

```
1  % Program for the control of spatiotemporal chaos using a wave of control
2
3  size1=256; % Size of the system
4  T=3000;  %Number of iterations
5  %Space and time steps
6  dx=0.5;dt=0.022;
7  %Boundary conditions - no flux
8  n=[1 1:(size1-1)];
9  s=[2:size1 size1];
10 e=n;w=s;
11 %Diffusion coeffecient
12 diffcoff = 1;
13
14 %Model parameters
15 x1=0.0026;x2=0.837;
16 y1=1.8;
17 silon1=1/75;silon2=1;silon3= 0.3;  %0.1<silon3<2
18 c1=20;c2=3;c3=15;
19 a=0.06;
```

```
20  k1=3.0;
21
22  %Control parameters
23  stim =3.8;%Strength of stimulation
24  dur=125; %Duration of control signal at every control point
25  c= 0.1586; % Velocity of the control wave
26  d = c*dur; % Width of the region over which control is applied
27  A=zeros(256,256);
28
29  %Defining the control grid over which the wave of control is applied
30  for i1=6:6:size1-1,
31      for j1=6:6:size1-1,
32          q=i1;q1=j1;
33          A(q,q1)=stim;A(q,q1+1)=stim;A(q+1,q1)=stim;
34          A(q,q1-1)=stim;A(q-1,q1)=stim;A(q+1,q1+1)=stim;
35          A(q-1,q1-1)=stim;A(q+1,q1-1)=stim;A(q-1,q1+1)=stim;
36      end;
37  end;
38  %Determining the shape and direction of the control wave
39  for i1= 1:256,
40      for j1= 1:256,
41          B(i1,j1)= i1*i1 + j1*j1;
42      end;
43  end;
44
45  %Initial conditions is obtained from running the program
46  %Panfilov_2d_spiralchaos.m
47  load Spiral_initial.mat
48
49  j1=1; %initialising the frames
50  %Time loop
51  for i1=1:T,
52      % Creating the wave of control
53      r2 = c*i1;
54      r1 = r2 - d;
55      istim = 0.25*(1+sign(B-(r1*r1))).*(1+sign((r2*r2)-B)).*A;
56      %Spatial term
57      spat=diffcoff*(u(n,:)+u(s,:)+u(:,e)+u(:,w)-4*u)/(dx*dx);
58      %Reaction terms for the fast and slow variables
59      del_u_1=-c1*0.5*u.*(1-sign(u-x1));
60      del_u_2=0.5*(1-sign(u-x1).*sign(u-x2)).*(c2*u-a);
61      del_u_3=-0.5*c3*(1-sign(x2-u)).*(u-1);
62      del_u=del_u_1+del_u_2+del_u_3-v;
63
64      del_v_1=silon3*0.5*(1+sign(sign(x1-u)+sign(y1-v)-0.5))...
65      .*(k1*u-v);
66      del_v_2=silon1*0.25*(1-sign(del_v_1-0.0001))...
67      .*(1-sign(u-x2)).*(k1*u-v);
68      del_v_3=silon2*0.5*(1-sign(x2-u)).*(k1*u-v);
69      del_v=del_v_1+del_v_2+del_v_3;
70
71      %Forward Euler
72      unew=u+dt*(spat+del_u+istim);
73      vnew=v+dt*del_v;
74
```

```
75      %Creating frames capturing the pseudocolor plots of fast variable
76        if(i1==100*(j1-1)+1)
77          imagesc(unew),caxis([-0.2 1.2])
78          urec1(j1)=getframe;%Stores the different frames
79          j1=j1+1;
80        end;
81        %Update the variables
82          u=unew;v=vnew;
83    end;
84    %Run the movie created by using the command
85    figure, movie(urec1)
```

Bibliography

J. A. Abildskov and R. L. Lux. Mechanisms in the interruption of reentrant tachycardia by pacing. *J. Electrocardiology*, 28:107, 1995.

K. Agladze, J. P. Keener, S. C. Muller, and A. V. Panfilov. Rotating spiral waves created by geometry. *Science*, 264:1746, 1994.

K. Agladze, M. W. Kay, V. Krinsky, and N. Sarvazyan. Interaction between spiral and paced waves in cardiac tissue. *Am. J. Physiol. Heart Circ. Physiol.*, 293:H503, 2007.

F. G. Akar and D. S. Rosenbaum. Transmural electrophysiological heterogeneities underlying arrhythmogenesis in heart failure. *Circ. Res.*, 93:638, 2003.

R. R. Aliev and A. V. Panfilov. A simple two-variable model of cardiac excitation. *Chaos, Solitons and Fractals*, 7:293, 1996.

M. Allessie, F. I. Bonke, and F. J. Schopman. Circus movement in rabbit atrial muscle as a mechanism of tachycardia. iii. the "leading circle" concept: a new model of circus movement in cardiac tissue without the involvement of an anatomical obstacle. *Circ. Res.*, 41:9, 1977.

S. Alonso, I. Sendina-Nadal, V. Pérez-Munuzuri, J. M. Sancho, and F. Sagués. Regular wave propagation out of noise in chemical active media. *Phys. Rev. Lett.*, 87:078302, 2001.

S. Alonso, F. Sagués, and A. S. Mikhailov. Taming Winfree turbulence of scroll waves in excitable media. *Science*, 299:1722, 2003.

S. Alonso, J. M. Sancho, and F. Sagués. Suppression of scroll wave turbulence by noise. *Phys. Rev. E*, 70:067201, 2004.

S. Alonso, F. Sagués, and A. S. Mikhailov. Periodic forcing of scroll rings and control of Winfree turbulence in excitable media. *Chaos*, 16:023124, 2006.

S. A. Alonso and A. V. Panfilov. Negative filament tension at high excitability in a model of cardiac tissue. *Phys. Rev. Lett.*, 100:218101, 2008.

C. Anderson and N. A. Trayanova. Success and failure of biphasic shocks: results of bidomain simulations. *Mathematical Biosciences*, 174(2):91, 2001.

I. S. Aranson and L. Kramer. The world of the complex Ginzburg–Landau equation. *Rev. Mod. Phys.*, 74:99, 2002.

A. Atri, J. Amundson, D. Clapham, and J. Sneyd. A single-pool model for in-tracellular calcium oscillations and waves in the *xenopus laevis* oocyte. *Bio-physical Journal*, 65(4):1727, 1993.

I. Banville and R. A. Gray. Effect of action potential duration and conduction velocity restitution and their spatial dispersion on alternans and the stability of arrhythmias. *J. Cardiovasc.Electrophysiol.*, 13:1141, 2002.

M. Bär, M. Falcke, and M. Or-Guil. Mechanisms of spiral breakup in chemi-cal and biological reaction-diffusion models. In S. C. Müller, J. Parisi, and W. Zimmermann, editors, *Transport and Structure: Their Competitive Roles in Biophysics and Chemistry*, pages 326–348. Springer, Berlin, 1999.

D. Barkley. Euclidean symmetry and the dynamics of rotating spiral waves. *Phys. Rev. Lett.*, 72:164, 1994.

D. Barkley, M. Kness, and L. S. Tuckerman. Spiral-wave dynamics in a simple model of excitable media: The transition from simple to compound rotation. *Phys. Rev. A*, 42:2489, 1990.

B. P. Belousov. A periodic reaction and its mechanism. *Compilation of Ab-stracts on Radiation Medicine*, 147(1), 1959.

B. Bengtsson, E. M. Chow, and J. M. Marshall. Activity of circular muscle of rat uterus at different times in pregnancy. *Am. J. Physiol. Cell. Physiol.*, 246(3):C216, 1984.

O. Berenfeld and A. Pertsov. Dynamics of intramural scroll waves in three-dimensional continuous myocardium with rotational anisotropy. *J. Theor. Biol.*, 199:383, 1999.

M. J. Berridge, M. D. Bootman, and H. L. Roderick. Calcium signalling: dynam-ics, homeostasis and remodelling. *Nature Reviews Molecular Cell Biology*, 4 (7):517, 2003.

D. M. Bers. Cardiac excitation contraction coupling. *Nature (London)*, 415 (6868):198, 2002.

V. N. Biktashev and J. Brindley. Phytoplankton blooms and fish recruitment rate: effects of spatial distribution. *Bulletin of Mathematical Biology*, 66(2): 233, 2004.

V. N. Biktashev and A. V. Holden. Reentrant waves and their elimination in a model of mammalian ventricular tissue. *Chaos*, 8:48, 1998.

V. N. Biktashev, A. V. Holden, and H. Zhang. Tension of organizing filaments of scroll waves. *Phil. Trans. Roy. Soc. Lond. A*, 347:611, 1994.

V. N. Biktashev, J. Brindley, A. V. Holden, and M. A. Tsyganov. Pursuit-evasion predator-prey waves in two spatial dimensions. *Chaos*, 14(4):988, 2004.

V. N. Biktashev, A. Arutunyan, and N. A. Sarvazyan. Generation and escape of local waves from the boundary of uncoupled cardiac tissue. *Biophys. J.*, 94:3726, 2008.

K. Binder and A. P. Young. Spin glasses: experimental facts, theoretical concepts, and open questions. *Rev. Mod. Phys.*, 58:801, 1986.

S. T. Blackburn. *Maternal, Fetal and Neonatal Physiology: A Clinical Perspective*. Saunders-Elsevier, St. Louis, MO, 2007.

G. Blatter, M. V. Feigel'man, V. B. Geshkenbein, A. I. Larkin, and V. M. Vinokur. Vortices in high-temperature superconductors. *Rev. Mod. Phys.*, 66:1125, 1994.

S. Boccaletti, C. Grebogi, Y-C Lai, H. Mancini, and D. Maza. The control of chaos: theory and applications. *Physics Reports*, 329(3):103, 2000.

M. C. Boerlijst and P. Hogeweg. Spatial gradients enhance persistence of hypercycles. *Physica D*, 88:29, 1995.

R. A. Brauch, M. Adnan El-Masri, J. C. Parker Jr., and R. S. El-Mallakh. Glial cell number and neuron/glial cell ratios in postmortem brains of bipolar individuals. *Journal of Affective Disorders*, 91:87, 2006.

W. C. Bray. A periodic reaction in homogeneous solution and its relations to catalysis. *J. Am. Chem. Soc.*, 43(6):1262, 1921.

J. Breuer and S. Sinha. Controlling spatiotemporal chaos in excitable media by local biphasic stimulation. eprint arXiv:nlin/0406047v1, 2004.

J. Breuer and S. Sinha. Death, dynamics and disorder: Terminating reentry in excitable media by dynamically-induced inhomogeneities. *Pramana*, 64:553, 2005.

J. Brindley, V. H. Biktashev, and M. A. Tsyganov. Invasion waves in populations with excitable dynamics. *Biological Invasions*, 7(5):807, 2005.

G. Bub, A. Shrier, and L. Glass. Spiral wave generation in heterogeneous excitable media. *Phys. Rev. Lett.*, 88:058101, 2002.

B. Burstein and S. Nattel. Atrial fibrosis: mechanisms and clinical relevance in atrial fibrillation. *Journal of the American College of Cardiology*, 51(8):802, 2008.

C. Cabo, A. M. Pertsov, W. T. Baxter, J. M. Davidenko, R. A. Gray, and J. Jalife. Wave-front curvature as a cause of slow conduction and block in isolated cardiac muscle. *Circ. Res.*, 75(6):1014, 1994.

J. M. Cao, Z. Qu, Y. H. Kim, T. J. Wu, A. Garfinkel, J. N. Weiss, and P. S. Chen. Spatiotemporal heterogeneity in the induction of ventricular fibrillation by rapid pacing importance of cardiac restitution properties. *Circ. Res.*, 84: 1318, 1999.

E. Carafoli. Calcium signaling: a tale for all seasons. *Proc. Natl. Acad. Sci. USA*, 99(3):1115, 2002.

E. Carafoli, L. Santella, D. Branca, and M. Brini. Generation, control, and processing of cellular calcium signals. *Critical Reviews in Biochemistry and Molecular Biology*, 36(2):107, 2001.

Cardiac Arrhythmia Suppression Trial (CAST). Investigators. preliminary report: Effect of encainide and flecainide on mortality in a randomized trial of arrhythmia suppression after myocardial infarction. *N. Engl. J. Med.*, 321 (6):406, 1989.

A. C. Charles. Glia-neuron intercellular calcium signaling. *Developmental Neuroscience*, 16:196, 1994.

P. S. Chen, P. D. Wolf, E. G. Dixon, N. D. Danieley, D. W. Frazier, W. M. Smith, and R. E. Ideker. Mechanism of ventricular vulnerability to single premature stimuli in open-chest dogs. *Circ. Res.*, 62:1191, 1988.

P. S. Chen, D. O. Walter, H. S. Karagueuzian, B. Kogan, S. J. Evans, M. Karpoukhin, C. Hwang, T. Uchida, M. Gotoh, O. Nwasokwa, P. Sager, and J. N. Weiss. Quasiperiodicity and chaos in cardiac fibrillation. *Journal of Clinical Investigation*, 99(2):305, 1997.

W. Chen, S. C. Cheng, E. Avalos, O. Drugova, G. Osipov, P. Y. Lai, and C. K. Chan. Synchronization in growing heterogeneous media. *Europhys. Lett.*, 86: 18001, 2009.

D. R. Chialvo. Generic excitable dynamics on a two-dimensional map. *Chaos, Solitons & Fractals*, 5(3):461, 1995.

T. R. Chigwada, P. Parmananda, and K. Showalter. Resonance pacemakers in excitable media. *Phys. Rev. Lett.*, 96:244101, 2006.

L. Chilton, W. R. Giles, and G. L. Smith. Evidence of intercellular coupling between co-cultured adult rabbit ventricular myocytes and myofibroblasts. *J. Physiology*, 583(1):225, 2007.

G. Christakos, R. A. Olea, and H-L Yu. Recent results on the spatiotemporal modelling and comparative analysis of black death and bubonic plague epidemics. *Public Health*, 121(9):700, 2007.

D. J. Christini and J. J. Collins. Using noise and chaos control to control nonchaotic systems. *Phys. Rev. E*, 52:5806, 1995.

D. J. Christini and J. J. Collins. Control of chaos in excitable physiological systems: a geometric analysis. *Chaos*, 7(4):544, 1997.

D. J. Christini and L. Glass. Introduction: Mapping and control of complex cardiac arrhythmias. *Chaos*, 12:732, 2002.

D. J. Christini, K. M. Stein, S. M. Markowitz, S. Mittal, D. J. Slotwiner, M. A. Scheiner, S. Iwai, and B. B. Lerman. Nonlinear-dynamical arrhythmia control in humans. *Proc. Natl. Acad. Sci. USA*, 98:5827, 2001.

D. J. Christini, M. L. Riccio, C. A. Culianu, J. J. Fox, A. Karma, and R. F. Gilmour. Control of electrical alternans in canine cardiac Purkinje fibers. *Phys. Rev. Lett.*, 96:104101, 2006.

R. H. Clayton and A. V. Panfilov. A guide to modelling cardiac electrical activity in anatomically detailed ventricles. *Progress in Biophysics and Molecular Biology*, 96(1):19, 2008.

R. H. Clayton, A. V. Holden, and W. C. Tong. Can endogenous, noise-triggered early after-depolarizations initiate reentry in a modified Luo–Rudy ventricular virtual tissue? *Int. J. Bif. Chaos*, 13:3835, 2003.

R. H. Clayton, O. Bernus, E. M. Cherry, H. Dierckx, F. H. Fenton, L. Mirabella, A. V. Panfilov, F. B. Sachse, G. Seemann, and H. Zhang. Models of cardiac tissue electrophysiology: progress, challenges and open questions. *Progress in Biophysics and Molecular Biology*, 104(1):22, 2011.

S. M. Cobbe and A. C. Rankin. Cardiac arrythmias. In D. A. Warrell, T. M. Cox, J. D. Firth, and E. J. Benz, editors, *Oxford Textbook of Medicine*, chapter 15. Oxford University Press, Oxford, 2005.

P. Comtois and A. Vinet. Curvature effects on activation speed and repolarization in an ionic model of cardiac myocytes. *Phys. Rev. E.*, 60:4619, 1999.

P. Comtois and A. Vinet. Resetting and annihilation of reentrant activity in a model of a one-dimensional loop of ventricular tissue. *Chaos*, 12:903, 2002.

J. Coromilas, A. E. Saltman, B. Waldecker, S. M. Dillon, and A. L. Wit. Electrophysiological effects of flecainide on anisotropic conduction and reentry in infarcted canine hearts. *Circulation*, 91(8):2245, 1995.

M. Courtmanche. Complex spiral wave dynamics in a spatially distributed ionic model of cardiac electrical activity. *Chaos*, 6:579, 1996.

M. Cross and H. Greenside. *Pattern Formation and Dynamics in Nonequilibrium Systems*. Cambridge University Press, Cambridge, 2009.

M. C. Cross and P. C. Hohenberg. Pattern formation outside of equilibrium. *Rev. Mod. Phys.*, 65:851, 1993.

I. A. Csapo and H. A. Kuriyama. Effects of ions and drugs on cell membrane activity and tension in the postpartum rat myometrium. *J. Physiol.*, 165: 575, 1963.

E. Cytrynbaum and J. P. Keener. Stability conditions for the traveling pulse: modifying the restitution hypothesis. *Chaos*, 12:788, 2002.

M. A. Dahlem, R. Graf, A. J. Strong, J. P. Dreier, Y. A. Dahlem, M. Sieber, W. Hanke, K. Podoll, and E. Schöll. Two-dimensional wave patterns of spreading depolarization: retracting, re-entrant, and stationary waves. *Physica D*, 239(11):889, 2010.

J. M. Davidenko, A. V. Pertsov, R. Salomonsz, W. Baxter, and J. Jalife. Stationary and drifting spiral waves of excitation in isolated cardiac muscle. *Nature*, 355:349, 1992.

J. M. Davidenko, R. Salomonsz, A. M. Pertsov, W. T. Baxter, and J. Jalife. Effects of pacing on stationary reentrant activity theoretical and experimental study. *Circ. Res.*, 77:1166, 1995.

C. de Chillou, D. Lacroix, D. Klug, I. Magnin-Poull, C. Marquie, M. Messier, M. Andronache, C. Kouakam, N. Sadoul, J. Chen, E. Aliot, and S. Kacet. Isthmus characteristics of reentrant ventricular tachycardia after myocardial infarction. *Circulation*, 105:726, 2002.

G. W. de Young and J. Keizer. A single-pool inositol 1, 4, 5-trisphosphate-receptor-based model for agonist-stimulated oscillations in ca2+ concentration. *Proc. Nat. Acad. Sci. USA*, 89(20):9895, 1992.

K. Devlin. *The Man of Numbers: Fibonacci's Arithmetic Revolution.* Bloomsbury Publishing, London, 2011.

W. L. Ditto, M. L. Spano, V. In, J. Neff, B. Meadows, J. J. Langberg, A. Bolmann, and K. McTeague. Control of human atrial fibrillation. *Int. J. Bif. Chaos.*, 10:593, 2000.

R. A. Duquette, A. Shmygol, C. Vaillant, A. Mobasheri, M. Pope, T. Burdyga, and S. Wray. Vimentin-positive, c-kit-negative interstitial cells in human and rat uterus: a role in pacemaking? *Biol. Reproduction*, 72:276, 2005.

B. Echebarria and A. Karma. Instability and spatiotemporal dynamics of alternans in paced cardiac tissue. *Phys. Rev. Lett.*, 88:208101, 2002a.

B. Echebarria and A. Karma. Spatiotemporal control of cardiac alternans. *Chaos*, 12:923, 2002b.

I. R. Efimov, V. Sidorov, Y. Cheng, and B. Wollenzier. Evidence of three-dimensional scroll waves with ribbon-shaped filament as a mechanism of ventricular tachycardia in the isolated rabbit heart. *J. Cardiovasc. Electrophysiol.*, 10:1452, 1999.

V. M. Eguiluz, D. R. Chialvo, G. A. Cecchi, M. Baliki, and A. V. Apkarian. Scale-free brain functional networks. *Phys. Rev. Lett.*, 94(1):018102, 2005.

N. El-Sherif, A. Smith, and K. Evans. Canine ventricular arrhythmias in the late myocardial infarction period. 8. epicardial mapping of reentrant circuits. *Circ. Res.*, 49:255, 1981.

I. R. Epstein and J. A. Pojman. *An Introduction to Nonlinear Chemical Dynamics: Oscillations, Waves, Patterns, and Chaos*. Oxford University Press, New York, 1998.

D. E. Euler. Cardiac alternans: mechanisms and pathophysiological significance. *Cardiovascular Research*, 42(3):583, 1999.

I. V. Everett, H. Thomas, and J. E. Olgin. Atrial fibrosis and the mechanisms of atrial fibrillation. *Heart Rhythm*, 4(3):S24, 2007.

M. Falcke and H. Engel. Influence of global coupling through the gas phase on the dynamics of co oxidation on Pt (110). *Phys. Rev. E.*, 50:1353, 1994.

V. G. Fast and A. G. Kleber. Role of wavefront curvature in propagation of cardiac impulse. *Cardiovasc. Res.*, 33:258, 1997.

V. G. Fast and A. M. Pertsov. Drift of vortex in the myocardium. *Biophysics*, 35:478, 1990.

H. Fei, M. S. Hanna, and L. H. Frame. Assessing the excitable gap in reentry by resetting implications for tachycardia termination by premature stimuli and antiarrhythmic drugs. *Circulation*, 94:2268, 1996.

F. H. Fenton and A. Karma. Vortex dynamics in three-dimensional continuous myocardium with fiber rotation: filament instability and fibrillation. *Chaos*, 8:20, 1998.

F. H. Fenton, S. J. Evans, and H. M. Hastings. Memory in an excitable medium: a mechanism for spiral wave breakup in the low-excitability limit. *Phys. Rev. Lett.*, 83(19):3964, 1999.

F. H. Fenton, E. M. Cherry, H. M. Hastings, and S. J. Evans. Multiple mechanisms of spiral wave breakup in a model of cardiac electrical activity. *Chaos*, 12:852, 2002.

F. H. Fenton, S. Luther, E. M. Cherry, N. F. Otani, V. Krinsky, A. Pumir, E. Bodenschatz, and R. F. Gilmour. Termination of atrial fibrillation using pulsed low-energy far-field stimulation. *Circulation*, 120:467, 2009.

L. Fibonacci. *Fibonacci's Liber abaci: A Translation into Modern English of Leonardo Pisano's Book of Calculation*. Springer, New York, 2003.

R. J. Field and R. M. Noyes. Oscillations in chemical systems. iv: Limit cycle behavior in a model of a real chemical reaction. *Journal of Chemical Physics*, 60(5):1877, 1974.

R. J. Field, E. Körös, and R. M. Noyes. Oscillations in chemical systems. ii: Thorough analysis of temporal oscillation in the bromate-cerium-malonic acid system. *Journal of the American Chemical Society*, 94(25):8649, 1972.

J. D. Fisher, R. Mehra, and S. Furman. Termination of ventricular tachycardia with bursts of rapid ventricular pacing. *Am. J. Card.*, 41:94, 1978.

R. FitzHugh. Thresholds and plateaus in the Hodgkin–Huxley nerve equations. *Journal of General Physiology*, 43(5):867, 1960.

R. FitzHugh. Impulses and physiological states in theoretical models of nerve membrane. *Biophysical Journal*, 1(6):445, 1961.

P. Foerster, S. C. Müller, and B. Hess. Curvature and propagation velocity of chemical waves. *Science*, 241:685, 1988.

P. Foerster, S. C. Müller, and B. Hess. Curvature and spiral geometry in aggregation patterns of *Dictyostelium discoideum*. *Development*, 109:11, 1990.

R. N. Fogoros. *Antiarrhythmic Drugs: A Practical Guide*. John Wiley & Sons, 2008.

J. J. Fox, E. Bodenschatz, and R. F. Gilmour. Period-doubling instability and memory in cardiac tissue. *Phys. Rev. Lett.*, 89(13):138101, 2002a.

J. J. Fox, R. F. Gilmour, and E. Bodenschatz. Conduction block in one-dimensional heart fibers. *Phys. Rev. Lett.*, 89:198101, 2002b.

J. J. Fox, M. L. Riccio, F. Hua, E. Bodenschatz, and R. F. Gilmour. Spatiotemporal transition to conduction block in canine ventricle. *Circ. Res.*, 90:289, 2002c.

A. L. Fradkov and A. Yu. Pogromsky. Nonlinear and adaptive control of chaos. In E. Schöll and H-G. Schuster, editors, *Handbook of Chaos Control*, chapter 7, page 129. Wiley-VCH Verlag, Weinham, 2 edition, 2007.

D. W. Frazier, P. D. Wolf, J. M. Wharton, A. S. L. Tang, W. M. Smith, and R. E. Ideker. Stimulus-induced critical point: mechanism for the electrical initiation of reentry in normal canine myocardium. *J. Clin. Invest.*, 83:1039, 1989.

J. García-Ojalvo and L. Schimansky-Geier. Noise-induced spiral dynamics in excitable media. *Europhysics Letters*, 47(3):298, 1999.

R. E. Garfield and W. L. Maner. Physiology and electrical activity of uterine contractions. *Sem. Cell Dev. Biol.*, 18:289, 2007a.

R. E. Garfield and W. L. Maner. Physiology and electrical activity of uterine contractions. *Sem. Cell. Dev. Biol.*, 18(3):289, 2007b.

R. E. Garfield, G. Saade, C. Buhimschi, I. Buhimschi, L. Shi, S. Q. Shi, and K. Chwalisz. Control and assessment of the uterus and cervix during pregnancy and labour. *Human Reproduction Update*, 4:673, 1998.

A. Garfinkel and Z. Qu. Nonlinear dynamics of excitation and propagation in cardiac muscle. cardiac electrophysiology. In D. P. Zipes and J. Jalife, editors, *Cardiac Electrophysiology: From Cell to Bedside*, chapter 36, page 327. Saunders, Philadelphia, 2004.

A. Garfinkel, M. L. Spano, W. L. Ditto, and J. N. Weiss. Controlling cardiac chaos. *Science*, 257:1230, 1992.

A. Garfinkel, Y. H. Kim, O. Voroshilovsky, Z. Qu, J. R. Kil, M. H. Lee, H. S. Karagueuzian, J. N. Weiss, and P. S. Chen. Preventing ventricular fibrillation by flattening cardiac restitution. *Proc. Nat. Acad. Sci. USA*, 97:6061, 2000.

D. J. Gauthier, G. M. Hall, R. A. Oliver, E. G. Dixon-Tulloch, P. D. Wolf, and S. Bahar. Progress toward controlling in vivo fibrillating sheep atria using a nonlinear-dynamics-based closed-loop feedback method. *Chaos*, 12:952, 2002.

L. A. Geddes, W. A. Tacker, J. Mcfarlane, and J. Bourland. Strength-duration curves for ventricular defibrillation in dogs. *Circ. Res.*, 27:551, 1970.

M. U. Gillette and T. J. Sejnowski. Biological clocks coordinately keep life on time. *Science*, 309:1196, 2005.

B. T. Ginn and O. Steinbock. Quantized spiral tip motion in excitable systems with periodic heterogeneities. *Phys. Rev. Lett.*, 93:158301, 2004.

S. D. Girouard and D. S. Rosenbaum. Role of wavelength adaptation in the initiation, maintenance, and pharmacologic suppression of reentry. *J. Cardiovasc. Electrophys.*, 12:697, 2001.

L. Glass. Patterns of supernumerary limb regeneration. *Science*, 198:321, 1977.

L. Glass. Synchronization and rhythmic processes in physiology. *Nature(London)*, 410:277, 2001.

L. Glass and M. E. Josephson. Resetting and annihilation of reentrant abnormally rapid heartbeat. *Phys. Rev. Lett.*, 75:2059, 1995.

L. Glass, Y. Nagai, K. Hall, M. Talajic, and S. Nattel. Predicting the entrainment of reentrant cardiac waves using phase resetting curves. *Physics Today*, 65:021908, 2002.

M. R. Gold and S. R. Shorofsky. Strength-duration relationship for human transvenous defibrillation. *Circulation*, 96:3517, 1997.

M. Golubitsky, I. Stewart, P.-L. Buono, and J. J. Collins. Symmetry in loco-motor central pattern generators and animal gaits. *Nature (London)*, 401: 693, 1999.

N. A. Gorelova and J. Bureš. Spiral waves of spreading depression in the isolated chicken retina. *Journal of Neurobiology*, 14(5):353, 1983.

G. Gottwald, A. Pumir, and V. Krinsky. Spiral wave drift induced by stimu-lating wave trains. *Chaos*, 11:487, 2001a.

G. Gottwald, A. Pumir, and V. Krinsky. Spiral wave drift induced by stimu-lating wave trains. *Chaos*, 11:487, 2001b.

R. A. Gray. Termination of spiral wave breakup in a FitzHugh–Nagumo model via short and long duration stimuli. *Chaos*, 12:941, 2002.

R. A. Gray and N. Chattipakorn. Termination of spiral waves during cardiac fibrillation via shock-induced phase resetting. *Proc. Natl. Acad. Sci. USA*, 102:4672, 2005.

R. A. Gray, J. Jalife, A. V. Panfilov, W. T. Baxter, C. Cabo, J. M. Davidenko, and A. M. Pertsov. Nonstationary vortexlike reentrant activity as a mech-anism of polymorphic ventricular tachycardia in the isolated rabbit heart. *Circulation*, 91:2454, 1995a.

R. A. Gray, J. Jalife, A. V. Panfilov, W. T. Baxter, C. Cabo, J. M. Davidenko, A. M. Pertsov, A. V. Panfiov, P. Hogeweg, and A. T. Winfree. Mechanisms of cardiac fibrillation. *Science*, 270:1222, 1995b.

R. A. Gray, A. M. Pertsov, and J. Jalife. Spatial and temporal organization during cardiac fibrillation. *Nature (Lond.)*, 392:75, 1998.

J. M. Greenberg, B. D. Hassard, and S. P. Hastings. Pattern formation and periodic structures in systems modeled by reaction-diffusion equations. *Bull. Amer. Math. Soc.*, 84(6):1296, 1978.

J. L. Greenstein and R. L. Winslow. An integrative model of the cardiac ventricular myocyte incorporating local control of ca2+ release. *Biophysical Journal*, 83(6):2918, 2002.

G. Grégoire and H. Chaté. Onset of collective and cohesive motion. *Phys. Rev. Lett.*, 92:025702, 2004.

B. T. Grenfell, O. N. Bjørnstad, and J. Kappey. Travelling waves and spatial hierarchies in measles epidemics. *Nature*, 414(6865):716–723, 2001.

A. C. Guyton and J. E. Hall. *Textbook of Medical Physiology*. Saunders Elsevier, Philadelphia, 12 edition, 2011.

V. Hakim and N. Brunel. Fast global oscillations in networks of integrate-and-fire neurons with low firing rates. *Neural. Comput.*, 11:1621, 1999.

V. Hakim and A. Karma. Theory of spiral wave dynamics in weakly excitable media: asymptotic reduction to a kinematic model and applications. *Phys. Rev. E*, 60:5073, 1999.

G. M. Hall and D. J. Gauthier. Experimental control of cardiac muscle alternans. *Phys. Rev. Lett.*, 88:198102, 2002.

K. Hall, D. J. Christini, M. Tremblay, J. J. Collins, L. Glass, and J. Billette. Dynamic control of cardiac alternans. *Phys. Rev. Lett.*, 75:4518, 1997.

M. E. Harris-White, S. A. Zanotti, S. A. Frautschy, and A. C. Charles. Spiral intercellular calcium waves in hippocampal slice cultures. *Journal of Neurophysiology*, 79(2):1045, 1998.

M. P. Hassell, H. N. Comins, and R. M. May. Spatial structure and chaos in insect population dynamics. *Nature (London)*, 353(6341):255, 1991a.

M. P. Hassell, R. M. May, S. W. Pacala, and P. L. Chesson. The persistence of host-parasitoid associations in patchy environments. i: A general criterion. *American Naturalist*, 138:568, 1991b.

D. He, G. Hu, M. Zhan, W. Ren, and Z. Gao. Pattern formation of spiral waves in an inhomogeneous medium with small-world connections. *Phys. Rev. E.*, 65:055204, 2002.

C. S. Henriquez. Simulating the electrical behavior of cardiac tissue using the bidomain model. *Critical Reviews in Biomedical Engineering*, 21(1):1, 1992.

H. Henry. Spiral wave drift in an electric field and scroll wave instabilities. *Phys. Rev. E*, 70:026204, 2004.

H. Henry and V. Hakim. Scroll waves in isotropic excitable media: Linear instabilities, bifurcations, and restabilized states. *Phys. Rev. E*, 65:046235, 2002.

M. Hildebrand, M. Bär, and M. Eiswirth. Statistics of topological defects and spatiotemporal chaos in a reaction-diffusion system. *Phys. Rev. Lett*, 75:1503, 1995.

D. L. Hill, R. B. Daroff, A. Ducros, N. J. Newman, and V. Biousse. Most cases labeled as "retinal migraine" are not migraine. *Journal of Neuro-Ophthalmology*, 27(1):3, 2007.

C. Hirth, U. Borchard, and D. Hafner. Effects of the calcium antagonist diltiazem on action potentials, slow response and force of contraction in different cardiac tissues. *J. Mol. Cell. Cardiol.*, 15:799, 1983.

A. L. Hodgkin and A. F. Huxley. Currents carried by sodium and potassium ions through the membrane of the giant axon of loligo. *J. Physiology*, 116 (4):449, 1952a.

A. L. Hodgkin and A. F. Huxley. The components of membrane conductance in the giant axon of loligo. *J. Physiology*, 116(4):473, 1952b.

A. L. Hodgkin and A. F. Huxley. The dual effect of membrane potential on sodium conductance in the giant axon of loligo. *J. Physiology*, 116(4):497, 1952c.

A. L. Hodgkin and A. F. Huxley. A quantitative description of membrane current and its application to conduction and excitation in nerve. *J. Physiology*, 117(4):500, 1952d.

A. L. Hodgkin, A. F. Huxley, and B. Katz. Measurement of current-voltage relations in the membrane of the giant axon of loligo. *J. Physiology*, 116(4): 424, 1952.

D. A. Hooks, K. A. Tomlinson, S. G. Marsden, I. J. LeGrice, B. H. Smaill, A. J. Pullan, and P. J. Hunter. Cardiac microstructure implications for electrical propagation and defibrillation in the heart. *Circ. Res.*, 91:331, 2002.

M. Hörning, A. Isomura, K. Agladze, and K. Yoshikawa. Liberation of a pinned spiral wave by a single stimulus in excitable media. *Phys. Rev. E*, 79:026218, 2009.

L. Horwood, S. VanRiper, and T. Davidson. Antitachycardia pacing: an overview. *Am. J. Crit. Care*, 4:397, 1995.

K. A. Hossmann. Periinfarct depolarizations. *Cerebrovascular and Brain Metabolism Reviews*, 8(3):195, 1995.

Z. Hou and H. Xin. Noise-sustained spiral waves: effect of spatial and temporal memory. *Phys. Rev. Lett.*, 89(28):280601, 2002.

X. Huang, W. C. Troy, Q. Yang, H. Ma, C. R. Laing, S. J. Schiff, and J-Y. Wu. Spiral waves in disinhibited mammalian neocortex. *J. Neuroscience*, 24(44): 9897, 2004.

S.-M. Hwang, T. Y. Kim, and K. J. Lee. Complex-periodic spiral waves in confluent cardiac cell cultures induced by localized inhomogeneities. *Proc. Nat. Acad. Sci. USA.*, 102:10363, 2005.

R. E. Ideker and J. M. Rogers. Human ventricular fibrillation wandering wavelets, mother rotors, or both? *Circulation*, 114:530, 2006.

T. Ikeda, M. Yashima, T. Uchida, D. Hough, M. C. Fishbein, W. J. Mandel, P.-S. Chen, and H. S. Karagueuzian. Attachment of meandering reentrant wave fronts to anatomic obstacles in the atrium role of the obstacle size. *Circ. Res.*, 81:753, 1997.

R. Imbihl and G. Ertl. Oscillatory kinetics in heterogeneous catalysis. *Chemical Reviews*, 95(3):697, 1995.

A. Isomura, M. Horning, K. Agladze, and K. Yoshikawa. Eliminating spiral waves pinned to an anatomical obstacle in cardiac myocytes by high-frequency stimuli. *Phys. Rev. E*, 78:066216, 2008.

H. Ito and L. Glass. Spiral breakup in a new model of discrete excitable media. *Phys. Rev. Lett.*, 66:671, 1991.

V. Jacquemet. Pacemaker activity resulting from the coupling with nonexcitable cells. *Phys. Rev. E.*, 74:011908, 2006.

V. Jacquemet and C. S. Henriquez. Loading effect of fibroblast-myocyte coupling on resting potential, impulse propagation and repolarization: insights from a microstructure model. *Am. Journ. Physiol. Heart and Circ. Physiol.*, 294(5):H2040, 2008.

S. Jakubith, H. H. Rotermund, W. Engel, A. von Oertzen, and G. Ertl. Spatiotemporal concentration patterns in a surface reaction: Propagating and standing waves, rotating spirals, and turbulence. *Phys. Rev. Lett.*, 65:3013, 1990.

J. Jalife. Ventricular fibrillation: mechanisms of initiation and maintenance. *Ann. Rev. Physiol.*, 62:25, 2000.

J. Jalife and M. Delmar. Ionic basis of the Wenckebach phenomenon. In P. Hunter L. Glass and A. McCulloch, editors, *Theory of Heart*, page 359. Springer-Verlag, New York, 1991.

J. Jalife, C. Antzelevitch, V. Lamanna, and G. K. Moe. Rate-dependent changes in excitability of depressed cardiac Purkinje fibers as a mechanism of intermittent bundle branch block. *Circulation*, 67:912, 1983.

J. Jalife, R. A. Gray, G. E. Morley, and J. M. Davidenko. Self-organization and the dynamical nature of ventricular fibrillation. *Chaos*, 8:57, 1998.

J. Jalife, J. M. Anumonwo, M. Delmar, and J. M. Davidenko. *Basic Cardiac Electrophysiology for the Clinician*. Futura Publishing, Armonk, 1999.

T. Jesan. *Systems Biology from Cell to Society: Transmission Dynamics in Complex Networks with Mesoscopic Organization*. PhD thesis, Homi Bhabha National Institute, September 2013.

X. Jie, B. Rodriguez, J. R. de Groot, R. Coronel, and N. Trayanova. Reentry in survived subepicardium coupled to depolarized and inexcitable midmyocardium: insights into arrhythmogenesis in ischemia phase 1b. *Heart Rhythm*, 5:1036, 2008.

Z. A. Jimenez and O. Steinbock. Scroll wave filaments self-wrap around unexcitable heterogeneities. *Phys. Rev. E.*, 86:036205, 2012.

Z. A. Jimenez, B. Marts, and O. Steinbock. Pinned scroll rings in an excitable system. *Phys. Rev. Lett.*, 102:244101, 2009.

N. P. Johnson and J. Mueller. Updating the accounts: global mortality of the 1918-1920 "Spanish" influenza pandemic. *Bulletin of the History of Medicine*, 76(1):105, 2002.

J. L. Jones and O. H. Tovar. Electrophysiology of ventricular fibrillation and defibrillation. *Critical Care Medicine*, 28(11):N219, 2000.

P. N. Jordan and D. J. Christini. Therapies for ventricular cardiac arrhythmias. *Critical Reviews in Biomedical Engineering*, 33(6), 2005.

M. E. Josephson. *Clinical Cardiac Electrophysiology: Techniques and Interpretation*. Lea Febiger, Philadelphia, 2nd edition edition, 1993.

P. Jung and G. Mayer-Kress. Noise controlled spiral growth in excitable media. *Chaos: An Interdisciplinary Journal of Nonlinear Science*, 5(2):458, 1995a.

P. Jung and G. Mayer-Kress. Spatiotemporal stochastic resonance in excitable media. *Phys. Rev. Lett.*, 74:2130, 1995b.

M. M. Kaplan and R. G. Webster. The epidemiology of influenza. *Scientific American*, 237(6):88, 1977.

A. Karma. Spiral breakup in model equations of action potential propagation in cardiac tissue. *Phys. Rev. Lett.*, 71:1103, 1993.

A. Karma. Electrical alternans and spiral wave breakup in cardiac tissue. *Chaos*, 4:461, 1994.

A. Karma and R. F. Gilmour. Nonlinear dynamics of heart rhythm disorders. *Phys. Today*, 60(3):51, 2007.

J. Keener and J. Sneyd. *Mathematical Physiology*. Springer-Verlag, New York, 1998.

J. P. Keener. The effects of discrete gap junction coupling on propagation in myocardium. *J. Theo. Biol.*, 148:49, 1991.

J. P. Keener and K. Bogar. A numerical method for the solution of the bidomain equations in cardiac tissue. *Chaos*, 8(1):234, 1998.

J. P. Keener and T. J. Lewis. The biphasic mystery: Why a biphasic shock is more effective than a monophasic shock for defibrillation. *Journal of Theoretical Biology*, 200(1):1, 1999.

J. P. Keener and A. V. Panfilov. A biophysical model for defibrillation of cardiac tissue. *Biophys. J.*, 71:1335, 1996.

J. P. Keener and A. V. Panfilov. The effects of geometry and fibre orientation on propagation and extracellular potentials in myocardium. In A. V. Panfilov and A. V. Holden, editors, *Computational Biology of the Heart*, page 235. Wiley, New York, 1997.

W. O. Kermack and A. G. McKendrick. A contribution to the mathematical theory of epidemics. *Proceedings of Royal Society A*, 115(772):700, 1927.

Y. H. Kim, A. Garfinkel, T. Ikeda, T-J Wu, C. A. Athill, J. N. Weiss, H. S. Karagueuzian, and P-S. Chen. Spatiotemporal complexity of ventricular fibrillation revealed by tissue mass reduction in isolated swine right ventricle. further evidence for the quasiperiodic route to chaos hypothesis. *J. Clin. Invest.*, 100:2486, 1997.

Y. H. Kim, F. Xie, M. Yashima, T.-J. Wu, M. Valderrabano, M.-H. Lee, T. Ohara, O. Voroshilovsky, R. N. Doshi, and M. C. Fishbein. Role of papillary muscle in the generation and maintenance of reentry during ventricular tachycardia and fibrillation in isolated swine right ventricle. *Circulation*, 100: 1450, 1999.

C. J. H. J. Kirchhof, F. Chorro, G. J. Scheffer, J. Brugada, K. Konings, Z. Zetelaki, and M. Allessie. Regional entrainment of atrial fibrillation studied by high-resolution mapping in open-chest dogs. *Circulation*, 88(2):736, 1993.

A. G. Kleber, C. B. Riegger, and M. J. Janse. Electrical uncoupling and increase of extracellular resistance after induction of ischemia in isolated, arterially perfused rabbit papillary muscle. *Circ. Res.*, 61:271, 1987.

J. Kockskämper and L. A. Blatter. Subcellular ca2+ alternans represents a novel mechanism for the generation of arrhythmogenic ca2+ waves in cat atrial myocytes. *J. Physiology*, 545(1):65, 2002.

B. Y. Kogan, W. J. Karplus, B. S. Billett, A. T. Pang, H. S. Karagueuzian, and S. S. Khan. The simplified Fitzhugh–Nagumo model with action potential duration restitution: effects on 2-d wave propagation. *Physica D*, 50:327, 1991.

P. Kohl, A. G. Kamkin, I. S. Kiseleva, and D. Noble. Mechanosensitive fibroblasts in the sino-atrial node region of rat heart: interaction with cardiomyocytes and possible role. *Exp. Physiol.*, 79:943, 1994.

M. L. Koller, M. L. Riccio, and R. F. Gilmour. Dynamic restitution of action potential duration during electrical alternans and ventricular fibrillation. *Am. J.Physiol. Heart Circ. Physiol.*, 275:H1635, 1998.

V. Krinsky. Fibrillation in excitable media. *Problemy Kibernetiki*, 2:59, 1968.

V. Krinsky and K. Agladze. Interaction of rotating waves in an active chemical medium. *Physica D*, 8:50, 1983.

V. Krinsky and A. Pumir. Models of defibrillation of cardiac tissue. *Chaos*, 8: 188, 1998.

V. Krinsky, F. Plaza, and V. Voignier. Quenching a rotating vortex in an excitable medium. *Phys. Rev. E*, 52(3):2458, 1995.

V. I. Krinsky, E. Hamm, and V. Voignier. Dense and sparse vortices in excitable media drift in opposite directions in electric field. *Phys. Rev. Lett.*, 76:3854, 1996.

A. K. Kryukov, V. S. Petrov, L. S. Averyanova, G. V. Osipov, W. Chen, O. Drugova, and C. K. Chan. Synchronization phenomena in mixed media of passive, excitable, and oscillatory cells. *Chaos*, 18:037129, 2008.

M. Kuperman and G. Abramson. Small world effect in an epidemiological model. *Phys. Rev. Lett.*, 86(13):2909, 2001.

Y.-C. Lai and R. L. Winslow. Fractal basin boundaries in coupled map lattices. *Phys. Rev. E.*, 50:3470, 1994.

W. J. Lammers, H. Mirghani, B. Stephen, S. Dhanasekaran, A. Wahab, M. A. Al Sultan, and F. Abazer. Patterns of electrical propagation in the intact pregnant guinea pig uterus. *Am. J. Physiol. Regul. Integr. Comp. Physiol.*, 294: R919, 2008a.

W. J. E. P. Lammers. Circulating excitations and re-entry in the pregnant uterus. *Pflügers Arch.*, 433:287, 1997.

W. J. E. P. Lammers. The electrical activities of the uterus during pregnancy. *Reprod. Sci.*, 20(2):182, 2013.

W. J. E. P. Lammers, H. Mirghani, B. Stephen, S. Dhanasekaran, A. Wahab, M. A. H. Al Sultan, and F. Abazer. Patterns of electrical propagation in the intact pregnant guinea pig uterus. *Am. J. Physiol. Regul. Integr. Comp. Physiol.*, 294(3):R919, 2008b.

J. F. Landa, C. T. West, and J. B. Thiersch. Relationships between contraction and membrane electrical activity in the isolated uterus of the pregnant rat. *Am. J. Physiol.*, 196(4):905, 1959.

M. Lauritzen. Pathophysiology of the migraine aura the spreading depression theory. *Brain*, 117(1):199, 1994.

A. A. P. Leao. Pial circulation and spreading depression of activity in the cerebral cortex. *Journal of Neurophysiology*, 7(6):391, 1944a.

A. A. P. Leao. Spreading depression of activity in the cerebral cortex. *Journal of Neurophysiology*, 7:359, 1944b.

A. A. P. Leao. Further observations on the spreading depression of activity in the cerebral cortex. *Journal of Neurophysiology*, 10(6):409, 1947.

J. Lechleiter, S. Girard, E. Peralta, and D. Clapham. Spiral calcium wave propagation and annihilation in *Xenopus laevis* oocytes. *Science*, 252:123, 1991.

K. J. Lee. Wave pattern selection in an excitable system. *Phys. Rev. Lett.*, 79 (15):2907, 1997.

K. J. Lee, E. C. Cox, and R. E. Goldstein. Competing patterns of signaling activity in *Dictyostelium discoideum*. *Phys. Rev. Lett.*, 76(7):1174, 1996.

K. J. Lee, R. E. Goldstein, and E. C. Cox. Resetting wave forms in *dictyostelium* territories. *Phys. Rev. Lett.*, 87:068101, 2001.

H. Levine, I. Aranson, L. Tsimring, and T. V. Truong. Positive genetic feedback governs camp spiral wave formation in *Dictyostelium*. *Proc. Natl. Acad. Sci. USA*, 93(13):6382, 1996.

T. Lewis. *The Mechanism and Graphic Registration of the Heart Beat*. Shaw, London, 1920.

G. Li, Q. Ouyang, V. Petrov, and H. L. Swinney. Transition from simple rotating chemical spirals to meandering and traveling spirals. *Phys. Rev. Lett.*, 77:2105, 1996.

Z. Y. Lim, B. Maskara, F. Aguel, R. Emokpae, and L. Tung. Spiral wave attachment to millimeter-sized obstacles. *Circulation*, 114:2113, 2006.

R. Loch-Caruso, K. Criswell, C. Grindatti, and K. Brant. Sustained inhibition of rat myometrial gap fjunctions and contractions by lindane. *Reproductive Biology and Endocrinology*, 1(1):62, 2003.

C. Luo and Y. Rudy. A dynamic model of the cardiac ventricular action potential. i. Simulations of ionic currents and concentration changes. *Circulation Research*, 74(6):1071, 1994.

C. H. Luo and Y. Rudy. A model of the ventricular cardiac action potential: Depolarization, repolarization, and their interaction. *Circ. Res*, 68:1501, 1991.

M. F. MacDorman, W. M. Callaghan, T. J. Mathews, D. L. Hoyert, and K. D. Kochanek. Trends in preterm-related infant mortality by race and ethnicity. *International Journal of Health Services*, 37:635, 2007.

D. L. Mackas and C. M. Boyd. Spectral analysis of zooplankton spatial hetero-geneity. *Science*, 204(4388):62, 1979.

R. Majumder, A. R. Nayak, and R. Pandit. Scroll-wave dynamics in human cardiac tissue: lessons from a mathematical model with inhomogeneities and fiber architecture. *PLoS*, 6:e18052, 2011.

H. Malchow, S. Petrovskii, and A. Medvinsky. Pattern formation in models of plankton dynamics: A synthesis. *Oceanologica Acta*, 24(5):479, 2001.

M. Markus, Zh. Nagy-Ungavarai, and B. Hess. Phototaxis of spiral waves. *Science*, 257:225, 1992.

J. A. Martin, B. E. Hamilton, P. D. Sutton, S. J. Ventura, F. Menacker, S. Kirmeyer, and T. J. Mathews. Births: preliminary data for 2009. *National Vital Statistics Reports*, 57:7, 2009a.

J. A. Martin, S. Kirmeyer, M. Osterman, and R. A. Shepherd. Born a bit too early: recent trends in late preterm births. Technical report, Centers for Disease Control and Prevention National Center for Health Statistics 3311 Toledo Road, Hyattsville, Maryland 20782, USA., 2009b.

H. Martins-Ferreira, M. Nedergaard, and C. Nicholson. Perspectives on spread-ing depression. *Brain Research Reviews*, 32(1):215, 2000.

S. A. Marvel, T. Martin, C. R. Doering, D. Lusseau, and M. E. J. Newman. The small-world effect is a modern phenomenon. arXiv:1310.2636v1, 2013.

N. McHale, M. Hollywood, G. Sergeant, and K. Thornbury. Origin of sponta-neous rhythmicity in smooth muscle. *J. Physiol.*, 570(1):23, 2006.

E. Meron. Pattern formation in excitable media. *Physics Reports*, 218:1, 1992.

S. Mesiano. Myometrial progesterone responsiveness and the control of human parturition. *J. Soc. Gynec. Invest.*, 11:193, 2004.

M. Mézard, M. A. Virasoro, and G. Parisi. *Spin Glass Theory and Beyond*. World Scientific, 1987.

A. S. Mikhailov and K. Showalter. Control of waves, patterns and turbulence in chemical systems. *Physics Reports*, 425(2):79, 2006.

A. S. Mikhailov, V. A. Davydov, and V. S. Zykov. Complex dynamics of spiral waves and motion of curves. *Physica D*, 70:1, 1994.

S. M. Miller, R. E. Garfield, and E. E. Daniel. Improved propagation in my-ometrium associated with gap junctions during parturition. *Am. J. Physiol. Cell. Physiol.*, 256:130, 1989.

G. R. Mines. On dynamic equilibrium in the heart. *J. Physiol.(London)*, 46: 349, 1913.

H. Miyoshi, M. B. Boyle, L. B. MacKay, and R. E. Garfield. Voltage-clamp studies of gap junctions between uterine muscle cells during term and preterm labor. *Biophys. J.*, 71:1324, 1996.

K. G. Moe, W. C. Rheinboldt, and J. A. Abildskov. A computer model of atrial fibrillation. *American Heart Journal*, 67(2):200, 1964.

N. Morita, W. J. Mandel, Y. Kobayashi, and H. S. Karagueuzian. Cardiac fibrosis as a determinant of ventricular tachyarrhythmias. *Journal of Arrhythmia*, 2014.

M. A. Moskowitz. The 2006 Thomas Willis lecture: The adventures of a translational researcher in stroke and migraine. *Stroke*, 38(5):1645, 2007.

A. P. Munuzuri, V. Pérez Villar, and M. Markus. Splitting of autowaves in an active medium. *Phys. Rev. Lett.*, 79:1941, 1997.

C. B. Muratov, E. Vanden-Eijnden, and E. Weinan. Noise can play an organizing role for the recurrent dynamics in excitable media. *Proc. Natl. Acad. Sci. USA*, 104:702, 2007.

C. J. L. Murray, L. C. Rosenfeld, S. S. Lim, K. G. Andrews, K. J. Foreman, D. Haring, N. Fullman, M. Naghavi, R. Lozano, and A. D. Lopez. Global malaria mortality between 1980 and 2010: a systematic analysis. *Lancet*, 379 (9814):413, 2012.

J. Murray. *Mathematical Biology*. Springer, New York, 2002.

S. Nadkarni and P. Jung. Spontaneous oscillations of dressed neurons: a new mechanism for epilepsy? *Phys. Rev. Lett.*, 91(26):268101, 2003.

M. P. Nash, C. P. Bradley, P. Sutton, M.Hayward, D. J. Paterson, and P. Taggart. Human hearts possess large regions of steep and flat APD restitution. *Europace*, 6:187, 2004.

M. P. Nash, C. P. Bradley, P. Sutton, M. Hayward, D. J. Paterson, and P. Taggart. Spatial heterogeneity of action potential duration restitution in humans. *Heart Rhythm*, 2:S216, 2005.

M. P. Nash, A. Mouradand, R. H. Clayton, P. M. Sutton, C. P. Bradley, M. Hayward, D. J. Paterson, and P. Taggart. Evidence for multiple mechanisms in human ventricular fibrillation. *Circulation*, 114:536, 2006.

A. R. Nayak, T. K. Shajahan, A. V. Panfilov, and R. Pandit. Spiral-wave dynamics in a mathematical model of human ventricular tissue with myocytes and fibroblasts. *PloS One*, 8(9), 2013.

M. Nelson and J. Rinzel. The Hodgkin–Huxley model. In *The Book of GENESIS*, page 29. Springer, 1998.

T. I. Netoff, R. Clewley, S. Arno, T. Keck, and J. A. White. Epilepsy in small-world networks. *The Journal of Neuroscience*, 24(37):8075, 2004.

A. J. Nicholson and V. A. Bailey. The balance of animal populations. Part i. In *Proceedings of the Zoological Society of London*, volume 105, page 551, 1935.

P. M. F. Nielsen, I. J. Le Grice, B. H. Smaill, and P. J. Hunter. Mathematical model of geometry and fibrous structure of the heart. *Am. J. Physiol.*, 260: 1365, 1991.

D. Noble. The relation of Rushton's liminal length for excitation to the resting and active conductances of excitable cells. *J. Physiol.*, 226:573, 1972.

J. B. Nolasco and R. W. Dahlen. A graphic method for the study of alternation in cardiac action potentials. *J. Appl. Physiol.*, 25(2):191–196, 1968.

T. Nomura and L. Glass. Entrainment and termination of reentrant wave propagation in a periodically stimulated ring of excitable media. *Phys. Rev. E*, 53:6353, 1996.

T. Ohara, K. Ohara, J.-M. Cao, M.-H. Lee, M. C. Fishbein, W. J. Mandel, P.-S. Chen, and H. S. Karagueuzian. Increased wave break during ventricular fibrillation in the epicardial border zone of hearts with healed myocardial infarction. *Circulation*, 103:1465, 2001.

G. V. Osipov and J. J. Collins. Using weak impulses to suppress traveling waves in excitable media. *Phys. Rev. E*, 60:54, 1999.

H. G. Othmer and Y. Tang. Oscillations and waves in a model of insp3-controlled calcium dynamics. In *Experimental and Theoretical Advances in Biological Pattern Formation*, page 277. Plenum Press, New York, 1993.

E. Ott, C. Grebogi, and J. A. Yorke. Controlling chaos. *Phys. Rev. Lett.*, 64: 1196, 1990.

Q. Ouyang, H. L. Swinney, and G. Li. Transition from spirals to defect-mediated turbulence driven by a Doppler instability. *Physical Review Letters*, 84(5): 1047, 2000.

S. W. Pacala, M. P. Hassell, and R. M. May. Symmetry breaking instabilities in dissipative systems. ii. *Nature*, 344:150, 1990.

R. Pandit, A. Pande, S. Sinha, and A. Sen. Spiral turbulence and spatiotemporal chaos: characterization and control in two excitable media. *Physica A*, 306:211, 2002.

A. Panfilov and A. Pertsov. Ventricular fibrillation: evolution of the multiple-wavelet hypothesis. *Phil. Trans. Roy. Soc. Lond. A*, 359(1783):1315, 2001.

A. V. Panfilov. Spiral breakup as a model of ventricular fibrillation. *Chaos*, 8: 57, 1998.

A. V. Panfilov. Spiral breakup in an array of coupled cells: the role of the intercellular conductance. *Phys. Rev. Lett.*, 88:118101, 2002.

A. V. Panfilov and P. Hogeweg. Spiral breakup in a modified FitzHugh–Nagumo model. *Phys. Lett. A*, 176:295, 1993.

A. V. Panfilov and A. V. Holden. Self-generation of turbulent vortices in a two-dimensional model of cardiac tissue. *Physics Letters A*, 151(1):23, 1990.

A. V. Panfilov and A. V. Holden. Spatiotemporal irregularity in a two-dimensional model of cardiac tissue. *International Journal of Bifurcation and Chaos*, 1(01):219, 1991.

A. V. Panfilov and J. P. Keener. Effects of high frequency stimulation on cardiac tissue with an inexcitable obstacle. *J. Theor. Biol.*, 163:439, 1993.

A. V. Panfilov and A. N. Rudenko. Two regimes of the scroll ring drift in the three-dimensional active media. *Physica*, 28D:215, 1987.

A. V. Panfilov and B. N. Vasiev. Vortex initiation in a heterogeneous excitable medium. *Physica D*, 49:107, 1991.

J. M. Pastore, S. D. Girouard, K. R. Laurita, F. G. Akar, and D. S. Rosenbaum. Mechanism linking t-wave alternans to the genesis of cardiac fibrillation. *Circulation*, 99(10):1385, 1999.

D. Pazo, L. Kramer, A. Pumir, S. Kanani, I. Efimov, and V. Krinsky. Pinning force in active media. *Phys. Rev. Lett.*, 93:168303, 2004.

B. Percha, R. Dzakpasu, M. Żochowski, and J. Parent. Transition from local to global phase synchrony in small world neural network and its possible implications for epilepsy. *Phys. Rev. E*, 72(3):031909, 2005.

A. Pertsov and M. Vinson. Dynamics of scroll waves in inhomogeneous excitable media. *Phil. Trans. R. Soc. Lond. A*, 347:687, 1994.

A. M. Pertsov, E.A. Ermakova, and A.V. Panfilov. Rotating spiral waves in a modified FitzHugh–Nagumo model. *Physica D*, 14:117, 1984.

A. M. Pertsov, R. R. Aliev, and V. I. Krinsky. Three-dimensional twisted vortices in an excitable chemical medium. *Nature*, 345:419, 1990.

A. M. Pertsov, J. M. Davidenko, R. Salomonsz, W. T. Baxter, and J. Jalife. Spiral waves of excitation underlie reentrant activity in isolated cardiac muscle. *Circ. Res.*, 72:631, 1993.

E. Pervolaraki and A. V. Holden. Human uterine excitation patterns leading to labour: Synchronization or propagation? In *Lecture Notes in Computer Science*, volume 7223, page 162. Springer, Berlin, Heidelberg, 2012.

N. S. Peters, J. Coromilas, N. J. Severs, and A. L. Wit. Disturbed connexin43 gap junction distribution correlates with the location of reentrant circuits in the epicardial border zone of healing canine infarcts that cause ventricular tachycardia. *Circulation*, 95:988, 1997.

N. S. Peters, J. Coromilas, M. S. Hanna, M. E. Josephson, C. Costeas, and A. L. Wit. Characteristics of the temporal and spatial excitable gap in anisotropic reentrant circuits causing sustained ventricular tachycardia. *Circ. Res.*, 82: 279, 1998.

A. Pikovsky, M. Rosenblum, and J. Kurths. *Synchronization*. Cambridge University Press, Cambridge, 2003.

A. S. Pikovsky and J. Kurths. Coherence resonance in a noise-driven excitable system. *Phys. Rev. Lett.*, 78:775, 1997.

R. D. Pingree, R. R. Pugh, P. M. Holligan, and G. R. Forster. Summer phytoplankton blooms and red tides along tidal fronts in the approaches to the English Channel. *Nature (London)*, 1975.

Y. Pomeau and P. Manneville. Intermittent transition to turbulence in dissipative dynamical systems. *Commun. Math. Phys.*, 74:189, 1980.

L. M. Popescu, S. M. Ciontea, and D. Cretoiu. Interstitial Cajal-like cells in human uterus and fallopian tube. *Ann. N. Y. Acad. Sci.*, 1101:139, 2007.

M. Potse, B. Dubé, J. Richer, A. Vinet, and R. M. Gulrajani. A comparison of monodomain and bidomain reaction-diffusion models for action potential propagation in the human heart. *IEE Transactions on Biomedical Engineering*, 53(12):2425, 2006.

C. W. Potter. A history of influenza. *J. Applied Microbiology*, 91:572, 2001.

W. H. Press, S. A. Teukolsky, W. T. Vetterling, and B. P. Flannery. *Numerical Recipes in C*. Cambridge University Press, Cambridge, 1995.

I. Prigogine and R. Lefever. Symmetry breaking instabilities in dissipative systems. ii. *The Journal of Chemical Physics*, 48(4):1695, 1968.

A. Pumir, A. Arutunyan, V. Krinsky, and N. Sarvazyan. Genesis of ectopic waves: role of coupling, automaticity, and heterogeneity. *Biophys. J.*, 89: 2332, 2005.

A. Pumir, V. Nikolski, M. Hörning, A. Isomura, K. Agladze, K. Yoshikawa, R. Gilmour, E. Bodenschatz, and V. Krinsky. Wave emission from heterogeneities opens a way to controlling chaos in the heart. *Phys. Rev. Lett.*, 99: 208101, 2007.

A. Pumir, S. Sinha, S. Sridhar, M. Argentina, M. Hörning, S. Filippi, C. Cherubini, S. Luther, and V. Krinsky. Wave-train-induced termination of weakly anchored vortices in excitable media. *Phys. Rev. E*, 81:010901(R), 2010.

K. Pyragas. Continuous control of chaos by self-controlling feedback. *Physics Letters A*, 170(6):421, 1992.

Z. Qu, J. N. Weiss, and A. Garfinkel. Spatiotemporal chaos in a simulated ring of cardiac cells. *Phys. Rev. Lett.*, 78:1387, 1997.

Z. Qu, J. N. Weiss, and A. Garfinkel. Cardiac electrical restitution properties and stability or reentrant spiral waves: a simulation study. *Am. J.Physiol.Heart Circ.Physiol.*, 276:H269, 1999.

Z. Qu, A. Garfinkel, P-S Chen, and J. N. Weiss. Mechanisms of discordant alternans and induction of reentry in simulated cardiac tissue. *Circulation*, 102:1664, 2000a.

Z. Qu, J. Kil, F. Xie, A. Garfinkel, and J. N. Weiss. Scroll wave dynamics in a three-dimensional cardiac tissue model: roles of restitution, thickness, and fiber rotation. *Biophysical Journal*, 78(6):2761, 2000b.

Z. Qu, J. N. Weiss, and A. Garfinkel. From local to global spatiotemporal chaos in a cardiac tissue model. *Phys. Rev. E.*, 61(1):727, 2000c.

Z. Qu, F. Xie, A. Garfinkel, and J. N. Weiss. Origins of spiral wave meander and breakup in a two-dimensional cardiac tissue model. *Annals of Biomedical Engineering*, 28(7):755, 2000d.

W. Quan and Y. Rudy. Unidirectional block and reentry of cardiac excitation: a model study. *Circ. Res.*, 66:367, 1990.

W-J. Rappel. Filament instability and rotational tissue anisotropy: a numerical study using detailed cardiac models. *Chaos*, 11(1):71, 2001.

W-J. Rappel, F. Fenton, and A. Karma. Spatiotemporal control of wave instabilities in cardiac tissue. *Phys. Rev. Lett.*, 83:456, 1999.

A. Reichenbach. Glia: neuron index: review and hypothesis to account for different values in various mammals. *Glia*, 2(2):71, 1989.

J. G. Restrepo and A. Karma. Spatiotemporal intracellular calcium dynamics during cardiac alternans. *Chaos*, 19(3):037115, 2009.

M. L. Riccio, M. L. Koller, and R. F. Gilmour. Electrical restitution and spatiotemporal organization during ventricular fibrillation. *Circulation Research*, 84(8):955, 1999.

S. Rohr, J. P. Kucera, and A. G. Kleber. Slow conduction in cardiac tissue, i: Effects of a reduction of excitability versus a reduction of electrical coupling on microconduction. *Circ. Res.*, 83:781, 1998.

D. S. Rosenbaum, L. E. Jackson, J. M. Smith, H. Garan, J. N. Ruskin, and R. J. Cohen. Electrical alternans and vulnerability to ventricular arrhythmias. *New England Journal of Medicine*, 330(4):235, 1994.

M. E. Rosenthal and M. E. Josephson. Current status of antitachycardia devices. *Circulation*, 82:1889, 1990.

H. H. Rotermund. Imaging of dynamic processes on surfaces by light. *Surface Science Reports*, 29(7):265, 1997.

H. H. Rotermund, W. Engel, M. Kordesch, and G. Ertl. Imaging of spatio-temporal pattern evolution during carbon monoxide oxidation on platinum. *Nature*, 343:355, 1990.

A. Roxin, H. Riecke, and S. A. Solla. Self-sustained activity in a small-world network of excitable neurons. *Phys. Rev. Lett.*, 92(19):198101, 2004.

A. N. Rudenko and A. V. Panfilov. Drift and interaction of vortices in two-dimensional heterogeneous active medium. *Studia Biophysica*, 98:183, 1983.

Y. Rudy. Reentry: Insights from theoretical simulations in a fixed pathway. *J. Cardiovasc. Electrophys.*, 6:294, 1995.

W. A. H. Rushton. Initiation of the propagated disturbance. *Proc. Royal Soc. London. B.*, 124:210, 1937.

F. B. Sachse, A. P. Moreno, and J. A. Abildskov. Electrophysiological modeling of fibroblasts and their interaction with myocytes. *Annals of Biomedical Engineering*, 36(1):41, 2008.

P. Sadeghi and H. H. Rotermund. Gradient induced spiral drift in heterogeneous excitable media. *Chaos*, 21:013125, 2011.

H. Sakaguchi and T. Fujimoto. Elimination of spiral chaos by periodic force for the Aliev–Panfilov model. *Phys. Rev. E*, 67:067202, 2003.

H. Sakaguchi and Y. Kido. Elimination of spiral chaos by pulse entrainment in the Aliev–Panfilov model. *Phys. Rev. E*, 71:052901, 2005.

F. H. Samie, O. Berenfeld, J. Anumonwo, S. F. Mironov, S. Udassi, J. Beaumont, S. Taffet, A. M. Pertsov, and J. Jalife. Rectification of the background potassium current: A determinant of rotor dynamics in ventricular fibrillation. *Circ. Res.*, 89:1216, 2001.

S. K. Sarna. Are interstitial cells of cajal plurifunction cells in the gut? *American Journal of Physiology-Gastrointestinal and Liver Physiology*, 294(2): G372, 2008.

S. Sawai, P. A. Thomason, and E. C. Cox. An autoregulatory circuit for long-range self-organization in *Dictyostelium* cell populations. *Nature (London)*, 433(7023):323, 2005.

G. Schram, M. Pourrier, P. Melnyk, and S. Nattel. Differential distribution of cardiac ion channel expression as a basis for regional specialization in electrical function. *Circ. Res.*, 90:939, 2002.

G. P. Sergeant, M. A. Hollywood, K. D. McCloskey, K. D. Thornbury, and N. G. McHale. Specialised pacemaking cells in the rabbit urethra. *J. Physiol.*, 526 (2):359, 2000.

T. K. Shajahan, S. Sinha, and R. Pandit. Ventricular fibrillation in a simple excitable medium model of cardiac tissue. *Int. J. Mod. Phys. B*, 17:5645, 2003.

T. K. Shajahan, S. Sinha, and R. Pandit. Spatiotemporal chaos and spiral turbulence in models of cardiac arrhythmias: an overview. *Proc. Indian Natl. Sci.*, 71:47, 2005.

T. K. Shajahan, S. Sinha, and R. Pandit. Spiral-wave dynamics depend sensitively on inhomogeneities in mathematical models of ventricular tissue. *Phys. Rev. E*, 75:011929, 2007.

T. K. Shajahan, A. R. Nayak, and R. Pandit. Spiral-wave turbulence and its control in the presence of inhomogeneities in four mathematical models of cardiac tissue. *PLoS One*, 4:e4738, 2009.

R. M. Shaw and Y. Rudy. Electrophysiologic effects of acute myocardial ischemia: A mechanistic investigation of action potential conduction and conduction failure. *Circ. Res.*, 80:124, 1997.

M. Shibata and J. Bureš. Reverberation of cortical spreading depression along closed-loop pathways in rat cerebral cortex. *Journal of Neurophysiology*, 35 (3):381, 1972.

Y. Shiferaw, D. Sato, and A. Karma. Coupled dynamics of voltage and calcium in paced cardiac cells. *Phys. Rev. E*, 71(2):021903, 2005.

A. Shmygol, A. M. Blanks, G. Bru-Mercier, J. E. Gullam, and S. Thornton. Control of uterine ca2+ by membrane voltage. *Ann. N. Y. Acad. Sci.*, 1101 (1):97, 2007.

R. Singh and S. Sinha. Spatiotemporal order, disorder, and propagating defects in homogeneous system of relaxation oscillators. *Physical Review E*, 87(1): 012907, 2013.

R. Singh, J. Xu, N. Garnier, A. Pumir, and S. Sinha. Self-organized transition to coherent activity in disordered media. *Physical Review Letters*, 108(6): 068102, 2012.

S. Sinha. Noise-free stochastic resonance in simple chaotic systems. *Physica A*, 270:204, 1999.

S. Sinha and J. Breuer. Terminating ventricular tachycardia by pacing induced dynamical inhomogeneities in the reentry circuit. arXiv:nlin/0406046v1, 2004.

S. Sinha and B. K. Chakrabarti. Deterministic stochastic resonance in a piece-wise linear chaotic map. *Phys. Rev. E*, 58:8009, 1998.

S. Sinha and D. J. Christini. Termination of reentry in an inhomogeneous ring of model cardiac cells. *Phys. Rev. E*, 66:061903, 2002.

S. Sinha and S. Sridhar. Controlling spatiotemporal chaos and spiral turbulence in excitable media. In E. Schöll and H-G. Schuster, editors, *Handbook of Chaos Control*, chapter 32, pages 703–718. Wiley-VCH, Weinheim, 2 edition, 2008.

S. Sinha, A. Pande, and R. Pandit. Defibrillation via the elimination of spiral turbulence in a model for ventricular fibrillation. *Phys. Rev. Lett.*, 86:3678, 2001.

S. Sinha, K. M. Stein, and D. J. Christini. Critical role of inhomogeneities in pacing termination of cardiac reentry. *Chaos*, 12:893, 2002.

S. Sinha, J. Saramäki, and K. Kaski. Emergence of self-sustained patterns in small-world excitable media. *Phys. Rev. E.*, 76:015101, 2007.

B. T. Skale, M. J. Kallok, E. N. Prystowsky, R. M. Gill, and D. P. Zipes. Inhibition of premature ventricular extrastimuli by subthreshold conditioning stimuli. *Circulation*, 68:707, 1983.

G. Somjen. Mechanisms of spreading depression and hypoxic spreading depression-like depolarization. *Physiological Reviews*, 81(3):1065, 2001.

G. G. Somjen. Aristides Leao's discovery of cortical spreading depression. *Journal of Neurophysiology*, 94(1):2, 2005.

J. C. Sommerer and E. A. Ott. A physical system with qualitatively uncertain dynamics. *Nature*, 365:138, 1993.

S. Sridhar and S. Sinha. Controlling spatiotemporal chaos in excitable media using an array of control points. *Europhys. Lett.*, 81:50002, 2008.

S. Sridhar and S. Sinha. Response to sub-threshold stimulus is enhanced by spatially heterogeneous activity. *EPL*, 92(6):60006, 2010.

S. Sridhar, S. Sinha, and A. V. Panfilov. Anomalous drift of spiral waves in heterogeneous excitable media. *Phys. Rev. E*, 82:051908, 2010.

S. Sridhar, A. Ghosh, and S. Sinha. Critical role of pinning defects in scroll-wave breakup in active media. *Europhysics Letters*, 103(5):50003, 2013a.

S. Sridhar, D-M Le, Y-C Mi, S. Sinha, P-Y Lai, and C. K. Chan. Suppression of cardiac alternans by alternating-period-feedback stimulations. *Phys. Rev. E.*, 87(4):042712, 2013b.

A. T. Stamp, G. V. Osipov, and J. J. Collins. Suppressing arrhythmias in cardiac models using overdrive pacing and calcium channel blockers. *Chaos*, 12:931, 2002.

J. M. Starobin and C. F. Starmer. Boundary-layer analysis of waves propagating in an excitable medium: Medium conditions for wave-front obstacle separation. *Phys. Rev. E*, 54:430, 1996.

J. H. Steele. In *Fisheries Mathematics*. Academic Press, London, 1977.

O. Steinbock, J. Schutze, and S. C. Muller. Electric-field-induced drift and deformation of spiral waves in an excitable medium. *Phys. Rev. Lett.*, 68: 248, 1992.

O. Steinbock, F. Siegert, S.C. Muller, and C.J. Weijer. Three-dimensional waves of excitation during *Dictyostelium* morphogenesis. *Proc. Natl. Acad. Sci. USA*, 90:7332, 1993.

O. Steinbock, P. Kettunen, and K. Showalter. Anisotropy and spiral organizing centers in patterned excitable media. *Science*, 269:1857, 1995a.

O. Steinbock, A. Tóth, and K. Showalter. Navigating complex labyrinths: optimal paths from chemical waves. *Science*, 267:868, 1995b.

B. Surawicz and B. Surawicz. *Electrophysiological Basis of ECG and Cardiac Arrhythmia*. Lippincott Williams and Wilkins, Philadelphia, 1995.

S. Takagi, A. Pumir, D. Pazo, I. Efimov, V. Nikolski, and V. Krinsky. Unpinning and removal of a rotating wave in cardiac muscle. *Phys. Rev. Lett.*, 93:058101, 2004.

K. H. W. J. ten Tusscher and A. V. Panfilov. Reentry in heterogeneous cardiac tissue described by the Luo–Rudy ventricular action potential model. *Am. J. Physiol. Heart Circ. Physiol*, 284:H542, 2003a.

K. H. W. J. ten Tusscher and A. V. Panfilov. Influence of nonexcitable cells on spiral breakup in two-dimensional and three-dimensional excitable media. *Phys. Rev. E.*, 68:062902, 2003b.

K. H. W. J. ten Tusscher and A. V. Panfilov. Wave propagation in excitable media with randomly distributed obstacles. *Multiscale Model. Simul.*, 3:265, 2005.

K. H. W. J. ten Tusscher and A. V. Panfilov. Influence of diffuse fibrosis on wave propagation in human ventricular tissue. *Europace*, 9:vi38, 2007.

K. H. W. J. ten Tusscher, D. Noble, P. J. Noble, and A. V. Panfilov. A model for human ventricular tissue. *Am. J. Physiol. Heart Circ. Physiol.*, 286:H1573, 2004.

M. Tinsley, J. Cui, F. V. Chirila, A. Taylor, S. Zhong, and K. Showalter. Spatiotemporal networks in addressable excitable media. *Phys. Rev. Lett.*, 95 (3):038306, 2005.

M. Toiya, V. K. Vanag, and I. R. Epstein. Diffusively coupled chemical oscillators in a microfluidic assembly. *Angewandte Chemie*, 120(40):7867–7869, 2008.

E. C. Tolkacheva, D. G. Schaeffer, D. J. Gauthier, and W. Krassowska. Condition for alternans and stability of the 1:1 response pattern in a "memory" model of paced cardiac dynamics. *Phys. Rev. E.*, 67:031904, 2003.

W.-C. Tong, C. Y. Choi, S. Karche, A. V. Holden, H. Zhang, and M. J. Taggart. A computational model of the ionic currents, ca2+ dynamics and action potentials underlying contraction of isolated uterine smooth muscle. *PLoS One*, 6:18685, 2011.

A. Tóth and K. Showalter. Logic gates in excitable media. *J. Chem. Phys.*, 103(6):2058, 1995.

N. Trayanova. Defibrillation of the heart: insights into mechanisms from modelling studies. *Experimental Physiology*, 91(2):323, 2006.

J. E. Truscott. Environmental forcing of simple plankton models. *Journal of Plankton Research*, 17(12):2207, 1995.

J. E. Truscott and J. Brindley. Ocean plankton populations as excitable media. *Bulletin of Mathematical Biology*, 56(5):981, 1994a.

J. E. Truscott and J. Brindley. Equilibria, stability and excitability in a general class of plankton population models. *Philosophical Transactions of the Royal Society of London: A*, 347(1685):703, 1994b.

M.-L. Tsai, K. Cesen-Cummings, R. C. Webb, and R. Loch-Caruso. Acute inhibition of spontaneous uterine contractions by an estrogenic polychlorinated biphenyl is associated with disruption of gap junctional communication. *Toxicol. Appl. Pharmacol.*, 152:18, 1998.

R. W. Tsien, R. S. Kass, and R. Weingart. Cellular and subcellular mechanisms of cardiac pacemaker oscillations. *J. Exp. Biol.*, 81:205, 1979.

A. M. Turing. The chemical basis of morphogenesis. *Phil. Trans. R. Soc. Lond. B.*, 237:37, 1952.

M. Valderrabano, Y-H. Kim, M. Yashima, and T-J Wu. Obstacle-induced transition from ventricular fibrillation to tachycardia in isolated swine right ventricles: Insights into the transition dynamics and implications for the critical mass. *Journal of the American College of Cardiology*, 36:2000, 2000.

M. Valderrabano, M.-H. Lee, T. Ohara, A. C. Lai, M. C. Fishbein, S.-F. Lin, H. S. Karagueuzian, and P. S. Chen. Dynamics of intramural and transmural reentry during ventricular fibrillation in isolated swine ventricles. *Circ. Res.*, 88:839, 2001.

M. Valderrabano, P. S. Chen, and S. F. Lin. Spatial distribution of phase singularities in ventricular fibrillation. *Circulation*, 108:354, 2003.

T. Vicsek, A. Czirók, E. Ben-Jacob, I. Cohen, and O. Shochet. Novel type of phase transition in a system of self-driven particles. *Phys. Rev. Lett.*, 75: 1226, 1995.

A. Vinet and F. A. Roberge. The dynamics of sustained reentry in a ring model of cardiac tissue. *Ann. Biomed. Eng.*, 22:568, 1994.

M. Vinson, A. Pertsov, and J. Jalife. Anchoring of vortex filaments in 3d excitable media. *Physica D*, 72:119, 1994.

M. Vinson, S. Mironov, S. Mulvey, and A. Pertsov. Control of spatial orientation and lifetime of scroll rings in excitable media. *Nature (London)*, 386:477, 1997a.

M. Vinson, S. Mironov, S. Mulvey, and A. Pertsov. Control of spatial orientation and lifetime of scroll rings in excitable media. *Nature*, 386:477, 1997b.

V. Voignier, E. Hamm, and V. Krinsky. Chirality dependent component of vortex advection in excitable media. *Chaos*, 9:238, 1999.

C-T. Wang and R. Loch-Caruso. Phospholipase-mediated inhibition of spontaneous oscillatory uterine contractions by lindane *in vitro*. *Toxicol. Appl. Pharmacol.*, 182(2):136, 2002.

S. Y. Wang, M. Yoshino, J. L. Sui, M. Wakui, P. N. Kao, and C. Y. Kao. Potassium currents in freshly dissociated uterine myocytes from nonpregnant and late-pregnant rats. *J. Gen. Physiol.*, 112(6):737, 1998.

Y. Wang and Y. Rudy. Action potential propagation in inhomogeneous cardiac tissue: safety factor considerations and ionic mechanism. *Am. J. Physiol. Heart Circ. Physiol.*, 278:H1019, 2000.

M. Watanabe, F. Fenton, S. Evans, H. Hastings, and A. Karma. Mechanisms of discordant alternans. *J. Cardvasc. Electrophys.*, 12:196, 2001.

M. A. Watanabe and M. L. Koller. Mathematical analysis of dynamics of cardiac memory and accommodation: theory and experiment. *American Journal of Physiology: Heart and Circulatory Physiology*, 282(4):H1534, 2002.

D. J. Watts and S. H. Strogatz. Collective dynamics of small-world networks. *Nature (London)*, 393(6684):440, 1998.

C. J. Weijer. Morphogenetic cell movement in *Dictyostelium*. *Sem. Cell. Dev. Biol.*, 10:609, 1999.

J. N. Weiss, P. S. Chen, Z. Qu, H. S. Karagueuzian, and A. Garfinkel. Ventricular fibrillation how do we stop the waves from breaking? *Circ. Res.*, 87: 1103, 2000.

J. N. Weiss, A. Karma, Y. Shiferaw, P-S Chen, A. Garfinkel, and Z. Qu. From pulsus to pulseless the saga of cardiac alternans. *Circ. Res.*, 98(10):1244, 2006.

B. J. Welsh, J. Gomatam, and A. E. Burgess. Three-dimensional chemical waves in the Belousov–Zhabotinskii reaction. *Nature (Lond.)*, 304:611, 1983.

N. Wiener and A. Rosenblueth. The mathematical formulation of the problem of conduction of impulse in a network of connected excitable element, specifically in cardiac muscle. *Arch. Inst. Cardiol. Mexico*, 16:205, 1946.

L. D. Wilson and D. S. Rosenbaum. Mechanisms of arrythmogenic cardiac alternans. *Europace*, 9:vi77, 2007.

A. T. Winfree. Spiral waves of chemical activity. *Science*, 175:634, 1972.

A. T. Winfree. *When Time Breaks Down*. Princeton University Press, Princeton, 1987.

A. T. Winfree. Electrical instability in cardiac muscle: Phase singularities and rotors. *J. Theo. Biol.*, 311:611, 1989.

A. T. Winfree. The electrical thresholds of ventricular myocardium. *J. Cardiovasc. Electrophys.*, 1:393, 1990.

A. T. Winfree. Varieties of spiral wave behavior: An experimentalist's approach to the theory of excitable media. *Chaos*, 1:303, 1991.

A. T. Winfree. Evolving perspectives during 12 years of electrical turbulence. *Chaos*, 8:1, 1998.

A. T. Winfree. *The Geometry of Biological Time*. Springer, New York, 2000.

A.T. Winfree. Scroll-shaped waves of chemical activity in three dimensions. *Science*, 181:937, 1973.

F. X. Witkowski, L. J. Leon, P. A. Penkoske, W. R. Giles, M. L. Spano, W. L. Ditto, and A. T. Winfree. Spatiotemporal evolution of ventricular fibrillation. *Nature (Lond.)*, 392:78, 1998.

M. Woltering and M. Markus. Control of spatiotemporal disorder in an excitable medium. *ScienceAsia*, 28:43, 2002.

S. Wray, S. Kupittayanant, A. Shmygol, R. D. Smith, and T. Burdyga. The physiological basis of uterine contractility: A short review. *Exp. Physiol.*, 86 (2):239, 2001.

T.-J. Wu, J. J. Ong, C. Hwang, J. J. Lee, M. C. Fishbein, L. Czer, A. Trento, C. Blanche, R. M. Kass, and W. J. Mandel. Characteristics of wave fronts during ventricular fibrillation in human hearts with dilated cardiomyopathy: role of increased fibrosis in the generation of reentry. *J. Am. Coll. Card.*, 32: 187, 1998.

T-J. Wu, S-F. Lin, A. Baher, Z. Qu, A. Garfinkel, J. Weiss, C-T. Ting, and P-S. Chen. Mother rotors and the mechanisms of d600-induced type 2 ventricular fibrillation. *Circulation*, 110(15):2110, 2004.

F. Xie, Z. Qu, and A. Garfinkel. Dynamics of reentry around a circular obstacle in cardiac tissue. *Phys. Rev. E*, 58:6355, 1998.

F. Xie, Z. Qu, J. N. Weiss, and A. Garfinkel. Interactions between stable spiral waves with different frequencies in cardiac tissue. *Phys. Rev. E*, 59(2):2203, 1999.

F. Xie, Z. Qu, A. Garfinkel, and J. N. Weiss. Electrophysiological heterogeneity and stability of reentry in simulated cardiac tissue. *Am. J. Physiol. Heart Circ. Physiol.*, 280:H535, 2001a.

F. Xie, Z. Qu, J. N. Weiss, and A. Garfinkel. Coexistence of multiple spiral waves with independent frequencies in a heterogeneous excitable medium. *Phys. Rev. E*, 63:031905, 2001b.

Y. Xie, A. Garfinkel, P. Camelliti, P. Kohl, J. N. Weiss, and Z. Qu. Effects of fibroblast-myocyte coupling on cardiac conduction and vulnerability to reentry: a computational study. *Heart Rhythm*, 6(11):1641, 2009.

M. Yoshino, S. Y. Wang, and C. Y. Kao. Sodium and calcium inward currents in freshly dissociated smooth myocytes of rat uterus. *J. Gen. Physiol.*, 110 (5):565, 1997.

R. C. Young. Myocytes, myometrium, and uterine contractions. *Ann. N. Y. Acad. Sci.*, 1101(1):72, 2007.

G. Yuan, G. Wang, and S. Chen. Control of spiral waves and spatiotemporal chaos by periodic perturbation near the boundary. *Europhys. Lett.*, 72:908, 2005.

A. N. Zaikin and A. M. Zhabotinsky. Concentration wave propagation in two-dimensional liquid-phase self-oscillating system. *Nature (Lond.)*, 225:535, 1970.

H. Zhang, R. W. Winslow, and A. V. Holden. Re-entrant excitation initiated in models of inhomogeneous atrial tissue. *J. Theor. Biol.*, 50:279, 1998.

H. Zhang, B. Hu, and G. Hu. Suppression of spiral waves and spatiotemporal chaos by generating target waves in excitable media. *Phys. Rev. E*, 68:026134, 2003.

H. Zhang, Z. Cao, N-J Wu, H-P Ying, and G. Hu. Suppress Winfree turbulence by local forcing excitable systems. *Phys. Rev. Lett.*, 94:188301, 2005a.

H-J. Zhang, P-Y. Wang, and Y-Y. Zhao. Light-gradient-induced spiral wave drifts in a Belousov–Zhabotinsky reaction. *Chin. Phys. Lett.*, 22:287, 2005b.

Lu Qun Zhou and Qi Ouyang. Experimental studies on long-wavelength instability and spiral breakup in a reaction-diffusion system. *Phys. Rev. Lett.*, 85: 8, 2000.

Lu Qun Zhou and Qi Ouyang. Spiral instabilities in a reaction-diffusion system. *J. Phys. Chem. A.*, 105:112, 2001.

D. P. Zipes and J. Jalife. *Cardiac Electrophysiology: From Cell to Bedside.* Saunders, Philadelphia, 2004.

V. S. Zykov. Cycloidal circulation of spiral waves in excitable medium. *Biofizika*, 175:634, 1986.

Index

Milton Keynes UK
Ingram Content Group UK Ltd.
UKHW051948071024
449327UK00026B/2213

9 780367 377984